The
Sociable Humanist

James Harris in 1745: engraving by Charles Bestland of Joseph
Highmore's portrait

The Sociable Humanist

The Life and Works of
James Harris 1709–1780

Provincial and Metropolitan Culture
in Eighteenth-century England

CLIVE T. PROBYN

CLARENDON PRESS · OXFORD
1991

Oxford University Press, Walton Street, Oxford OX2 6DP

Oxford New York Toronto
Delhi Bombay Calcutta Madras Karachi
Petaling Jaya Singapore Hong Kong Tokyo
Nairobi Dar es Salaam Cape Town
Melbourne Auckland

and associated companies in
Berlin Ibadan

Oxford is a trade mark of Oxford University Press

Published in the United States
by Oxford University Press, New York

© *Clive T. Probyn 1991*

British Library Cataloguing in Publication Data
Probyn, Clive T.
The sociable humanist: the life and works of James Harris
1709–1780: provincial and metropolitan culture in
eighteenth-century England.
1. England. Social life, 1714–1820—Biographies
I. Title   942.07092

ISBN 0–19–818563–4

Library of Congress Cataloging in Publication Data
Probyn, Clive T.
The sociable humanist: the life and works of James Harris,
1709–1780: provincial and metropolitan culture in eighteenth-
century England/Clive T. Probyn.
Includes bibliographical references and index.
1. Harris, James, 1709–1780.   2. Salisbury (England)—Intellectual
life—18th century.   3. London (England)—Intellectual life—18th
century.   4. England—Civilization—18th century.   5. Politicians—
England—Biography.   6. Impresarios—England—Biography.
7. Humanists—England—Biography.   I. Title.
DA483.H34P76   1991
942.3'19—dc20   90–38552
ISBN 0–19–818563–4

Typeset by Cambrian Typesetters, Frimley, Surrey
Printed and bound in Great Britain by
Biddles Ltd, Guildford and King's Lynn

For Meg

Acknowledgements

THE biographical materials in this book, and much else, are very largely based on previously unpublished manuscripts. Some of these are accessible in public archives and libraries, but most are not. It could not have been carried through without the help of many people, and I now know that it could not have been written at all without the signal generosity and co-operation of the Earl and Countess of Malmesbury, who allowed me repeated and unrestricted access to the Harris papers in their private possession, and graciously allowed me to quote from them. I hope that what follows is at least in part a recompense for the trouble which they kindly took on my behalf, as well as a corrective to previous neglect of their distinguished ancestor.

I also wish to record my gratitude to those friends and colleagues who responded to my requests and tolerated my demands: Martin and Ruthe Battestin, in particular; Gavin Betts; Jeremy Black; Mr and Mrs John Cordle (of Malmesbury House); Clifford Davies (Keeper of the Archives, Wadham College); Michael Crump (British Library); Bertrand Goldgar; Dr S. J. Goss (Keeper of Pictures, Wadham College); Jocelyn Harris; B. L. Harrison (Assistant Archivist, West Yorkshire Archive Service); Roger Highfield (Librarian, Merton College); John Jacob; Michael Jubb (Public Record Office); Francis King; Dr Harold Love; Dr Barbara Krettek (University Library, Göttingen); Tom Lockwood; Maureen Mann; Marjorie Paskin, MBE; Miss Penelope Rundle (Wiltshire Record Office); Frank Ryder; Peter Sabor; Derek Shorrocks (Somerset County Archivist); and Lars Troide. My parents happily shared my interest in both Salisbury and the Harris connection.

Miss Rosemary Dunhill, Hampshire County Archivist and Diocesan Record Officer, enabled my work in ways I could hardly have expected from someone less dedicated to her profession.

I am grateful for the assistance of the library staff at Monash University, Victoria, the State Library of Victoria, the Baillieu Library, University of Melbourne, Merton and Wadham Colleges, Oxford, the British Library and the Colindale Newspaper Library, Salisbury Public Library, the Public

Record Offices at Kew and Chancery Lane, Leicester University Library, the County Record Offices in Winchester and Trowbridge, the Alderman Library of the University of Virginia, the Library of Congress, and the Osborn Collection at Yale University. I am also grateful to Sheila Wilson for her keen eye and expert typing, and to Tony Miller and Miss Sheila Steffens for photographic assistance.

All dates are given in New Style. Punctuation has been modernized only minimally. A short section in Chapter 3, comparing Harris's dialogue 'On Happiness' with Johnson's *Rasselas*, has previously appeared in *The Modern Language Review*, 73 (April 1978), 261–6. I am grateful for the editor's permission to reproduce a modified version. I am happy to acknowledge financial support for my research from the Monash University Special Research Fund and from the Australian Academy of the Humanities.

C.T.P.

Contents

List of Figures

List of Abbreviations

EHR	*English Historical Review*
ELH	*Journal of English Literary History*
Gertrude Harris, 'Memoir'	'Memoir of J. Harris Author of Hermes' by [Katherine] Gertrude Robinson [née Harris]: PRO Kew 30/43/1/4 (Lowry Cole Papers)
Harris Papers	Papers belonging to the Harris family: in private possession, and catalogued by the first earl in manuscript
Harris, *Works*	*The Works of James Harris, Esq.*, ed. James Harris, first earl of Malmesbury (Oxford, 1841)
HRO	Hampshire Record Office (Winchester)
Malmesbury, 'Memoir'	'Memoirs of the Harris Family', by James, first earl of Malmesbury (1800): PRO Kew 30/43/1/2 (Lowry Cole Papers)
Malmesbury Papers	As Harris Papers above, but selected, referenced, and photo-copied by the History of Parliament Trust.
MLR	*Modern Language Review*
MP	*Modern Philology*
N&Q	*Notes and Queries*
PMLA	*Publications of the Modern Language Association of America*
PQ	*Philological Quarterly*
PRO	Public Record Office (Kew or Chancery Lane)
Series of Letters	*A Series of Letters of the First Earl of Malmesbury His Family and Friends from 1745 to 1820*, ed. the earl of Malmesbury, 2 vols. (London, 1870)

SRO	Somerset Record Office
TSLL	*Texas Studies in Literature and Language*
WANHS	*Wiltshire Archaeological and Natural History Magazine*
WRO	Wiltshire Record Office (Trowbridge)

Praxid	Sicell	Thomas	Lucy	John	Robert
bapt.	bapt.	bapt.	bapt.	bapt.	bapt.
24 Aug. 1589	27 Aug. 1592	10 May 1595	2 Aug. 1597	25 Mar. 1600	20 Feb. 1602

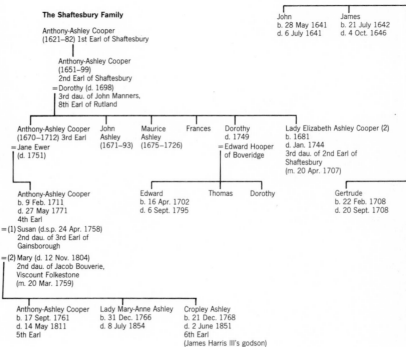

The Shaftesbury Family

John
b. 28 May 1641
d. 6 July 1641

James
b. 21 July 1642
d. 4 Oct. 1646

Anthony-Ashley Cooper
(1621–82) 1st Earl of Shaftesbury

Anthony-Ashley Cooper
(1651–99)
2nd Earl of Shaftesbury
= Dorothy (d. 1698)
3rd dau. of John Manners,
8th Earl of Rutland

Anthony-Ashley Cooper
(1670–1712) 3rd Earl
= Jane Ewer
(d. 1751)

John
Ashley
(1671–93)

Maurice
Ashley
(1675–1726)

Frances

Dorothy
d. 1749
= Edward Hooper
of Boveridge

Lady Elizabeth Ashley Cooper (2)
b. 1681
d. Jan. 1744
3rd dau. of 2nd Earl of
Shaftesbury
(m. 20 Apr. 1707)

Anthony-Ashley Cooper
b. 9 Feb. 1711
d. 27 May 1771
4th Earl
= (1) Susan (d.s.p. 24 Apr. 1758)
2nd dau. of 3rd Earl of
Gainsborough
= (2) Mary (d. 12 Nov. 1804)
2nd dau. of Jacob Bouverie,
Viscount Folkestone
(m. 20 Mar. 1759)

Edward
b. 16 Apr. 1702
d. 6 Sept. 1795

Thomas

Dorothy

Gertrude
b. 22 Feb. 1708
d. 20 Sept. 1708

Anthony-Ashley Cooper
b. 17 Sept. 1761
d. 14 May 1811
5th Earl

Lady Mary-Anne Ashley
b. 31 Dec. 1766
d. 8 July 1854

Cropley Ashley
b. 21 Dec. 1768
d. 2 June 1851
6th Earl
(James Harris III's godson)

THE HARRIS FAMILY

William Harris
of Orcheston St George
co. Wilts. d. 1570
= Cecily Sherne widow
(m. 7 July 1561)

Thomas Harris
of Orcheston St George
bapt. 15 Mar. 1565
d. 24 Aug. 1605
= Praxid Perry
dau. of Robert Perry

JAMES HARRIS I
of the Close of Sarum
bapt. 6 Oct. 1605
d. Aug. 1679
= Gertrude (1592–1678)
dau. of Robert Tounson,
Bishop of Salisbury

| Thomas Harris of the Close of Sarum b. 22 Feb. 1643 d. 13 Jan. 1679 = (1) Dorothy d.s.p. 25 Mar. 1672 dau. of George Cary, DD Dean of Exeter = (2) Joan (1650–1734) m. 21 July 1673 dau. of Sir Wadham Wyndham of Norrington, Wilts. | Gertrude b. 4 May 1645 d. Jan. 1646 | Jane b. 18 June 1646 d. 30 Apr. 1647 | Margaret b. 25 Apr. 1647 d. 23 June 1679 = Capt. Gabriel Ashley (1646–1702) (m. 8 July 1669) | Praxid b. 26 Dec. 1648 d. 22 Oct. 1649 |

| = JAMES HARRIS II of the Close of Sarum b. 17 Apr. 1674 d. 26 Aug. 1731 | = (1) Catherine (1681–13 June 1705) dau. of Charles Cocks MP of Worcester, and Maria, sister and co-heir of John, Lord Somers (m. 3 Aug. 1704) | Wadham b. 13 Dec. 1678 d. 30 Mar. 1685 |

| JAMES HARRIS III of the Close of Sarum b. 25 July 1709 d. 22 Dec. 1780 =Elizabeth b. 1722 d. 16 Oct. 1781 dau. of John Clarke of Sandford, Somerset, MP and heir of her brother John Clarke of Sandford (m. 8 July 1745) | Thomas b. 1 Jan. 1712 d.s.p. 21 Feb. 1785 Master in Chancery =Catharine (1710–8 June 1796) dau. of Sir Edward Knatchbull, 4th Bt. and sister of Sir Wyndham Knatchbull Wyndham, 5th Bt. | George William b. 19 Sept. 1714 d. 23 Aug. 1777 Prebendary of Sarum and Rector of Eccliffe, Co. Durham | Catherine b. 13 June 1705 d. 8 Jan. 1741 = Sir Wyndham Knatchbull Wyndham, 5th Bt. (d. 23 July 1747) (3 children) |

| JAMES HARRIS IV b. 9 Apr. 1746 d. 21 Nov. 1820 (Created Baron of Malmesbury 19 Sept. 1788; Viscount Fitzharris and Earl of Malmesbury 29 Dec. 1800) =Harriot-Mary b. 2 May 1762 d. 20 Aug. 1830 dau. of Sir George Amyand and sister of Sir George Cornwall, Bt. | Elizabeth b. July 1747 d. 13 Apr. 1749 | Katherine Gertrude b. 18 Apr. 1750 d.s.p. 8 June 1834 = Hon. Frederick Robinson b. 11 Oct. 1746 d. 28 Dec. 1792 2nd son of Thomas, 1st Lord Grantham (m. 13 June 1785) | John Thomas b. June 1751 d. 9 Dec. 1752 | Louisa Margaret b. 11 Jan. 1753 d. 26 May 1826 |

(5 children)

Introduction

'Who is this Harris?' (Charles Townshend, 1777)

ON 26 February 1781 His Britannic Majesty's Envoy Extraordinary and Minister Plenipotentiary to the Empress of all the Russias wrote a letter to one of his two sisters in England. The subject of this letter was the recent death of their father, James Harris:

the Life of my Father would be written by Doctr. Kippis . . . a Life ought to be written . . . it will be requisite to furnish Doctr. Kippis with such papers as may be necessary, but I should wish to see what he has written before it be published.

Such a biography was neither published nor composed. Indeed, Andrew Kippis reached no further than the letter *F* in his *Biographica Britannica* before he died in 1795, and the Harris papers were kept in the possession of the family, where they still remain to this day. For a variety of reasons, the life of James Harris has remained sketchy at best, misinterpreted at worst, and generally ignored. There is no book on James Harris, least of all a full-scale biography.

As custodian of his father's papers and heir to his magnificent library, the British Envoy and first Earl of Malmesbury faithfully arranged his father's correspondence, edited the major published works, and was in turn similarly served by his grandson, the third earl, in 1844 and 1870. Other than a brief memoir of James Harris prefixed to an edition of his *Works* in 1801, the published record of his life is bare.

The few public facts are well-enough known. When Harris died, on 22 December 1780, *The Gentleman's Magazine* reported the outlines: Fellow of the Royal Society, Trustee of the British Museum, Member of Parliament for Christchurch, Hampshire (from 1761 to 1780), a Lord Commissioner of the Admiralty, a member of the Board of Treasury, comptroller to Queen Charlotte, nephew of Anthony, third Earl of Shaftesbury 'whose refinement of taste and manners Mr Harris inherited', and a man with few equals in the theory and practice of music. As an author, he had made a mark with four theoretical books dealing with art, music, painting, poetry, happiness, universal grammar, Aristotelian philosophy, and (soon to appear) a work on the history of European medieval literature and literary criticism.

Harris's career touched almost every aspect of Augustan culture and successfully bridged its provincial and metropolitan contexts. His friends and acquaintances included his cousin the fourth Earl of Shaftesbury, Henry Fielding, John Hoadly, George Lyttelton, George Frideric Handel, Samuel Richardson, Sir Joshua Reynolds, Samuel Johnson, Joseph and Thomas Warton, Lord Kames, and Lord Monboddo. Harris's son, created first Earl of Malmesbury in 1800 for his brilliant diplomatic successes in Madrid, Berlin, St Petersburg, The Hague, Paris, Lille, and Brunswick, has been the subject of two scholarly books, and makes two appearances in fiction (in Thackeray's *Vanity Fair*, and, more substantially, in Dennis Wheatley's *The Shadow of Tyburn Tree*, 1948, as a member of 'a brilliant trio at Merton', with Fox and William Eden). James Harris has inspired no biographer, although he too has appeared in fiction, in Thomas Love Peacock's *Melincourt* (1817: chapter 21), and in Edward Rutherfurd's epic novel on the history of Salisbury, *Sarum* (1987). He is better known in France and Germany than in his own country.

This neglect of Harris is largely due to the fact that, until very recently, the unpublished materials for a biographical and critical enquiry had never been made accessible. But it is also true that the current versions of the intellectual history of eighteenth-century England would discourage such an undertaking in the first place. Harris's broad array of interests does not fit the generally familiar categories. He is unpopular with philosophers for his attacks on Locke (though Locke is almost never mentioned by name); with critics because of his apparently low opinion of contemporary literature (though his admiration for Henry Fielding's work was boundless, and he read Richardson and Pope with warm admiration); with scientists for his dismay at the contemporary fashion for empirical and materialist philosophies (though he was enthralled by Hunter's lectures on comparative anatomy); and with the general reader because of his daunting classical scholarship, even though his books are specifically written to display Europe's classical inheritance in the most accessible manner to English readers. In his own words, from the dedication to his last book, Harris wished to lead his readers 'to inspect authors far superior to myself, many of whose works (like hidden treasures) have lain for years out of sight'. He succeeded too well in acting as the enabling medium for the ideas of others, and his modesty has concealed his own originality.

His reputation, unluckily, was largely determined at an early stage by the disapproval of Samuel Johnson, his friend and exact contemporary. Rebounding from an intemperate attack on Mrs Elizabeth Montagu's pretensions to literary scholarship and classical learning, Johnson turned to Boswell's defence of Harris, one of Montagu's supporters, in order to win his point: 'Harris is a sound sullen scholar; he does not like interlopers. Harris . . . is a prig, and a bad prig' (Boswell, *Life of Johnson*, 7 April 1778). Boswell never understood Johnson's judgement on 'the amiable philosopher of Salisbury'. If there was any real substance to Johnson's remark, it was, as Boswell suggests, Harris's *manner* as a writer to which Johnson objected. On 14 July 1763 Boswell had been alleviating his dissipation by rereading Hume, and James Harris's 'very sensible and accurate' 'Concerning Happiness: A Dialogue', and although he objected to the latter's 'somewhat aukward' dialogue form, its 'frequent repetition of "said he" and "I replied" ', he nevertheless 'rose from his Dialogue happier and more disposed to follow virtue'. Twelve years later, in a letter to William Johnson Temple, Boswell's 'oldest and dearest friend', Boswell remarked:

I am invited to a dinner on the banks of the Thames, at Richard Owen Cambridge's where we are to be Reynolds, Johnson, and Hermes Harris. 'Do you think so? said he. Most certainly, said I.' Do you remember how I used to laugh at his style when we were in the Temple? He thinks himself an ancient Greek from these little peculiarities.

Schoolboys are prone to tell such stories about their masters, and there is in Johnson's hostility to Harris, one suspects, an element of resentment at being instructed. Even so, Johnson's oft-cited, dismissive, and enigmatic verdict on Harris is better remembered than the object of its provocation.

There was nothing in Harris's *life* of which Johnson could disapprove, but a good deal that he might have coveted. When Johnson imagined the scholar's life he projected a life constrained by Toil, Envy, Want, the Garret, the Gaol, or the Patron. As a leading member of the county *haute bourgeoisie*, Harris had both the economic means and an almost missionary desire to commit himself to a lifetime of study. Remarkably free with his hospitality, Harris's own meat and drink was literary scholarship. Almost universally known as 'the amiable Mr Harris', he spoke enviably of no one nor ill of anyone except bungling scholars, tyrants in Church, and State, and eschewed (to his biographer's chagrin) what his daughter was to call 'the tittle tattle Gossip of the day'. For patrons, at least in Johnson's sense, Harris had not the slightest need. His role model as a writer, foolishly again as far as Johnson was concerned, was the third Earl of Shaftesbury. As for the gaol, Harris's only experience of that lay in sending malefactors there as a justice of the peace.

Harris and Johnson were frequently in each other's company, but in some important ways each inhabited competing, and eventually divergent, intellectual worlds. The former has been placed, if at all, in the vanguard of cultural resistance to the new, the latter, indisputably, is seen as one of the great makers of his age's character.

Thus James Harris has come down to us, so far, like a piece of unsorted intellectual debris, *disjecta membra* from a larger creation. His letters appear in biographies of Handel, Richardson, and Fielding; his conversations in Boswell's *Life of Johnson*; quotations from his work crop up in Reynolds's *Discourses on Art*; books on Wessex theatre history commemorate a brilliant provincial musical tradition which he stage-managed for more than thirty years in Salisbury; his political diaries compiled under George Grenville's administration in the 1760s are recognized as the best and most detailed to have survived; his longest poem, *Concord*, has been claimed as the best poetic version of Shaftesbury's aesthetic theories; and his close association with the Burney family has ensured that his extraordinary passion for music (both theory and practice) has not been wholly forgotten. Henry Fielding regarded Harris as 'the man whom I esteem most of any person in this World', and may have commemorated their close friendship in the surname of the heroine in Fielding's last and most autobiographical novel, *Amelia*. One of Handel's librettists, Thomas Morell, wrote about a performance of Handel's final masterpiece in Salisbury cathedral in 1764: 'My own favourite is *Jephtha*, which I wrote in 1751, and in composing of which Mr Handel fell blind. I had the pleasure to hear it finely perform'd at Salisbury under Mr Harris.' Harris's diaries record such comments as this: 'Mr Boswell, the Corsican writer told me that David Hume was vain, but a very good natur'd man . . . that Dr Samuel Johnson (now 68) was much afraid of death'; and his daughter records that 'Mr Harris said he knew no two men speak so exactly as they wrote as Sterne and Johnson'.

Except for his continuing prominence as a linguistic theorist, as for example in Noam Chomsky's *Cartesian Linguistics* (1966), where he is one of only two English linguists discussed and cited at any length, Harris has always remained in the background. To a large extent this was Harris's choice. Towards those who asked for his help he was both generous and self-effacing. Among his beneficiaries, and to varying extents, there were Thomas Birch, John Upton, Jonathan Toup, William Young ('Parson Adams'), Henry and Sarah Fielding, Jane Collier, Mrs Elizabeth Carter, Handel's first biographer Mainwaring, Joseph Warton, Samuel Richardson, Monboddo, and many more. Harris preferred to keep his name out of others' publications, but it now seems only just that *his* work and influence should finally be recognized.

In 1769 the fourth Earl of Shaftesbury gave Harris the third earl's own annotated copy of *The Sociable Enthusiast* (a part of *The Moralists*). The title of the present book, *The Sociable Humanist*, is chosen to reflect not only the pervasive influence on Harris of Shaftesbury's writings but also to emphasize a characteristic not shared by the valetudinarian third earl himself, a love of company and of society in general. As Harris's daughter Gertrude put it: 'His love of Society was invariable, and his House was always open to his friends and

Acquaintance; and I might add, to their Acquaintance also
. . . there was seldom a day without his having *some person* to
breakfast dine or sup with him. His hospitality extended to all
strangers who were recommended to him.' Whether in
provincial Salisbury or in metropolitan London, Harris
sought out people and events to gratify that most characteristic
and generous of Augustan impulses—as well as the fulfilment
of its duty to itself—conversation.

Salisbury Origins

FOR anyone with a sense of history, present-day Salisbury and its environs offer an almost dizzying richness in every period of the past. Tracking around the points of the compass today will yield the co-ordinates of several thousand years of a nation's history. There is Avebury, Stonehenge, and Vespasian's camp to the north, the Roman road to Silchester to the north-east, the ruins of Henry II's Clarendon Palace to the east, Grim's ditch to the south, the seat of the Pembroke family at Wilton in the west, the vast expanse of Salisbury Plain with its artillery ranges in the north-west, all surrounded by the inscrutable silence of prehistoric barrows, earthworks, and tumuli. At the centre soars the magnificent cathedral, containing, appropriately enough, the works of the oldest mechanical clock in England.

In this part of Wiltshire, all roads lead to Salisbury, a relatively modern city built after Old Sarum, fortified by Celts, Saxons, and Normans in turn, was abandoned in the early twelfth century. Old Sarum cathedral stood derelict for many years until its final demolition in 1331, when its stones were transported to the new city in order to build the walls of its Close. Old Sarum lived on as a political fantasy, for here was the classic and almost entirely depopulated rotten borough, whose successive owners (notably the Pitt family) returned two members of parliament to Westminster until its abolition under the Reform Bill of 1832.

The roll-call of personalities who left their marks on this landscape is, similarly, like a shortened version of the nation's biography. Since Salisbury was founded by and for a religious purpose, clerics dominate its early history. There was St Augustine, who met the British priests at Cricklade; Bishop Poore, who laid the foundation stone of the new cathedral (in 1220), and whose descendant Edward Poore died in the same year as James Harris (1780); there was St Osmond, who completed Poore's works; Lord Herbert of Cherbury, the founder of deism, and his brother, the rector of nearby Bemerton, George Herbert; Bishop Ken, one of the original seven Non-Jurors; the charitable Bishop Seth Ward, who had been Professor of Astronomy at Oxford and founder member of the Royal Society before his appointment as Bishop of Salisbury in 1667; Gilbert Burnet, historian, architect of the Bill of Rights in 1689, and the man who performed the marriage of James Harris's father in 1707; the incendiary preacher of non-resistance Dr Henry Sacheverell (born at Marlborough); and George Crabbe, the rector of Trowbridge. Among other poets, there was Sir Philip Sidney, who wrote part of his *Arcadia* at Wilton, and Sir John Davies, born at Tisbury. The dramatist Philip Massinger was born in Salisbury, and John Dryden was living at Charlton during the Plague and Fire of London, writing *Annus Mirabilis* there. There is also the historian of the Rebellion, Edward Hyde, Earl of Clarendon, Lord High Chancellor, born at Dinton, 10 miles west of Salisbury, for whom one of the Harrises was to act as a land-agent; the 'philosopher of Malmesbury' Thomas Hobbes; the ecclesiastical historian Richard Hooker (rector of East Boscombe for 4 years); the novelist Henry Fielding; William Beckford, the eccentric author of *Vathek*; the brilliant satirist Sydney Smith (curate of Netheravon for a time); and the critic William Hazlitt, who spent much of his time at Winterslow. The musicians Henry and William Lawes, sons of one of the lay vicars, Thomas Lawes, probably learnt their art as choristers in the cathedral; Sir Christopher Wren was born at East Knoyle and was later employed at Longleat as well as on a major survey of Salisbury's cathedral. There was William Pitt, Lord Chatham, and Lady Mary Wortley Montagu: each claimed Salisbury origins. Salisbury is the focus, the social, cultural, and spiritual centre-point, an assertion of the spiritual and architectural brilliance of the so-called Dark Ages, the setting for William Golding's novel, *The Spire*, of 1964.

Like the history of New Sarum itself, the biography of any family begins in obscurity, with few 'facts' beyond dates of births, marriages, and deaths, the legal fingerprints left behind as evidence of whole lives whose character we can never fully recreate. But a process is clear: across the generations of the Harris family from the early sixteenth to the late eighteenth century a rural Wiltshire obscurity is steadily displaced by increasing national, and eventually international, prominence. The blood of an obscure Wiltshire country yeoman was to run in the veins of an accomplished scholar and philosopher who enjoyed a European reputation and, in the next generation, in one of Britain's most accomplished European ambassadors. For virtually the whole of his life James Harris moved only between Salisbury and London; his son was to spend most of his life entirely out of England, leaving from Salisbury for Madrid, Berlin, and St Petersburg. But it was the father who made the furthest intellectual journey. His interests lay not only in the classical roots of Western culture, in Aristotle, Plato, and Homer, as well as in the English 'classical' tradition of Chaucer, Shakespeare, and Milton, but also in something overarching all this, something more ambitious. Beyond speculation in literary, linguistic, and aesthetic theories in general, he sought evidence of the ultimate and universal operations of the mind itself.

I

The origins of the Harris family may be traced to the small village of Orcheston St George, about 11 miles north of Salisbury, in the fertile centre of Salisbury Plain. Here William Harris married Cecily Sherne on 7 July 1561, and died in 1570. That is all we know about him. Thomas Harris, William's son, also of Orcheston St George, was baptized on 15 March 1565, married Praxid Perry, and, apart from raising seven children and paying £10 composition tax to Charles I's commissioners because he had at least £40 annual income from freehold lands in 1631, he also slips into obscurity.[1] One

---

[1] E. A. Fry, 'Knighthood Compositions for Wiltshire', *Wiltshire Notes and Queries*, 1 (1893–6), 108. The biographical outlines for this chapter are taken from James, earl of Malmesbury, 'Memoirs of the Harris Family', Lowry Cole Papers, PRO Kew 30/43/1/2, a 40 page MS dated 1800, not in Malmesbury's hand but a fair copy prepared for the press. Unlike the printed 'Memoir' prefixed to the edition of his father's *Works* in 1801 and reprinted in 1841, a somewhat stiff though undoubtedly sincere act of filial piety, the unpublished draft is more detailed on the family background and full of personal anecdote. Although it is incomplete (e.g. the account of 'Hermes' Harris's musical activities in Salisbury and London on p. 20 is completely blank, James having no ear for and less interest in music), it is nevertheless the best memoir of

of Thomas's four sons was christened James, the first in a long line. Born in the year of the Gunpowder Plot (1605), he settled in the Close of Sarum around 1625, and died in the year that Gilbert Burnet, Bishop of Salisbury from 1689 to 1715, published the first part of his *History of the Reformation of the Church of England* (1679). It was the move to Salisbury that lifted the Harris family from a doubtless worthy but nevertheless obscure provincial existence to increasing involvement in national affairs, first as witnesses, and then, in the next century, as participants and directors.

The first James Harris was a successful country lawyer with an eye to social advancement. In 1640 he married Gertrude Tounson, one of the fifteen children of Robert Tounson, who had been installed as bishop of Salisbury on 9 July 1620. Tounson died on 15 May 1621, however, and was buried in Westminster Abbey. He left his large family poor, so much so that King James expressly recommended that Tounson's successor should remain celibate, 'because his Predecessor had left so large a Family meanly provided for, though not altogether destitute'.[2] In the fifteenth year of the Harris's marriage Salisbury was singed by what Clarendon was to call 'a little fire, which might have kindled and inflamed all the kingdom'.[3] This was the royalist and disaffected republican rising against Cromwell on 11 March 1655, led by Sir Joseph Wagstaff, Colonel Penruddock, and John Mompesson. During the assizes in Salisbury the town was occupied by up to 200 armed men, both judges and the sheriff were seized, the gaol emptied of its prisoners, and the King proclaimed at two o'clock in the afternoon. This was the only overt uprising in a

Harris that we have. All unacknowledged references, and those to Malmesbury's 'Memoir', are to this MS, corrected and supplemented from standard references such as Burke's *Peerage and Baronetage* (s.v. Malmesbury), Collin's *Peerage of England* (s.v. Harris), and from the MS pedigree deposited in Wadham College described below. In addition, the 5th Earl of Malmesbury published 'Some Anecdotes of the Harris Family' in *The Ancestor*, 1 (Apr. 1902), 1–27, which includes reproductions of family portraits.

[2] Anon., *Magna Britannia et Hibernia: A New Survey of Great Britain*, 6 vols. (1720–31), vi. 178. This passage was written in 1728 (see p. 183).

[3] *History of the Rebellion*, Bk. IV. See also C. H. Firth, 'Cromwell and the Insurrection of 1655', *EHR* 3 (1888), 323–50, and 4 (1888), 110–31; and *A Collection of the State Papers of John Thurloe*, 7 vols. (1742), iii. 295, 370–95.

planned national rebellion, and it petered out in a humiliating failure, having failed to gain any popular support. Its leaders were executed at Exeter and others were put to death in Salisbury itself. There is no mention of Harris in any of the records.

A family anecdote suggests that he *may* have been of republican persuasion.[4] It is possible that the Whig loyalties of subsequent generations of the Harris family began with the first James Harris, and also possible that their subsequent friendship with the Collier family of Steeple Langford stemmed from the 1650s. What is certain is that two members of the Collier family, the brothers Henry and Joseph, were transported to the West Indies for their part in the Salisbury uprising.[5]

If the first James Harris nurtured some misgivings about the return of Charles and the bishops, he kept them to himself. He was well advised to do so, for in the early days of the Restoration a close watch was kept on such things. During the election of Members of Parliament for London in mid-March 1661, many letters were being intercepted at the Post Office, including one from John Davenant to Harris on 18 March. This described the preparations for the coronation, the arrival of the French ambassador, and the selection of four candidates by the Common Council for the next parliament, each of whom was regarded as either Presbyterian or Independent and unanimously anti-episcopal. Davenant clearly assumes that Harris would share his and the 'honest party's' forebodings at such a choice of candidates, and this is perhaps the clearest evidence so far of Harris's own political alignment.[6] More conclusive evidence is suggested by Harris's move to the Cathedral Close in Salisbury, for the Bishop, through the

---

[4] Malmesbury, 'Some Anecdotes', p. 4, who suggests that the 'high-crowned headpiece' possibly worn by Harris in 1643 betokened puritan sympathies.

[5] *Thurloe State Papers*, iii. 368–9, 394–5; and W. W. Ravenhill, 'Records of the Rising in the West, A.D. 1655', *WANHS* 14:40 (1874), 46–9. Arthur Collier was the nephew of the transported brothers Joseph and Henry.

[6] *Calendar of State Papers, Domestic Series, of the Reign of Charles II, 1660–1661*, ed. M. A. E. Green (1860), 535–6. Green surmises that this letter and others of opposite political persuasion were intercepted to prevent news of the Court party's defeat in the London elections from prejudicing those in the country. The writer of this letter may have been the father of John Davenant of Sarum: see J. Foster, *Alumni Oxonienses*, 1 (Early Series: Kraus repr., 1968), 375.

Dean and Chapter, let him rent a house inside the Close at the clearly nominal rent of 27s. 4d. a year.

They would not have done this for a tenant of anti-episcopal views, even had his wife not been a bishop's daughter. At the Restoration, the house was an odd accretion of medieval and Jacobean elements, constructed on the basis of three separate tenements which had not been combined until 1583. It was described in 1660 as a house with 'hall, parlour, kytchen, larder, buttery and dyning room above stairs, three chambers and two garrets', with a separate 'shopp, chamber, cockloft, and a stable of two bays', with 'a little green courtyard before the house and two gardens and an orchard well planted'.[7] Today, Malmesbury House is one of the most elegant houses in the Close, and its Queen Anne facade screens a combination of rooms remodelled with Augustan symmetry. Its rear wall is still an integral part of the fourteenth-century Close wall built from the ruins of the old Norman cathedral of Old Sarum. The Close itself combined private comfort and public squalor throughout most of the eighteenth century. An observer remarked in 1782:

The Close is comfortable, and the divines well seated, but the house of God is kept in sad order, to the disgrace of our Church, and of Christianity . . . The churchyard is like a cow-common, as dirty and as neglected, and thro' the centre stagnates a boggy ditch. I wonder that the residents do not subscribe to plant near, and rowl the walks, and cleanse the ditch, which might be made a handsome canal . . . when the new bishop arrives . . . he will be shock'd at the dilapidations of the beautiful old Chapter house; and the Cloisters; thro' the rubbish of which they are now making a passage for his new Lordship's installation.[8]

The first James Harris was a prominent member of the local legal profession. Two letters to him from Lord Chancellor Clarendon, virtually the head of Charles II's Government at the time, indicate that he was regarded as a skilled, discreet, and responsible land-agent who could be given wide powers of

---

[7] G. Jackson-Stops, *Malmesbury House* [Salisbury, 1971], 1–2. The house was built on the site of an earlier canonical house known as Cole Abbey or Copt Hall. See C. R. Everett, 'Notes on the Decanal and other Houses in the Close of Sarum', *WANHS* 50: 181 (Dec. 1944), 432–3.

[8] Hon. John Byng, cited in D. H. Robertson, *Sarum Close* (1938), 249.

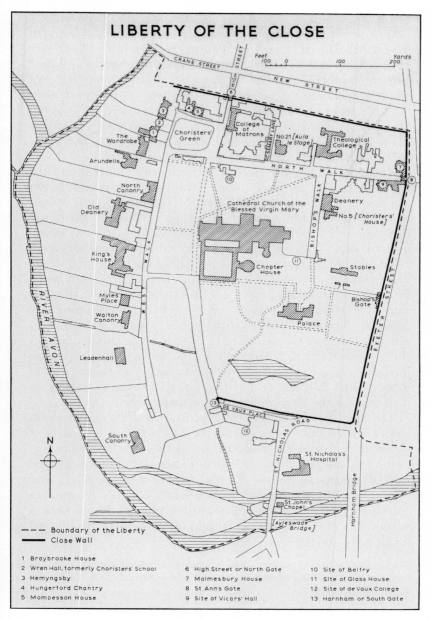

LIBERTY OF THE CLOSE

CRANE STREET

HIGH STREET

NEW STREET

Feet 100 0 100   Yards 200

Choristers' Green

College of Matrons

ROSEMARY LANE

No 21 [Aula le Stage]

Theological College

The Wardrobe

Arundells

NORTH WALK

Deanery

BISHOP'S WALK

No.5 [Choristers' House]

North Canonry

Old Deanery

Cathedral Church of the Blessed Virgin Mary

WEST WALK

King's House

Chapter House

Stables

EXETER STREET

Myles Place

Bishop's Gate

Walton Canonry

Palace

Leadenhall

RIVER AVON

South Canonry

DE VAUX PLACE

ST NICHOLAS ROAD

St Nicholas's Hospital

Harnham Bridge

St John's Chapel

[Ayleswade Bridge]

N

– – – Boundary of the Liberty
—— Close Wall

1 Braybrooke House
2 Wren Hall, formerly Choristers' School
3 Hemyngsby
4 Hungerford Chantry
5 Mompesson House

6 High Street or North Gate
7 Malmesbury House
8 St. Ann's Gate
9 Site of Vicars' Hall

10 Site of Belfry
11 Site of Glass House
12 Site of de Vaux College
13 Harnham or South Gate

1. Liberty of the Close, Salisbury.

negotiation and management relating to the Chancellor's estates at nearby Clarendon and Christchurch. In a letter to 'Good Mr Harris' dated 22 February 1664 from Worcester House, Clarendon remarks that 'I cannot rely upon any Mans Judgment more than I doe upon yours'. A subsequent letter of 28 July expresses Clarendon's complete satisfaction with Harris's lease of tenancies and goes on to instruct his agent about the collection of rents, adding, 'I would have you call to the fellow at Salisbury that took away the Topps of the Trees with the Kings officers, and make him pay for them, or I will sue him for it'. For his 'fidelity and integrity' Clarendon then formally appointed Harris one of three commissioners given the power of managing the Clarendon estates (27 June 1664), with 'full power and authority to lett and lett all or any of my messuages lands Tenements meadowes pastures woods underwoods Coppices Tithes and hereditaments whatsoever . . . And to make contracts and Bargaines for the letting of the same premises . . . as they shall think requisite.'[9] Here is the first indication of the family's legal prominence and the first sign that their connections were to encompass figures of national prominence. By 1673 James Harris was able to purchase land in his own right in the manor of Dibden, and in the records of the 1677 Heraldic Visitation to Wiltshire, whose purpose was to list the pedigrees and arms of all those entitled to be called Esquire or Gentleman, a 'Harris of Sarum' is included for the first time since the visitation of 1533.

The first James Harris died in August 1679, eight years before King James's Declaration of Toleration and Liberty of Conscience forced the remodelling of the Salisbury corporation. Only seven of the sixty Wiltshire magistrates interrogated proved amenable to the king's catholicizing policies, and although a 'James Harris' consequently lost his position of alderman on the Salisbury corporation, this must have been a namesake, even though we might infer similar disaffected loyalties in the Salisbury lawyer.[10]

---

[9] Malmesbury Papers, box 1:1, 32, 33, 34. See also *Calendar of the Clarendon State Papers*, ed. F. J. Routledge (Oxford, 1970), v. 386, for additional evidence of Harris's activities for Clarendon.

[10] See Sir George Duckett, 'Proposed Repeal of the Test and Penal Statutes by King James II in 1688', *WANHS* 18: 54 (1879), 366. Also displaced as a 'common-councell' man was the poetaster Dr Walter Pope, a pensioner of Dr Seth Ward, the

The only son of James and Gertrude to survive infancy was their first-born, Thomas (1643–78). Their daughter, Margaret (1647–79), married Captain Gabriel Ashley of the Close (1646–1702). Thomas was probably the first of the Harrises to attend Wadham College. He is recorded as a commoner on 13 April 1660, and as matriculated on 6 June of the same year.[11] He followed his father's professional and ecclesiastical preferences, becoming a barrister at law in the Middle Temple, and, in 1671, marrying Dorothy Cary, the daughter of the Dean of Exeter. When Dorothy died (25 March 1672) he married Joan Wyndham (1647–1734), the daughter of Sir Wadham Wyndham of Norrington, Wiltshire, one of the Justices of the King's Bench and a descendant of the founder of Wadham College, Nicholas Wadham. This was to be a connection of particular value to the Harrises' ambitions.

By this second marriage Thomas had two sons, a *second* James Harris (1674–1731), and the doubtless proudly named Wadham, who died at the age of 6 in 1685. After only 6 years of marriage to Thomas, Joan was widowed. She was to outlive both her sons and a daughter-in-law, but in 1679 she was left in sole management of the 5-year-old James and the Harris estate. She was, fortunately, the first in a long line of capable Harris women and steered her son's education purposefully. Her great-grandson was to write of her in 1800: 'she appears to have been a shrewd notable Woman, and altho she wrote a vile hand and spelt intolerably, to have been very careful during the minority of 17 years of my grandfather's property and education, she spared neither pains nor expence in this last as appears from her accounts.'[12]

The second James Harris was educated at Winchester and matriculated at Wadham on 8 July 1691 as a Gentleman Commoner. On 5 October he was received as a Fellow Commoner, and was also described as a student of Lincoln's Inn. He left Wadham without taking a degree in 1693.[13] For the next seven years he practised law, but at the age of 26 he

Bishop of Salisbury. For Pope's poetical tribute to Old and New Sarum, and to his patron, see *The Salisbury Ballad* (1713; repr. in Salisbury by E. Easton, 1770).

[11] R. B. Gardiner, ed. *The Registers of Wadham College, Oxford, part 1, 1613–1719* (1889), i. 234.   [12] Malmesbury, 'Memoir', p. 2.

[13] Gardiner, *Registers of Wadham College*, i. 372.

declared his long-held ambition for a political career. In retrospect, Harris's decision to stand for parliament seems premature. In 1701 his influential contacts were too few, and it was not until his second marriage in 1707 (by which time his political ambitions had ceased) that a network of political patrons became available to him.

For the election of November 1701, influence was more than ever a vital prerequisite. Following the death of ex-King James II, and after almost a year of a Tory administration opposed to William's policies, the Whigs moved swiftly and energetically to mobilize their candidates for the new parliament due to meet on 30 December. In the western counties their party manager was Thomas, Lord Wharton, and through his agent, Walter White (MP for Chippenham since 1695), direct contacts were made primarily with the 'Knights of the Shire'. One of them was Maurice Ashley Cooper, whose sister Harris was to marry just three years hence. Ashley (1675–1726) had served as MP for Weymouth and Melcombe Regis from 1695 to 1698, enjoyed the strong support of Burnet, and, with Sir William Ashe (MP for Heytesbury between 1689 and January 1701), was returned in place of the two sitting Tory members, Sir George Hungerford and Sir Richard How. For the first time in a thirty-year period, Wiltshire was now represented by two Whigs.

Harris harboured no illusions about the difficulties which faced an inexperienced and London-based outsider trying to penetrate the intense network of local loyalties. They were complex enough in his home town of Salisbury, but positively Byzantine in the constituency of his first choice, the classic rotten borough of Old Sarum. From his Lincoln's Inn chambers, on 11 November Harris wrote to his uncle, George Wyndham (1666–1746, the youngest son of Sir Wadham Wyndham), to announce the forthcoming dissolution of parliament and his intention of standing as a candidate:

I am now about . . . solliciting an interest in the next Election. Old Sarum is in short ye only place, where, from the small number of Electors (whereof some you know are my relations & others my neighbours and acquaintances), from Mr Harveys constant residence at a great distance, and that disposition you have sometimes said they shewed in favour of me at the last Election, I can conceive any likelyhood of succeeding in an attempt of this Nature. Yet after all I have still that check upon me from my apprehensions of not prevailing, that I scarce dare trust my intentions beyond your self unless upon a very probable likelyhood of carrying my point. And thereupon desire you would as from your own good will to me (which I have still experienced and shall always acknowledge) make such discoveries if possibly you can, either by positively engaging, or at least diving into the dispositions of the persons concerned . . . I have let fall some words formerly to this effect to Mr Mompesson, and have no reason to doubt of his assistance, however I think it best his mind were more fully known by your discovering him.[14]

The Pitt family had owned the almost wholly depopulated site of Old Sarum since 1692, and it was the Pitt family who determined its political character, either by electing members of their own family, or by ensuring that other successful candidates were in the Pitt interest. Voting rights were vested in burgage holders; in 1728 there were three voters, in 1734 there were five, and this situation remained until the borough was abolished by the Reform Bill of 1832.[15] Harris believed his prospects for election were good, an opinion strengthened by assurances and encouragement from his uncles George and Ashley, and from his mother Joan Wyndham. He accordingly drew up a sketch of the Old Sarum constituency in order to identify those with voting rights, and several times made a list of their likely preferences and interests. He came to the conclusion that land had been improperly and secretly split in order to create new voters. Harris's initial problem lay in gaining the controlling influence over the ten 'old votes' exercised by Mrs Pitt (whose husband was then at Fort St George in the East Indies). He then had to persuade Harvey's supporters to switch their allegiances without alienating Harvey's parliamentary colleague (and Harris's friend and neighbour in the Close), Charles Mompesson (MP for Old Sarum, 1698–1705, and for Wilton, 1708 and 1710). Mrs Pitt

[14] Malmesbury Papers, vol. 2, part 2, p. 1.

[15] See R. Sedgwick, *The House of Commons, 1715–54*, part 1. (1970), 350–1. The Pitts controlled Old Sarum from 1692 to 1751. New Sarum (Salisbury) had 54 voters. Anthony Duncombe, later Lord Feversham, established his Whig interest in 1721; from 1714 the interest was Tory, represented by Sir Edward Bouverie. From 1734 to 1747 both seats were held by Tories, and in 1753 William Beckford said of the Salisbury corporation: 'I scarcely know a more disinterested set of men in the kingdom.'

would give no unequivocal support to Harris, although she heartily disliked Mompesson, and the latter thought of Harris as an unwelcome competitor for his own seat. George Wyndham wrote to his nephew Harris, 18 October 1701, of Mrs Pitt's 'great respect for you and the Family', and that 'she did give me all hope except an absolute promise for yr. interest'. Mixing encouragement and dismay, Wyndham also added that 'there is nothing wanting but your personal appearance . . . but there are so many intricate turnings and windings in the management of the Votes that I think, after so great a disposition as People are in for you and against Mr Harvey, it wants the finishing Stroak'.[16] Harris's own detailed notes on the election of 24 November 1701 list *fourteen* voters (four of these being 'new votes' controlled by Mompesson and Harvey's united interests) and clarifies at least some of the intricacies in this hotbed of parliamentary vote-rigging. Wadham Wyndham, Harris's uncle, was against Harvey on the grounds that he was a stranger to the county, but was predetermined by the interest of Mrs Pitt, and would have supported Harris had he not been absent on election day because his wife was expecting a child; Charles Mompesson supported Harvey; Thomas Curgenven, the rector of Orcheston St George since 1690, and Governor Pitt's brother-in-law, would vote for Harris if instructed to do so by Mrs Pitt, and was against Mompesson; Francis Swanton, JP promised support for Harris but would also take instructions from Mrs Pitt; William Gerson's brother was a client of Harvey, was pre-engaged to Mompesson, but would (if a free agent) vote for Harris; William Harvey was for Mompesson; Mr Lydcott's son-in-law was for Harvey; Mr Elliott was for Mompesson, and therefore probably disposed by Mompesson's interest also to support Harvey; the vicar of Box, John Phillips, by Mrs Pitt's direction, was for both Harvey and Mompesson, but was also well disposed towards Harris; John Payne was for Mompesson and would only support Harris if the two of them were not in direct competition for the same seat. John Gauntlett, the MP for Wilton in 1695, 1698, 1701, 1702, and 1705, had given his support to Harris and Mompesson after Harris himself had directly solicited the

[16] Malmesbury Papers, box 1:2.

Earl of Pembroke's influence; William Hearst would probably not vote at all; John Brooks was Mompesson's man; and Charles Thompson, curate of Stratford-sub-Castle, was Harvey's client. Harris suspected that the 'new' votes of either Elliott or Gauntlett, Payne, and Thompson had been created by land-splitting, the latter arranged by Sir Elliot Harvey and Sir Thomas Mompesson. In spite of family connections, the support of his uncle George Wyndham (who also stood as a candidate), and the efforts of Wadham Wyndham to sound out likely supporters, Harris lost the election. He had few regrets, since there was a distinct possibility that if he had won, Mrs Pitt would have used him as her 'Champion' in her battle against Mompesson, by petitioning parliament against the 'new votes'. As he pointed out in a diary of the whole episode, this would have left him 'without any hopes of improving my own interests in the Burrough other than very precarious on Mr Pitt's return from the Indies'.

He also failed to be elected for New Sarum, an independent borough invariably represented by local men selected and supported by a corporation of small gentry and tradesmen who countenanced no outside interference in their selection procedures. Once given, their support was unwavering. The Bouverie family were to hold one of the seats without a break from 1741 to 1835. Harris's parliamentary ambition was clearly not matched by sufficient influence: the family was not yet strong enough to contest, least of all to depose, the established interests in the tightly knit political world of Salisbury. Even so, Harris's political instinct told him that the path to a successful political career lay through accommodation to and exploitation of the established social network. When Harris's eldest son entered parliament sixty years later, almost effortlessly and at the urging of the electorate, it was not to represent Old or New Sarum but Christchurch, and through the interest of a highly expert political manager, local landowner and Commissioner of Customs who was also a cousin, Edward Hooper.

Three years after the frustration of his parliamentary ambitions, James Harris married, on 3 August 1704. As for his father and grandfather before him, marriage was a social elevator. His first wife was Catherine Cocks, eldest daughter of Charles Cocks, JP for the county of Worcester, MP for Worcester in 1692, and for Droitwich in seven parliaments thereafter. Catherine's mother Maria was the sister and co-heir of John, Lord Somers, Lord Chancellor of Great Britain. Catherine was the eldest daughter and her younger sister Margaret married Philip Yorke, created first earl of Hardwicke in 1736, and later appointed Lord Chancellor. For a family of provincial lawyers, the connection with the two great chancellors Somers and Hardwicke was a dynastic coup with long-term though discreetly acknowledged repercussions. In 1751 the third James Harris was to dedicate his second book, *Hermes*, to Philip Yorke, Baron Hardwicke (1690–1764), Lord Chancellor from 1737 to 1756, and Harris's half-sister Catherine's uncle. When the second edition of *Hermes* appeared in 1765 the dedication was repeated, proof that respect and gratitude were the motive, not flattery of the powerful and influential. One of the third James Harris's closest friends, Henry Fielding, was to pay frequent and better known compliments to Hardwicke as 'this Great Man': in *Tom Jones* (Bk. IV, ch. 6) Hardwicke is to the social operation of merit and justice as the conscience is to the individual's mind and moral conduct.

Catherine Harris died on 13 June 1705 at the age of 24 giving birth to a daughter, also named Catherine. The well-connected widower then sought a second wife and stepmother for his only child. He chose Lady Elizabeth Ashley (1681–1743), the third daughter of Anthony, second Earl of Shaftesbury (1651–99: Dryden's target in *Absalom and Achitophel*) and Lady Dorothy Manners (1642–98). Elizabeth's brother, born at most 6 months after his parents' wedding, became the third earl (1671–1713) and wrote the *Characteristics*. The sceptical and asthmatic third earl had received two proposals of marriage for his younger sister by 1707, and their sibling relationship was not cordial. He complained that she cared little for 'me and mine, St Giles [the family seat at Wimborne St Giles], and all else, when made the finest for her and to please and winn her sisters'.[17] The ungrateful Elizabeth did

---

[17] Cited in R. Voitle, *The Third Earl of Shaftesbury, 1671–1713* (Baton Rouge and London, 1984), 262. Malmesbury's 'Memoir' states that the marriage took place in Ely Chapel, Holborn, not Hurn Court.

not suffer from her valetudinarian brother's allergy to the smoke-laden air of London and preferred the excitement of London to the life of provincial Dorsetshire. Even after the earl's return to England from Holland in the autumn of 1707 she chose to live with her younger brother Maurice Ashley. Their relationship seems to have mellowed somewhat in the years before the earl's death in 1713, but in 1707 the third earl had mixed feelings in approving the suit of James Harris. His first offer was rejected, but eventually Shaftesbury's steward called on Harris to inform him that £3,000 of Elizabeth's dowry had been made ready, and the marriage was performed by Gilbert Burnet, Bishop of Salisbury, on 20 April 1707 in Ely Chapel, Holborn. In the late autumn, Shaftesbury himself called on the newlyweds to deliver a set of silver plate worth £100.[18] For a family of lawyers, the connection with two lord chancellors was a distinct professional advantage; for the intellectual development of the third James Harris, however, the Shaftesbury connection was to be both seminal and definitive.

James and Elizabeth spent almost all their lives at their home in the Close. He consolidated and increased his inheritance by buying land, including an estate at Great Durnford in 1713 which his son was to use for extensive periods of study and writing, often for 10 days at a time.[19] Around 1724 he made further investments in real estate in the parish of Winterbourne, where his father had bought land before him, and at Hurdcott (west of Wilton) where his mother Joan had also invested. Between 1707 and 1719 he spent a good deal of money remodelling his house. Its west side was rebuilt to incorporate its present 'Queen Anne' façade, a hall with the two-storey staircase, a drawing-room, a dining-room, a bedchamber, and a magnificent library (which

---

[18] Ibid. 263.

[19] Records of Harris's property dealings are deposited in the Wiltshire Record Office, Trowbridge. See e.g. WRO 8M51 coffer 52, 14 Apr. 1702, a contract between Harris and Richard King of Dibden, Southampton. An attested copy of 'Hermes' Harris's release and marriage settlement (WRO 212B/5950) lists lands, tenements, and rents acquired by his father around 1724, by his grandfather before him at Winterbourne Dauntsey, Winterbourne Earls, Laverstock, Winterbourne Ford, and Hurdcott. For the younger Harris's property dealings between 1742 and 1754, see WRO 212B/203, B st J 26, 212B/191 (the last relating to Dr Arthur Collier, 8 Dec. 1742).

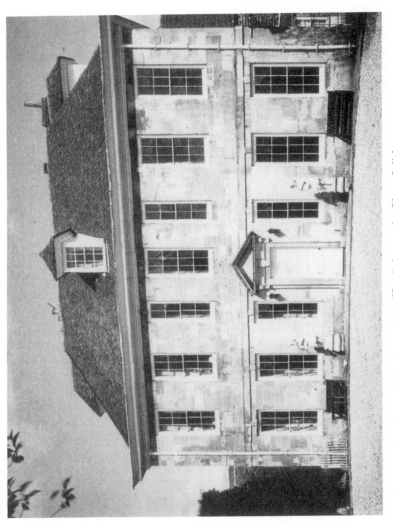

2. West front of the Harris home, the Close, Salisbury.

his son was to remodel in the latest Gothic style) on the first floor. This is the first allusion to the growing collection of books which by 1780 was to be described in *The Gentleman's Magazine* as one of the best private collections in Europe.[20] Neither the architect nor the craftsmen who executed the designs are known for certain, but it is possible that the Fort family of Salisbury masons carried out the whole project. Alexander Fort had been master mason in Sir Christopher Wren's Office of Works, and his younger brother, perhaps working from ready-made designs, carried on the business from their Salisbury office.[21] Whoever it was who completed the work, the credit for the transformation must go to the second James Harris. The splendid interior decoration, especially the fine plasterwork, well beyond the skills of a jobbing provincial plasterer, was completed in the 1760s, under the supervision of the third James Harris.

Harris's somewhat depleted fortune was augmented by some shrewd and timely speculation. Before the South Sea Bubble finally burst in 1720 he had sold his stock for enormous profits. On one occasion he recouped £2,250 on an initial investment of £500 at 450 per cent, and on another £4,400 on £1,000 at 440 per cent. Such a solid citizen may have been presented to the king on his 3-day visit to the Bishop of Salisbury on 29 August 1722, we might surmise. What is certain is that he took the oath of allegiance to King George II on 17 December 1723, in the company of William Wyndham of Dinton, Messrs Hersent, Graham (curate of Bishopston), Pitts of Burcombe, Eghill, and the Revd Arthur Collier (1680–1732), the rector of Langford Magna, nephew of the transported brothers, and brother-in-law of Richard Hele, prebendary of the cathedral and master of the Grammar School.[22] In general, the Collier family was to bring the Harris family more pain than pleasure, as will become clear later, but at this stage the rector of Langford Magna and two other contemporary philosophers in Salisbury provide a

---

[20] *The Gentleman's Magazine*, 51 (1781), 24, which makes particular mention of 'the Greek MSS'. [21] Jackson-Stops, *Malmesbury House*, p. 3.
[22] R. Benson, *Memoirs of the Life and Writings of the Rev. Arthur Collier . . . with Some Account of his Family* (1837), 215. A copy of Collier's *Clavis Universalis* was item 692 in the Hurn Court library sale of the Harris library in 1950.

crucial link between the interior, fragmentary, and domestic narrative of the Harris family, and the public, coherent, and national context of which Salisbury itself was an important part.

II

Life in a provincial city in the eighteenth century may have been more interrelated and familial than life in the metropolis. Salisbury experienced no significant growth in population during Harris's lifetime. Excluding the Close, its number of inhabitants remained at just under 7,000 from 1695 to 1775, its flourishing cultural life sustained by a relatively small and stable community. We might assume that most of Salisbury's residents, ecclesiastical and secular, were on nodding terms with many of their fellows; but we need also to remember that a small town in eighteenth-century England was probably more rather than less 'classed by subordination' (in Johnson's phrase) and that the large clerical community of Salisbury in particular would have *intensified* national sectarianism rather more than the vastness of London. The dangers of social presumption were very real, and subject to minute discrimina- tions which we easily overlook today. As a Low-Church Whig, Bishop Burnet's frequent battles with the intransigent, independent-minded, and Tory Salisbury corporation, not to mention those with the clerical and High Church members of his own diocese, indicate that boundaries were no less rigid for being provincial. For example, Burnet told the subdean of Salisbury, Thomas Naish, in 1696, that 'he loved him as he did his eyes'; but in the next few years this favoured protégé found preferment blocked, his rectory of St Edmund's taken away, and his brother dragged through the court and prosecuted for *scandalum magnatum* (i.e. for accusing Burnet of telling lies). In 1708 Naish concluded that Burnet had striven for the previous three years 'to crush me'. Naish had alienated Burnet by voting the wrong way in the election of 1701, supporting the Tory instead of the Whig candidates. It was an irreparable lapse of a subordinate's duty, not an exercise of his 'democratic' rights, and Naish's career was abruptly halted.

Thomas Naish (1699–1753) kept a diary for thirty years, and its survival enables us to reconstruct some of the links between the key families surrounding the Harrises at this time and during subsequent years.[23] Back in 1679, and for an unspecified period after that, Thomas Naish's father had been employed by the widowed Joan Harris to help manage the estate of the 5-year-old James Harris at Potterne and Dibden. She called him 'cousin' Naish, suggesting some kind of blood relationship. In 1691 the Tory rector John Norris arrived to take up residence at Bemerton, and two years later the younger Thomas Naish took up residence with him for half a year, recording that 'his conversation was very advantageous to me'. Just over a year later Naish became the godfather to Norris's second son. For some time before February 1696 the peripatetic Naish also lived with Colonel John Wyndham, the eldest son of Sir Wadham Wyndham, the judge who lived in the Salisbury house known as St Edmund's College, and the second James Harris's uncle. From there Naish moved to John Fielding's house in the Close, Fielding being a canon residentiary of the cathedral until his death on 31 January 1698. This was Henry Fielding's paternal grandfather. In 1700 it was Norris who 'overlooked . . . and corrected' Naish's sermon to the Society of Lovers of Musick in Salisbury before publication the following year. Naish's father, who shared his son's musical skills and also his High Church and Tory views, was described by Burnet in 1702 as 'the perfect incendiary of the place where he lives'. Having already alienated Burnet by insisting on exercising his own political judgement in 1701, the younger Naish again fell foul of his bishop in July 1706. Burnet had prepared an address of congratulation to the queen on the recent British victory at Ramillies (23 May 1706), but many of his subordinate clergy took objection to one of its clauses and refused to sign it. Thomas Naish thereupon drew up an alternative address after petitioning the Grand Jury for permission, personally travelled the diocese to obtain signatories, and then carried it to London himself, where he presented it to the Earl of Pembroke. Amongst many other names, there are those of John Norris and Arthur Collier, the latter at this very moment at work on his critique of Norris's idealist philosophy. Naish also continued the

---

[23] Information on Thomas Naish is here condensed from D. Slatter (ed.), *The Diary of Thomas Naish* (Devizes, 1965), vol. 20 of *WANHS* (1964).

family tradition of working for the Harrises. In January 1710 he notes in his diary: 'I hope I have got a tenant for Mr. Harris for his estate at Crowthorne in my parish of Corton.'

In the often vague, frequently subterranean, and always complex process known as the history of ideas, there are sometimes episodes of startling clarity, when connections are real and links concrete. The English tradition of neo-Platonic philosophy, revived by the Cambridge Platonists Henry More and Ralph Cudworth and continuing through Coleridge, Carlyle, and Bradley, has its early eighteenth-century phase in Salisbury. If John Norris was the English Malebranche, Arthur Collier, it has been said, was the English Spinoza. That both were to be largely neglected by subsequent historians of philosophy is not surprising, since both wrote against the received philosophical currents of the time, and both were in some ways subsumed in the reputations of greater figures (Norris by his mentor Malebranche, and Collier by Berkeley). Nevertheless, during the childhood and early manhood of 'Hermes' Harris the intellectual atmosphere of Salisbury was thick with Idealism. When James Harris took the oath of allegiance with Arthur Collier in 1723, Collier's *magnum opus* was already 10 years old, and he was only one of three speculative thinkers in Salisbury whose concerns were decidedly immaterial and other-worldly.

Salisbury was not London, but with Gilbert Burnet as its bishop from 1689 to 1715 neither was it an intellectual backwater. Preferment could only be obtained by obedience, and Burnet was an unusually interventionist bishop, a stickler for his prerogative powers, easily offended, and on occasions unforgiving to those who crossed him. His own Cambridge Platonism, his toleration of Protestant dissent, and his interests in philosophy, algebra, and chemistry, provided a context in which the moderate rationalist theologians of the diocese could contribute to the rising tide of deism which was to overflow in Matthew Tindal's *Christianity as old as the Creation* (1730). In the careers of John Norris (1657–1712), Arthur Collier (1680–1732), and Thomas Chubb (1679–1747)—two High Churchmen and a deist respectively—there is the measure of Salisbury's speculative freedom. Only the first lacks a demonstrable connection with the Harris family, and the Collier family was to remain close to the Harris family for two generations.

In 1691, and with John Locke's recommendation, Norris was presented to the benefice of Bemerton by the Earl of Pembroke, the dedicatee of Locke's *Essay Concerning Human Understanding* (1690). The rectory at Bemerton had been built and occupied between 1630 and 1633 by George Herbert, the younger brother of Edward, Lord Herbert of Cherbury the author of *De Veritate* (1624), a work commonly described as the earliest purely metaphysical work by an Englishman, and the foundation-stone of deism.[24] Norris himself is known as the sole English disciple of Malebranche, whose *La Recherche de la vérité* (1674–5), he called 'one of the best Books that is in the World'.[25] Next to his admiration for Platonic philosophy came Descartes and the Cambridge Platonist Henry More, whose importance he described in one of his many poems: 'Adam himself came short of thee,/He tasted of the Fruit, thou bear'st away the Tree.'[26] Norris's philosophical position might be described as a God-centred idealism in reaction against Hobbes's materialism. In *Reflections upon the Conduct of Human Life* (1690), for example, he described the philosopher of Malmesbury's definition of Reason as 'nothing else but . . . a well-order'd Train of Words', an unprecedented and gross 'piece of Idolatry', a plain proof of the 'great degeneracy of Mankind' (*Treatises*, p. 185). His most important work, a lengthy elaboration of ideas found throughout his poetry and prose, was the *Essay towards the Theory of the Ideal or Intelligible World*, published in two parts in 1701 and 1704. Although begun in Latin when he was still a Fellow of All Souls Oxford, the book was written anew and completed in English in the 'Solitude and Retirement' of Newton St Loe, Somerset, which he described to Lady Damaris Masham (Ralph Cudworth's

---

[24] M. H. Carré (trans.), Edward, Lord Herbert of Cherbury, *De Veritate* (Bristol, 1937), 5. *De Veritate* 1st pub, in Latin in Paris (1624), in London (1625).

[25] *Spiritual Counsel: or, The Father's Advice to his Children* (1694), in J. Norris's *Treatises upon Several Subjects* (1698), 501. All subsequent references to Norris's work are from this collection and are cited in the text as *Treatises*.

[26] *A Collection of Miscellanies* (2nd edn., 1692), 90, entitled 'To Dr. More'. Subsequent textual references are given in the text as *Miscellanies*. For a detailed discussion of Norris and his thought, see R. Acworth, *The Philosophy of John Norris of Bemerton (1657–1712)* (Hildesheim and New York, 1979).

daughter) as 'a little Corner of the World, where I must be more Company to my self than I have been ever yet' (*Treatises*, p. 259). Briefly, Norris's *Essay* argues that more than enough had been heard from the natural scientists about the material world and about what Norris elsewhere termed those 'unnecessary Curiosities . . . Quadrants, and Telescopes, Furnaces, Syphons, and Air-Pumps' (*Treatises*, p. 238), and that the time had come for a re-examination of 'the Ideal World which is within us', that '*Terra Incognita, a mere intellectual America*', which Malebranche ('the great *Gallileo* of the Intellectual World') had partly revealed.[27] Drawing mostly from scholastic sources, Norris argued that the Fall had separated the Soul's power of abstract perception from the Reason's sensory role, and that 'Sin . . . fortifies our Union with the *Sensible*, and weakens that Union which we have with the *Intelligible* World, which estranges us from the Divine *Light*, and indisposes us from being willing to have any Communion or Fellowship with it' (*Essay*, p. 5). His theory of knowledge states that man is in immediate contact with divine Ideas, so that Truth, Justice, and Beauty are 'immediately present' to the mind of everyone. In recuperating the world of Ideas, however, Norris left some of his readers (Lady Masham included) with the feeling that the world in which we actually live is an unlovable place. God was entitled to love, fellow human beings only to a disinterested benevolence, the latter being 'Prints and Impressions . . . the Shadows' of higher affection which is 'necessary, permanent and immutable' (*Essay*, p. 7). It is easy to misunderstand Norris, but the implied message that ordinary social affection is a distraction and an obstacle to the higher love of God seems inescapable.

More narrowly, Norris's attacks on the Quakers in *The Charge of Schism continued* (1691) and *Two Treatises concerning Divine Light* (1698) had, it seems, put an end to his prospects for promotion under Burnet's episcopal rule. In fact, Norris had attacked the Quakers not for their doctrine but for their presumption. He entirely shared their belief in an 'inner light': this was indeed the corner-stone of his philosophical position;

---

[27] *An Essay Towards the Theory of the Ideal or Intelligible World*, part 1 (1701), 4, 3. Subsequent textual references are to *Essay*.

but his argument stated that this was a *natural* as well as a spiritual means of perception, and that it was the property of *all* men, the 'Natural and Ordinary way of Understanding':

Man cannot be his own Light, or a Light to himself in the acquirement of Knowledge, and therefore . . . there must of necessity be some other Principle of Light in him distinct from his own Rational Nature . . . the *Divine Idea*.[28]

To argue that 'Divine ideas are the immediate objects of our thought in the perception of things' was one thing: to attack the post-revolutionary policy of toleration under the very nose of its chief Whig architect was another, and there is no doubt that Norris's opposition to the Toleration Act, as well as his view of Protestant nonconformists as schismatics (in *The Charge of Schism*) ensured a long residence at Bemerton. In 1707 Norris was informed that Burnet was 'absolutely resolved I shall never have anything here'.[29] He was free to admire the magnificent cathedral, but only from the outside.

Norris's High-Church and Tory sympathies met Burnet's disapproval, but his writings were discussed by Leibniz in his letters, recommended for family reading in Richardson's *Clarissa* (1747–8), and his *Reflections on the Conduct of Human Life* (1690) was described by John Wesley as 'that masterpiece of reason and religion . . . every paragraph of which must stand unshaken (with or without the Bible) till we are no longer mortal'.[30] Norris's single greatest effect, however, was immediate and local.

Arthur Collier, the rector of Langford Magna until financial and legal difficulties forced him to take lodgings in Salisbury, took ten years to complete his alarmingly subtitled *Clavis Universalis: Or, A New Inquiry after Truth. Being a Demonstration of the Non-Existence, or Impossibility of an External World* (1713).

---

[28] The second of *Two Treatises Concerning Divine Light*, in *Treatises*, p. 434.

[29] Norris to Dr Charlett, Master of University College, Oxford, 9 Apr. 1707; cited in Acworth, *Philosophy of John Norris*, pp. 313–14.

[30] Wesley to Samuel Farley, 14 Mar. 1756; cited in Acworth, *Philosophy of John Norris*, p. 301. Samuel Johnson's copy of Norris's *Miscellanies* (3rd edn., 1699) carefully marked passages for subsequent use in the *Dictionary*: see J. D. Fleeman, *The Sale Catalogue of Samuel Johnson's Library* (University of Victoria, 1975), 111. See also J. Harrison and P. Laslett, *The Library of John Locke* (2nd edn., Oxford, 1971), under Norris.

This book appeared four years after Berkeley's better-known but identical argument, *An Essay towards a New Theory of Vision*, and in the same year as Berkeley's popularized version, *Three Dialogues between Hylas and Philonus*. Collier and Samuel Clarke have been described as the only 'Absolute Idealists' in the history of English philosophy.[31] Collier's Idealism aimed to disprove completely the existence of matter as defined by Descartes, Malebranche, and most particularly by his highly esteemed but nevertheless misguided neighbour John Norris, whose words Collier cites at respectful length. Collier argues that 'all matter which exists, exists in, or dependently on, mind . . . there is no such thing as an external world . . . no visible object is, or can be, external'.[32] Like Norris, who provides Collier's starting-point and focus, Collier depends heavily on scholastic commentaries on Greek philosophy, and also like Norris he draws upon Descartes and Malebranche. Unlike Norris, he raises, without developing them, some theoretical consequences of his Idealism for language itself, specifically the relationship, if any, between words and things. Collier dodges the implications of his own theory by saying that 'we are at liberty, and also in some measure, obliged to use the common language of the world, notwithstanding that it proceeds almost wholly on the supposition of an external world: for, first, language is a creature of God, and therefore good, viz. for use, notwithstanding this *essential* vanity which belongs to it'.[33] When the third James Harris turned to such issues half a century later in *Hermes*, he wisely avoided mingling linguistic theory and theology, yet there was a clear stimulus for Harris's own brand of anti-materialism in Collier's theory that 'no matter is altogether external, but necessarily exists in some mind or other'. In Harris's later formulation, this was to become a dictum: *Nil est in sensu quod*

---

[31] Sir William Hamilton, 'Idealism; With Reference to the Scheme of Arthur Collier', *Discussions on Philosophy and Literature* (New York, 1868), 194. Ignored in England, Collier's *Clavis Universalis* was trans. into German by Eschenbach and not reprinted in English until Samuel Parr's *Metaphysical Tracts of the Eighteenth Century* (1837), which also includes *True Philosophy* and an abstract from Collier's *Treatise on the Logos.*

[32] *Clavis Universalis*, ed. E. Bowman (Chicago and London, 1909), 12.

[33] Ibid. 121. Unlike Norris, Collier was permitted to preach in Salisbury cathedral. He published his sermons of 29 May 1713 and 8 July 1716. The Salisbury printer Charles Hooten published his *Specimen of True Philosophy* in 1730.

non prius fuit in intellectu. His friendship with Collier's son, also Arthur, was to bring difficulties of an entirely materialistic kind, however.

The third member of this group of Salisbury metaphysicians was Thomas Chubb, as sociable as Norris was solitary, as tolerant as Norris was dogmatic, and as homely as Norris was rigidly intellectualist. Chubb caught Burnet's approving eye. A native of East Harnham, he spent most of his life in Salisbury. He was self-educated, a tallow-chandler's assistant before his writing attracted the financial support of Burnet's friend Sir Joseph Jekyll, Whig Master of the Rolls, and of William Cheselden, the brilliant surgeon and friend of Pope. Chubb's popular *Collection of Tracts* (1730) was dedicated to Burnet and gathered together his writings from 1715 onwards. They are dotted with references to Salisbury life and gained a measure of metropolitan fame. While at work on his *Essay on Man* (composed 1730–2; published 1733–4) Pope read Chubb's tracts and on 23 October 1730 wrote to John Gay to ask whether he had seen or talked to this 'wonderful phenomenon of Wiltshire'.[34] Though not wholly in agreement with Chubb's deist doctrines, Pope was nevertheless attracted to Chubb's rationalist tolerance and his optimistic arguments for social affection. Both were autodidacts, and both concluded their respective essays with benevolist arguments.

There is no mention of Shaftesbury in Chubb's tracts, but this did not prevent Johnson from linking the two of them (to Mrs Thrale) as 'wicked writers' whom it was dangerous to read. A passage such as the following could spell tolerance and moderation to a Burnet, a Shaftesbury, or a Pope, and anarchy to a Norris or a Johnson:

as religion is purely *personal*, and every man must be answerable for himself to God; so every man must, in reason, have a *right* to judge for himself, in all matters pertaining thereto; and, consequently, it must be *just* and *reasonable* in our governours, to *indulge* their subjects in the enjoyment of that right.[35]

[34] *The Correspondence of Alexander Pope*, ed. G. Sherburn, 5 vols. (Oxford, 1956), iii. 143: 'I have read thro' his whole volume with admiration of the writer: tho' not always with approbation of the doctrine.'

[35] T. Chubb, *A Collection of Tracts* (1730), treatise XXXI, p. 422. For Salisbury references, see pp. 47, 59, 66 ff. In an unpublished MS entitled 'A Short Map of Human Knowledge' prepared for an unnamed correspondent, Harris recommended

Chubb presided over a club in Salisbury convened by himself
in order to discuss his theories, and among those who are said
to have read over his manuscripts before publication ('but
never to have corrected them') were Samuel Clarke, Bishop
Hoadly, Dr John Hoadly, Archdeacon Rolleston, and James
Harris. If this were the young Harris, rather than the father
(for whom there is no surviving evidence of theological
interests), then this is the first sign of Harris's interest in either
deist thought or the business of writing for publication.
Helping others into print was to be a hallmark of James
Harris's later career: by contrast, Arthur Collier made a
collection of Chubb's business letters specifically to expose his
lack of learning.[36]

The three figures briefly discussed above provide, in
retrospect, more than a general background to the future
interests of the third James Harris. By comparison with the
work of the third Earl of Shaftesbury, whose influence on
'Hermes' Harris was to be pervasive, overt, and continuous,
they are indicative rather than instrumental influences. They
are, as it were, an alternative background which might have
been invoked if the third earl had never existed. Even so, they
indicate a Salisbury context of considerable intellectual depth,
and collectively establish the philosophical credentials of
Salisbury at a formative period in Harris's career. There is no
evidence to suggest that the second James Harris cultivated
Salisbury's literary men, but for the third James Harris the
existence of a lively, scholastic, Platonic, tolerant, and anti-
materialistic context of thought could have been nothing but a
stimulus, even though he was to avoid theological speculation

Chubb's *The Previous Question with regard to Religion* (1725) in the category of moral
philosophy (along with Aristotle, Plato, Xenophon, Tully, Pufendorf, Grotius,
Shaftesbury, Wollaston, and Harris's own 'Essay on Happiness'): Harris Papers, vol.
11, item 17.

[36] *The County Magazine* (Oct. 1786), 149, citing the anonymous *Memoirs of the Life
and Writings of Mr Thomas Chubb* (by 'Anti-Chubbius', i.e. Joseph Horler) (1747),
20, states that Collier 'made a large collection of his . . . business letters to show how
rude and ungrammatical they were . . . exhibited upon proper Occasions by Mr.
Collier to his friends'. For an epitaph on Chubb ('Born to be gen'rous, kind and
good'), see *The British Magazine* (Mar. 1747), 129; and for a sardonic memorial ('Woe
to the world, Tom Chubb cou'd read and write!'), see *The Gentleman's Magazine*, 17
(Mar. 1747), 148, and 18 (Apr. 1747), 193. Chubb's deist works are discussed at
length in J. Leland's *A View of the Principal Deistical Writers* (1754); see the 5th edn., 2
vols. (1798), i. 214–83. See also Fielding's *Covent-Garden Journal*, 13 (1752).

like the plague. The younger Arthur Collier was to remain Harris's close friend throughout his life, and Chubb's ideas, if not Thomas Chubb specifically, were to reappear in the figure of Square in his intimate friend's greatest novel, *Tom Jones* (1749).[37] But more important for Harris's future interests was the interest in all three writers in what was to be called the philosophy of mind. In his third book, *Philosophical Arrangements* (1775), Harris was to cover the same broad territory as Norris and Collier, but without their theological motives. When Harris speaks of that 'perceptive power, unmixed and pure intelligence' which is the 'part of our animating Form, that we must look [to] for the Immortal and Divine; 'tis this indeed is all . . . that a rational Man would wish to preserve, when he would be thankful to find his Passions and his Appetites extinct', he speaks the language of Norris and Collier. Unfortunately for the biographer bent on proving local influence, Harris was intent on rescuing the *classical* tradition, not on praising his contemporaries: thus Aristotle and Xenophon, in this case, are given as the sources for Harris's remark. Whatever Harris knew of Norris, Collier, and Chubb is subsumed in an infinitely more distinguished tradition. If Harris looked into Norris's work, he was more likely to see Cicero and Aristotle, just as looking at Pope's poetry revealed its Latin original in Horace. But if there is one message that Norris and Collier could have transmitted to Harris, of fundamental importance to each of them, it was, in Norris's words, that 'tho' the Natural World be the object of *Sense*, yet the Ideal World is the proper object of *Knowledge*, as well as Intelligible Vision or Perception'.[38] Although *this* idea could be traced all the way back through Christian theology, Greek philosophy, and eventually to the shadowy intellectual world of Hermes Trismegistus and Gnostic theosophy, its provenance in Salisbury in 1701 was a peculiarly appropriate beginning to the century.

---

[37] See M. Battestin's commentary in his edn. of Fielding's *Tom Jones*, 2 vols. (Oxford, 1974), i. 123–8.

[38] *Essay Towards the Theory of the Ideal or Intelligible World*, p. 61. Norris and Harris both wrote an essay on Happiness (the former in 1683, repr. in the *Miscellanies* of 1687, 1692, 1699, 1706, 1710, and 1730), and Norris also trans. Xenophon's *Cyropaedia*, anticipating M. Ashley's translation of 1728, dedicated to Harris's mother (see sect. IV pp. 37–40 below).

III

Inside the walls of the Harris home, meanwhile, a lively and varied literary and musical culture flourished. The second James Harris was an accomplished musician, a flautist who both composed and performed his own songs, and who wrote competent verse for the family's amusement. A regular exercise involved Latin poems composed by James Harris senior being translated into English by each of his sons. Twenty poems by Harris senior, three by his son Thomas, and seven by James Harris junior survive among the family manuscripts. The following caustic rebuttal to a poem by Lady Elizabeth Harris's uncle, the ninth Duke of Rutland, in praise of Caesar (both written in 1715), indicates Harris senior's scepticism of heroic myths as well as his Stuart loyalties:

> Caesar wd. be Glad, and Bless his Stars,
> If, after many hundred years,
> The Poets would be pleased at last
> To lett his Bones at Quiett Rest.
>
> What though he made his Country Bleed
> His Lust of Tyranny to Feed?
> What though he made a Rape on Rome
> Triumphant then, & in her Bloom?
> Must Rhiming Sotts for ever Be
> Reviving Tales of Cruelty,
> This wretched Peice of History?
>
> And yet, poor Fools, They mean him well
> They dress him up, & make him swell
> In every Epithet that's Great;
> To make his Character compleat,
> And fitt for Man to Imitate.
>
> Vain Man, whose Pride delights to tread
> Upon his Fellow Creature's Head:
> Whilst Those Alone can justly claime
> A Title to ye Hero's Name
> Who build on Liberty their fame.
> Of these But few as yet are Reckon'd
> William's ye First, & Anne ye Second.[39]

[39] Harris Papers, vol. 2, part 1, fo. 1, 'Family Pieces—Poetry and Music'. This is the source for the MS poems cited below.

Father and sons combined their talents in poems on the
Aurora Borealis of 5 January 1726 (regarded by Harris senior
as a celestial honour paid to George II's arrival in England),
and each contributed to a series of Aenigmas (the younger
James's being a riddle on a piano). Thomas turns an elegant
Latin poem on the death of Charles I, and competes with his
father in celebrating Isaac Newton's genius and memory
(1730 and 1731 respectively). The younger James is represented by a translation from the seventeenth-century French of
Jean-François Sarasin's satirical lines on Adam and Eve, a
brief translation from *Aeneid*, Bk. IV, a few lines in the
'sublime' genre of pictorial landscape, and seven lines entitled
'Socrates and Xantippe'. The father's poetry dominates both
in range and quality, however, including an epilogue to *The
Beggar's Opera* (28 January 1731), making the most of its anti-
conjugal ending, a verse paraphrase of Psalm 139, a translation of Jacopo Sannazzaro's Neapolitan Latin epigram on the
city of Venice, four drinking songs (two of them scored for
voice and flute), satirical poems on local personalities, and
'On a Spring Sun-Shine Day' (1729), which shows the
unsurprising influence of Pope:

> Let there be Light, sd. God; & it was so:
> Nor need ye great Creator more to doe;
> For Light is Life. Then came ye firstborn Spring,
> And Nature's Gen'rall Voice began to laugh and sing.
>   The Must had now her Birth too; Ev'ry Tongue
> Gave forth it's Sound, & that Sound was a song.
>   God saw his Works were good; then bid ye Sun
> Stand in ye Center, while ye Earth roll'd on.
> And Thou, said he, Thus plac'd wth. Guardian Eye,
> Watch o'er ye Whole, preserve its Harmony,
> And wth thy geniall Heat refresh it least it dye.
>   Hence 'tis, when drooping Nature seems as dead,
> His Quickning Spirit makes her raise her Head.
> Hence 'tis ye sprightly Lark salutes ye Day,
> When a clear Sky ye morning Beams display.
> And hence it is that Man grown old and dull
> When now his cup of Time is almost full,
> If haply, as ye vernall Sun draws nigh,
> His Thoughts revive to youthfull Poetry
> He acts thro' thee, O Sun; of such a one am I.

The most intriguing poem in the collection belongs to a minor
poetic industry in Salisbury's late 1720s: praise-poems to the
Cradock sisters. Since this particular poem lacks a designated
authorship, it is not possible to identify with any certainty
either its date or its author. Even so, some time between 1728
and 1730 (the dates of surrounding poems in 'Family Pieces—
Poetry and Musick') one of the Harris versifiers was evidently
casting a lascivious eye upon Catherine ('Kitty') Cradock, the
most sprightly of the three famous Salisbury beauties and
sister of Charlotte Cradock, whom Henry Fielding was
courting around 1730 and whom he was to marry in
November 1734. Fielding's own complimentary poem about a
beauty competition between the Cradock sisters ('The Queen
of Beauty', *c.*1729–30, in *Miscellanies*, 'writ when the Author
was very young'), has Jove saying to Venus:

> And can you, Daughter, doubt to whom
> (He cry'd) belongs the happy Doom,
> While C——cks yet make bless'd the Earth,
> C——cks, whom long before their Birth,
> I, by your own Petition mov'd,
> Decreed to be by all belov'd.
> C——cks, to whose celestial Dower;
> I gave all Beauties in my Power;
> To form whose lovely Minds and Faces,
> I stript half Heaven of its Graces.
> Oh let them bear an equal Sway,
> So shall Mankind well-pleas'd obey.[40]

In marked contrast to Fielding's gracefully indecisive court-
liness, the Harris poem to Catherine Cradock seems at times
almost a libidinous dissection, the fantasy of an elderly man
(Harris senior would be about 55), or of a very young one
(Thomas was 18 and James 20 in 1729). The first four stanzas
of the eight-stanza poem are as follows:

> Fine Belles & nice Beauxs
> Owe All to their Cloaths;
> But 'tis Pity to us makes her be well clad:
> Were She naked, I fear,
> No Mortal could bear
> A Sight so divine as Dear Kitty Crad—

---

[40] H. K. Miller (ed.), *Miscellanies by Henry Fielding, Esq.*, i (Oxford 1972), 80:
[Untitled] 'The Queen of Beauty'.

A very good Skin
For a Cat, or a Queen
Is enough; where no more can be had:
But the Critick must goe,
Much deeper than so,
That would come at the Best of my Dr Kitty Crad——

From her Eyes, Mouth, & Nose
Quite down to her Toes,
There's a Million of Beauties, I gad,
Wch lye here and there,
I won't tell you where;
For some Things are Secrets in my Dr Kitty Crad——

Three Goddesses chose
Their Charms to expose
To the Judgment of Paris, a Lusty Young Lad;
But were I to be
A Judge of those three
I'd lay them all by for my Dear Kitty Crad——

and the last stanza concludes:

Man's Life is a Race,
Which moves on apace;
Some gallop, some trot, few ride on a Pad.
Ah! how blessed were my Lott,
If, Instead of a Trott,
I could amble on gently with Dear Kitty Crad——.

It can only be speculation to suggest that Fielding and one of the Harris brothers were competing for each of the Cradock sisters at the same time. The fact that Fielding eventually chose Charlotte around 1730 need not imply that he did so out of deference because Catherine was bespoke by Thomas or James Harris. In fact, Kitty Cradock died many miles from Salisbury in the village of Codicote, Hertfordshire, a spinster, and under somewhat obscure circumstances. She was buried there on 24 April 1735.

IV

The second James Harris stimulated his son's interest in music and poetry, but it was Elizabeth Harris who transmitted to him her family's interest in philosophy. Her father, the second Earl of Shaftesbury, had commissioned four 'elemental' portraits of his daughters, and Elizabeth was depicted, appropriately, as 'Air'.[41] She read widely in English and French and habitually annotated the many books she read. Malmesbury's memoir remarks that there was 'no trace of a highly finished education in any of the four sisters except L[ad]y Elizabeth'. Largely self-educated, and in later life rather than in her childhood and early adulthood, orphaned at the age of 18, and a trial to her older brother, she felt a natural affinity with her younger brother Maurice Ashley [Cooper: 1675–1726], the third son.

Ashley served as MP for Weymouth and Melcombe Regis from 1695 to 1698, and in the autumn of 1701, as we have seen, he was elected to represent Wiltshire, enjoying the strong support of Burnet and, according to Thomas Naish, 'under great mistrusts of favouring dissenters and making alterations in the Church'.[42] Ashley served for Wiltshire until 1702, and again for his former constituency from 1705 to 1713, whereupon he seems to have retired to a life of scholarship. Certainly, by 1715, he was well into his translation of Xenophon's *Cyropaedia* and had completed half of it by 1717. He wrote to his sister Elizabeth about its slow progress, complaining of the tedium of making a fair copy: he was 'so cloyed with his own Part of the Performance, that he would sett his Papers on fire'.[43] Elizabeth Harris was consulted for her approval of the translation, and when it finally appeared, posthumously, in 1728, it was dedicated to her. Its preface contains a unique insight into the 'Disposition of Mind' of Elizabeth, undoubtedly the earliest and most formative influence on the young James, her son. As Malmesbury was to point out, this was a woman who had clearly 'accustomed her

[41] J. Hutchins, *The History and Antiquities of the County of Dorset*, 4 vols. (3rd edn., 1861–74), iii. 599. Appropriate quotations from the *Aeneid* accompanied each portrait.

[42] *The Diary of Thomas Naish*, p. 46. For Ashley's letter soliciting support in the election (17 Nov.), see Marquess of Lansdowne, 'Wiltshire Politicians (*c*.1700)', *WANHS* 46: 157 (Dec. 1932), 78–80.

[43] Harris Papers, vol. 2, part 10, fo.7, 'Shaftesbury Papers' (n.d.). The full title of Ashley's trans. is: *Cyropaedia: or, the Institution of Cyrus, By Xenophon. Translated from the Greek. By the late Honourable Maurice Ashley Cooper, Esq; to which is prefixed, A Preface, by Way of Dedication, to the Right Honourable, the Lady Elizabeth Harris*, 2 vols. (Dublin, 1728). All quotations are from this edn. and are given in the text below. Maurice died in Bedford Row in 1726. Perhaps Elizabeth herself saw the book through the press.

mind to serious thinking, and been in the habits of discoursing seriously with him [i.e. Ashley]'.

It was indeed serious thinking, and an anticipation of James Harris's own philosophical position in later years. Ashley's preface is sternly anti-Hobbesian and rigorously anti-Epicurean. Hobbes, Ashley argued, had copied the picture of his own times too well and projected his own troubled mind on to the world at large (Tom Jones's charge against the Man of the Hill). His pessimistic view of man in his natural state of 'War and Enmity' was therefore the specific product of particular and unrepresentative historical circumstances. Equally, Epicurean philosophy had erred in separating morality from theology, thereby reducing the former to a 'System of Manners . . . an Art of settling certain Rules of Behaviour upon a Principle of Interest, Convenience, or Pleasure' (p. 2). Ashley argues neither for a religion established by and serving the political purposes of the State (Sacheverell's 'national Establishment' in his example), nor for a religion so abstract that it is 'unconcern'd in the Administration of the lower World'. His notion of religion is that it is 'a mental Thing . . . absolutely independent, and has nothing to do with the Magistrate . . . a thing of nobler Nature, and its Truths are yet less subject to political Jurisdiction, than Mathematical Truths' (pp. 10–11). The natural enemies of such an idea are the princes of both Church and State, i.e. all those who would control the 'civil Rights of Men'.

Ashley's argument is classic deism, offering a synthesis of metaphysical, social, and ethical arguments based on a benevolist view of human psychology. Since man is 'a mild, gentle, sociable, and compassionate Creature' (p. 9), there is everything to be gained by free thinking:

As Passion is a domestick Oppressor of Liberty of Mind, so there are a Sort of foreign Oppressors of it. These are the *Hobbists* and the Favourers of Ecclesiastical Tyranny. No real Religion in the World, say these Men; no Rule of Right, or public Good in the State; no Virtue in Man; but all depends upon Tales authorized as Laws imposed by Power and Will. Now true Freedom of Thought here is to assert a Providence, Wisdom and Intelligence in the World; a Rule of Order in Societies of Men upon the Bottom of Public Good; Virtue and Worth in Man; and a Rule of Truth in all Things, which

to discover is Man's Wisdom; and to follow it is his Virtue,
Freedom, and Happiness. (p. 24)

In addressing his only book to Elizabeth Harris, whose 'chaste
and single' mind will surely respond to the 'Plainness and
Simplicity in this piece of *Xenophon*', Ashley assumes both her
understanding and approval of the preface's intellectual
position. Certainly, this is the most striking evidence of the
seminal influence of the *third* earl's philosophy on the Harris
family. When Ashley's book appeared, Elizabeth Harris's
intellectual character was, at the age of 47, no doubt settled.
James was 19, a student at Oxford, and responsive to such
Shaftesburian transmissions.

Ashley's version of Xenophon's didactic work on the ideal
form of government, and the ideal (masculine) ruler, was thus
addressed to a woman. However notable for her patriotism,
independent-mindedness, and horror of all kinds of tyranny,
Ashley was clearly aware of the reception this unusual gesture
might meet. He thus ends with a sardonic reflection on a
contemporary civic education which was based on perverse
notions of gender-roles:

I cannot but believe that even the States-Man, the Soldier, the
Divine, and the Learned in the Law of the present Age, would
readily excuse the addressing these Matters to a Lady . . . They will
be little concerned that such an Author should recommend the
Sciences and Arts of War and Government, of Justice and Religion,
to the Study of the Gentleman. For by means of Ignorance in these
Things, the Gentleman is render'd incapable of judging whether the
Mercenary in these Professions do their Duty for their Money; the
noblest Arts are thus left to the mercenary alone, and they become
the Guides and Governors of the World.

If this was an unusually philosophical mother for James
Harris, then he was fortunate. He was to breathe such
philosophical air for the whole of his adult life, and in many
ways the Shaftesbury connection was to provide the spur, the
confidence, and the context for his analysis of universal causes
in the areas of aesthetic, linguistic, epistemological, and
historical cultures. His debt was to be acknowledged in
specific ways: his longest and only notable poem, *Concord*, is
'the closest rendering' of the third earl's moral and aesthetic

theories,[44] and his first book, *Three Treatises*, was to be dedicated to the fourth earl.

Less obviously, Ashley's *Cyropaedia* signalled the family's view that classical literature was a proper study, and in the admired Xenophon Ashley had offered a cultural role-model of peculiar seductiveness: 'an ingenious, noble gentleman-like sort, not sedentary, not pedantick, and not servile, as all Learning may justly be called that is not acquir'd to get Money or Maintenance by' (p. 22). Unlike his friends Fielding and Johnson, James Harris was never to write for money, and neither had he need of subscribers to pay for his publications. He already possessed a laced waistcoat, and was born free to indulge in literary speculations. But it was the Shaftesbury connection, of which Harris was justifiably proud, that was to be used against him. Johnson chose to turn an acknowledged intellectual asset into an exercise of slavish and uncritical deference:

Is not young Rose Fuller a foolish fellow said I [Mrs Thrale] to keep from Church because old Rose is an Infidel?
—— Is it not foolisher says Johnson in James Harris to be an Infidel, because Lord Shaftesbury wrote the Characteristics?[45]

V

The second James Harris died in his fifty-seventh year, on 26 August 1731, leaving his widow Elizabeth (then 50 years old, and due to live another twelve years before her death in 1744), his mother Joan, who was to live for another three years, and his three sons. George William was the youngest, then almost 17, Thomas was 20, and the eldest son, the third James Harris, was 22. Their father's will, witnessed by William Burnet and the Salisbury surgeon Edward Goldwyre, had been drawn up on 14 February 1726 and proved on 5 October

[44] C. A. Moore, 'Shaftesbury and the Ethical Poets in England, 1700–1760', *PMLA* 24, (1916), 293. *Concord* appeared in F. Fawkes and W. Woty, *The Poetical Calendar* (1763), xii. 53–9.

[45] *Thraliana: The Diary of Mrs Hester Lynch Thrale (Later Mrs Piozzi) 1776–1809*, ed. K. C. Balderston, 2 vols. (Oxford, 1942), i. 35 (entry dated 28 May 1777).

1731.[46] It appointed two trustees, a kinsman William Harris of the Close, and the second James Harris's nephew Edward Hooper of Hurn Court, who were to oversee the family's affairs until each of the three sons reached their majority. James had already done so, and therefore immediately assumed the principal responsibility for managing the family's fortune and properties. The second James Harris left £5,000 capital stock in the Orphans Fund to be divided between Thomas and George at their majority, the interest meanwhile to provide for their education. A diamond necklace passed to his daughter by his first marriage, Catherine, already its *de facto* owner; and his house, books, and specific sums of money were left to his mother, wife, and daughter. Everything else passed to his son James, who inherited an estate charged with two jointures at the age of 22.

Harris's background lacked none of the constituents which were to make him famous. In broad terms the relationship is one of competence and performance, between untested, traditional convictions and skills on the one hand, and overt skills which were deliberately cultivated and modified by public responsibilities on the other. His daughter was to record in her 'Memoir' that when Harris 'was young he had been inclined to *Whiggism* but as he advanced in Life he felt the necessity of supporting Authority and Government'.[47] Salisbury intellectual life, markedly Platonist in the works of Norris and Collier, sustained rather than stifled metaphysics, and was the natural complement to the more powerful and immediate Shaftesburian model of philosophical and aesthetic theorizing. An ancient and still vital musical tradition in Salisbury regularly turned the amateurs who sustained it into 'professionals', and it was soon to be the turn of the Harris family.

Indeed, the life of James Harris seems, in retrospect, to be predictable from the beginning. There was the tradition of a Salisbury schooling, followed by Wadham; an hereditary

---

[46] PRO Chancery Lane, PROB 11 647, 5 Oct. 1731.

[47] 'Memoir of J. Harris Author of Hermes', by Mrs R[obinson]. Harris's daughter Catherine Gertrude married Hon. Frederick Robinson, the 2nd son of the 1st Lord Grantham, on 11 June 1785. This is an undated and unfinished draft MS, PRO Kew 30/43/1/4. Harris's remark was made to William Bowles.

disposition to the law; an as yet unrealized ambition for a career in national politics; some classical scholarship and music for respectable hobbies; the politics of an independent country Whig, and the religious views of an undisturbing and reasonable Anglicanism—tolerant, with deistical leanings. Apart from its steady upward mobility into the professions, its periodic links with lawyers and intellectual aristocrats, and a tradition of learned women, there was little to indicate that the third James Harris would amount to much more than a worthy member of the cultured landed gentry deploring foreign wars, religious disputes, and high taxation, in about equal proportions.

2

Of Books and Men

THE third James Harris was born on 25 July 1709. His was the only baptism performed in Salisbury cathedral on the following 10 August.[1] He was sent to the Grammar School in the Close 'under a Master at that time in high reputation', Richard Hele. Harris left no record of his schooldays, but doubtless he there acquired the usual competence in Latin (thereby learning whatever he could about English grammar), and Greek, using the famous Eton Latin and Greek grammars which were still in use at the school more than a century later.[2] Harris himself was to describe the syllabus at Winchester and Oxford in the reign of Edward II as grammar in the first followed by Aristotle and the rhetoric of Cicero in the second, 'a plan of education which still exists [and] which is not easy to be mended'.[3] The implications of this immovable school and university syllabus for Harris's subsequent career will be discussed later, but at this point it is useful to notice that, in Sir William Hamilton's judgement (1833), Harris's later proficiency in classical philology and philosophy was the single exception to two centuries of dullness and inertia in English study of logic, and that his expertise in grammar owed nothing whatsoever to school or university textbooks of the time.[4]

I

Whether Richard Hele was the model for Fielding's Thwackum in *Tom Jones*, there is no doubt that the Harrises,

---

[1] Salisbury Cathedral Register, WRO, part 1.

[2] N. Carlile, *Description of Endowed Grammar Schools* (1818), cited by Robertson, *Sarum Close*, pp. 266–7. Only 2 men, Edward Hardwick and Richard Hele, ran the school from 1673 to 1756. Hele married Amy Collier, sister of Arthur and William Collier of Steeple Langford (Robertson, *Sarum Close*, p. 225).

[3] *Philological Inquiries, Works* (1841), 501.

[4] Hamilton, *Discussions on Philosophy and Literature*, p. 713, app. 3, 'Oxford as it Might Be'.

both father and son, experienced his pedagogic tyranny. James Harris entered his son (later the first Earl of Malmesbury) in Hele's school, and then transferred him to Winchester at the age of 11, where he spent the next six years under the benevolent eye of Dr Joseph Warton. The contrast between the two pedagogues was remarkable. Although the young Harris says in his memoir of his father that Hele was 'long known and respected in the West of England as an instructor of youth', a different assessment is given later on in the unpublished manuscript. Hele was 'a dry pedantick scholar, without taste or fancy', who had once thrown a book at the young Harris for mispronunciation, cutting him over the eye.[5] After Harris senior intervened, Hele grovelled repentantly; but when the parent departed, the harsh treatment of the son was resumed with a vengeance. The following comparison of the two headmasters tells us as much as we can find out about the father and son's very different school experiences:

Doctor Warton was on the contrary all good humour and pleasantness, he had a great fund of elegant learning, but without a particle of pedagogism, full of fire and imagination, he made us, as we repeated our lessons in the Classicks to him partake of his enthusiasm; he would forgive us grammatical errors, or incorrect construing, but used to sigh and push at us if we read five lines with insipidity or passed by them unnoticed—He conversed with familiarity out of school, made companions of us, and I, in common with many of my school fellows, conceived a lasting affection and friendship for him which only ended at his death.[6]

Warton was all that Hele was not, communicating to his charges the passionate imagination of a scholar for whom classical literature was not a platform for rote learning but a living imaginative delight. Unlike his son, Harris senior knew Warton only as an adult, so we might infer that he was relieved to escape Hele's academy for Wadham College in his sixteenth year.

---

[5] Malmesbury, 'Memoir', p. 38.
[6] Ibid. In the 3rd earl's edn. of Malmesbury's *Diaries and Correspondence*, 4 vols. (1844), i, pp. x–xii, a letter of 1800 expresses a less encomiastic view of Warton's tutorial influence: 'I left Winchester in September 1762, I had been indulged there too much; Dr Warton erred in the contrary extreme from Mr Hale, I did nearly what I liked, and as boys always wish to be men, I thought myself a man too soon.'

The record of Harris's life at Wadham is scanty, and again we are obliged to fall back on his son's later experience. All we know is that a sum paid in advance as security for Harris's good behaviour ('caution money') was paid in the second half of 1725; that he matriculated eleven days before his seventeenth birthday on 14 July 1726; that he entered as a Fellow Commoner two days later; and that his caution money was returned, presumably on his departure, on 9 August 1729.[7] In general, the status of Fellow Commoner was not difficult to achieve for members of well-to-do families. They paid more for the privilege of learning less, although if they happened to be interested in learning as well as socializing with the Fellows, their status also bestowed upon them the privilege of access to the college library. Harris, like Gibbon at Magdalen, would have leapt at the opportunity, for otherwise the Fellow Commoner's life was not intellectually inspiring: 'the two years of my life I look back to', wrote Harris's son, 'as the most unprofitably spent were those I spent at Merton . . . a gentleman Commoner was under no restraint, and never called upon to attend either lectures, or chapel, or hall . . . My tutor gave himself no concern about his pupils.'[8]

On 10 June 1728 Harris's father (the second James Harris) drew up a family pedigree in support of his son Thomas's subsequently successful election to a Fellowship at Wadham. James, the older brother, was already in his penultimate year at Wadham as a Fellow Commoner, and Thomas's case depended on the privilege of 'Founder's Kin'. The college statutes allowed preference in elections to no more than three Scholars who could demonstrate kinship to Nicholas Wadham and Dorothy, his wife. Three generations of the Harris family had already passed through Wadham, thus on 12 July 1729 Harris received a bill for £9. 9s. 4d. from William Thomas, this being his charge for preparing and passing the pedigree under the common seal. The original draft survives, as well as the final copy kept at Wadham, and each traces the family back through the line of Wyndhams (Joan being Thomas's grandmother) to John Wyndham of Orchard in Somerset, brother of the founder Nicholas Wadham, who died without

[7] Gardiner, *Registers of Wadham College*, ii. 21–2.
[8] Malmesbury, 'Memoir', p. 40.

issue on 22 October 1609. Evidently, the Warden of Wadham had insisted on a properly attested pedigree, and perhaps the reason is hinted in William Thomas's letter to Harris senior (22 May 1729), reassuring him of his son's academic competence, but nevertheless pointing out that Thomas's 'greatest defect is in composition, wch I hope will soon be redressed'.[9] No question was ever raised about the scholarly ability of his older brother, and in the end Thomas Harris was elected a Fellow of Wadham, and then of Merton in 1734, before undertaking legal training at Lincoln's Inn. Unlike his father and brother before him, he stayed in the law, and was called to the Bar, rode the Western Circuit, and was eventually appointed Master in Chancery in 1754.

Thomas and his half-sister Catherine each married into the family of Sir Edward Knatchbull, the fourth baronet. Catherine married Sir Wyndham Knatchbull Wyndham, the fifth baronet, on 23 June 1730, when she was 25 years old. Catherine died in childbirth on 8 January 1741, and on Wyndham's death six years later James Harris became one of the guardians of their three orphaned children. On 11 June 1754, and at the comparatively advanced age of 43, Thomas married Catherine Knatchbull, Sir Edward Knatchbull's daughter. There were no children of this marriage. Thomas was a fine cellist, and the first extant Handel letter in English is to Harris's brother-in-law, Sir Wyndham Knatchbull (27 August 1734), probably a reply to Knatchbull's invitation to Handel to visit his country seat at Mersham le Hatch, near Ashford in Kent. It was Thomas, rather more than George William Harris, who was to keep his older brother in touch with Handel's London activities during the 1740s, and it was Thomas who was to be left £300 in Handel's will. Thomas also witnessed its first codicil and inherited Philip Mercier's magnificent portrait of Handel (executed around 1730). He also became a close colleague of his fellow lawyer, Henry Fielding, as well as a go-between for Fielding and James Harris in several literary projects.[10]

One further and important aspect of James Harris's time at Oxford should be clarified. His son's memoir states that his

[9] Harris Papers, box 18. 3.
[10] O. E. Deutsch, *Handel: A Documentary Biography* (1955), 369, 776.

father had not begun to study 'Aristotle and his commentators
. . . until many years after his retirement from London
[August 1731]. He had imbibed a prejudice, very common at
that time even among scholars, that Aristotle was an obscure
and unprofitable author, whose philosophy had been
deservedly superseded by that of Mr Locke.' Harris would
have studied the common syllabus of classical grammar,
rhetoric, logic, and metaphysics, the last two being Aristotelian.
It would have been literally possible for him not to have
studied Aristotle's actual works, but impossible to have
escaped the study of versions of Aristotle through the standard
university compilations such as Robert Sanderson's *Logicae
Artis Compendium* (Oxford, 1615) and Dean Henry Aldrich's
Artis Logicae Compendium (1691), i.e. simplified epitomes of the
Organon which Hamilton later scathingly described as having
'furnished, for above a century, the little all of logic doled out
. . . by the University of Bradwardine and Scotus'.[11]

It is a pity that more is not known about Harris's time at
Oxford. Later in life he was to complain about the anti-
Aristotelian bias of his pro-Lockian tutors, and his only
recorded comment on Oxford's intellectual provision was
made in a necessarily deferential context where plain speaking
was completely inappropriate. He sent a presentation copy of
his own book on Aristotelian logic to Lord North in 1775,
remarking that Oxford was 'a University where I was so
happy to be educated'.[12] Lord North had been the university's
chancellor for the previous three years, and he was neither the
appropriate recipient nor was this the occasion for proposals
on educational reform. Eventually, Harris left the university
without taking a degree (like his father before him), and in
1742 subscribed 3 guineas to the repair of the east window of
the college chapel.

Whatever the dissatisfactions may have been with the
teaching at Wadham, the college itself was an appropriate and
inevitable seed-bed for Harris's interests. Apart from its

---

[11] Hamilton, *Discussions on Philosophy and Literature*, p. 127.
[12] Harris Papers, vol. 14. part 2, 'Copies of my 3d Volume'. For the most recent
discussions of the Oxford curriculum, see L. S. Sutherland, 'The Curriculum', and J.
Yolton, 'Schoolmen, Logic and Philosophy', in L. S. Sutherland and L. G. Mitchell
(eds.), *The History of the University of Oxford* (Oxford, 1986), v. 469–91, 565–91.

distinctly Whiggish character, the family tradition of sending its sons there, and their 'Founder's Kin' privileges, Wadham had attracted some of the best minds of the Restoration and post-Restoration periods, many of whom had been the key figures in the new sciences. Cromwell's brother-in-law, John Wilkins, appointed warden in 1648 and at the centre of the brilliant 'Invisible' college, had published the single most important work on universal grammar before Harris's own, *An Essay towards a Real Character and a Philosophical Language* (1668). A presentation copy of this book was in Harris's library. A few years before Harris went up to Wadham, Joseph Trapp had been Oxford's first Professor of Poetry (1708–18).[13] The first edition of his *Lectures on Poetry Read in the Schools of Natural Philosophy* came out in Latin in 1711, the second in 1722, and then in Clarke and Bowyer's English translation in 1742, two years before Harris's own *Three Treatises*. Trapp's second and third lectures compared painting and poetry in a manner that certainly anticipated Harris's *Three Treatises*. He argued, for example, that poetry is supreme not only because it satisfies man's 'Passion for Harmony', but also because only poetry can delineate 'the inward Springs and Movements of the Soul, the Actions, Passions, Manners, the distinguishing Tempers and Natures of Men . . . the whole Circle of Learning enters into its Composition'. We cannot be sure that Harris knew Trapp's book, but there is no doubt at all that he came to know each of Trapp's successors in the post of Professor of Poetry. The second incumbent was the elder Thomas Warton (also vicar of nearby Basingstoke); Joseph Spence was the third (1728–38) and Harris was to subscribe to his *Polymetis* in 1747; the post was held by the younger Thomas Warton from 1756, and in the following year Joseph was to be Harris's son's headmaster at Winchester, as well as one of Harris's own close scholarly friends thereafter. There was also Robert Lowth, the future Bishop of London. Lowth was at New College in Harris's last year at Oxford, already a

---

[13] Trapp's *Praelectiones Poeticae: In Schola Naturalis Philosophiae Oxon. Habitae* (1711), in the English trans. of W. Bowyer and W. Clarke (1742), 30, 18, 23. For a review of Trapp's Latin trans. of Milton, see [H. Fielding], *The History of our Own Times*, 1 (Jan. 1741), 18–22. For further information on the Oxford professorship, see B. Hepworth, *Robert Lowth* (Boston, 1978), 47–54. Hepworth seems unaware of Harris's significance, giving him only 2 scant mentions.

published poet, and became Professor of Poetry in 1741. His *Short Introduction to English Grammar* of 1762—perhaps the most influential English grammar in Europe—was to be based on Harris's *Hermes*. The poet Edward Young was a Fellow of All Souls until 1730, and there were, of course, greater illuminati at a distance who had already brought extraordinary distinction to Oxford: Sprat, Boyle, Petty, and John Wallis, the latter second only to Newton in mathematics, Sir Christopher Wren (whose survey of Salisbury cathedral had been carried out in 1668), and Seth Ward (Bishop of Salisbury and founder of the Matron's College in the Close in 1685).

From what we know of Harris's subsequent reputation as one of the leading Hellenists of the period, a great deal of his time at Oxford must have been spent in private study, haunting the Bodleian library, scouring the bookshops in an attempt to build his own library, perhaps even trying out his own ideas in poetry and prose. Among his contemporaries at Wadham (from 31 May 1727 to 1733), there was Floyer Sydenham (1710–87), a scholar who was to devote his life to translating and editing Plato. In 1744 Harris dedicated the third of his *Three Treatises* to Sydenham.[14] Indeed, Harris seems to have regarded Oxford as the closest thing England could offer to the Greek academy: 'Open air, shade, water, and pleasant walks . . . in and about Oxford, may teach my own countrymen . . . and best explain how Horace lived, while a student at Athens.'[15]

After Oxford, Lincoln's Inn and the law detained him for exactly two years, the minimum period required for qualification. Whatever plans he may have made were cut short by his father's death on 26 August 1731. Harris immediately abandoned his chambers and returned directly to Salisbury to

[14] For Sydenham, see K. Raine and G. M. Harper (eds.), *Thomas Taylor the Platonist: Selected Writings* (Princeton, 1969). From Doctor's Commons on 10 June 1758 Arthur Collier informed Harris about the progress of Floyer Sydenham's project to translate Plato's *Dialogues*, to be printed by Richardson. The first 6 had already been translated, and were the *joint* effort of Sydenham and Collier: *Io, Greater Hippias, Symposium, Meno, Minos*, and *Philebus*. Harris was asked for a subscription, and sent 5 gns. a fortnight later. In Jan. Harris had sent a bank draft to Elizabeth Carter, and a list of subscribers' names, in support of her trans. of Epictetus, 1758 (Harris Papers, Letter Book, vol. 31, part 1, pp. 99–100 and 98).

[15] *Philological Inquiries*, p. 462.

take charge of the family. His legal training was to stand him in good stead in his subsequent career both in Salisbury and as George Grenville's aide after 1761, but for the moment, and at the age of 22, he returned to a substantial house, a comfortable fortune, a well-stocked library, and personal independence. He might have settled immediately to a comfortable life of provincial obscurity, but within eight months he was on the move, undertaking a modest version of the Grand Tour.

II

At 4 o'clock in the afternoon of 15 June 1732, accompanied by his friend John Warner and a servant James Gibbs, and armed with a pro forma letter in French inviting whomsoever to dine with them, Harris embarked from Dover. As an extension of, or, as some contemporaries believed of the Grand Tour, a compensation for, time spent unprofitably at the university, Harris was undertaking his only recorded European tour. He was to visit twenty-six towns and cities in Holland, Germany, Belgium, and France. If this was Harris's only European tour, with time and financial means now freely at his own disposal, then it is remarkable for a man soon to be known as the leading classicist of his day that he seems not to have visited Italy as well. Greece, of course, was still part of the Turkish empire and was not yet a common part of the Grand Tour. Two of the earliest exceptions, both members of the recently established Society of Dilettanti, were those two pioneers of classical archaeology, Nicholas Revett and James 'Athenian' Stuart (the latter Harris's illustrator for the revised edition of *Three Treatises* and *Hermes*). These were among the first to visit Greece in 1751, publishing their seminal *Antiquities of Athens* in 1762.[16]

There is just a possibility that Harris may have visited

---

[16] Revett and Stuart's book, and conversations with Stuart himself, who lived in Athens for 3 years, were Harris's primary sources for his knowledge of contemporary Athenian culture. It was Jacob Schnebbelie (1760–92), draughtsman to the Society of Antiquaries, who transcribed the Harris funerary inscriptions and engraved the Harris monument in Salisbury cathedral for publication in *The Gentleman's Magazine*, 62: 2 (1792), 817–18, 62: 1 (1792), 189–90.

Italy. Joseph Spence met a man denominated 'Harris' in Florence some time between late 1732 and early 1733, although there is no extant record in James Harris's papers to corroborate this.[17] Neither is there evidence from this period that he visited Paris, Versailles, or Fontainebleu. Yet when his son wrote to him from Paris in 1768 about the French capital's narrow and dirty streets, 'the singular way of lighting them, the height of their houses', he added that such details were 'all sufficiently known to you already'.[18] The squalor of the French capital had often been remarked by English travellers, and Malmesbury need not be implying that his father had experienced Paris at first hand.

The first month of Harris's roughly circular tour was spent in Holland. He arrived at Helvoetsluys and then travelled on to Rotterdam (where he copied down the funerary inscriptions to Erasmus and Grotius), Delft, The Hague, Leyden (where Fielding had been a student from March 1728 to August 1729), Haarlem ('ye neatest place in Holland'), and then turned south along the Zuider Zee through Medenblik, Enkhuizen, Monnickendam, and halted at Amsterdam. Here he selected the Ratshaus, the Physic Garden, the prison (with its array of torture instruments), its 'Old Church', and the Dutch Admiralty (with its forty ships), for particular scrutiny. For entertainment, and quite apart from a Lutheran church in which 'Whores sings Psalms', he chose Shakespeare, and describes a bizarre performance of *Titus Andronicus* in Amsterdam on 1 August:

a bloody and very dolefull Performance of thirteen Characters, by Water, Fire, Swords and Halter. Only one was surviving at the end of the Play. One was burnt alive in Sight of the Audience, not to mention Heads and Hands brought in innumerable. And yet, I could not perceive that they any way affected the Spectators, but by an occasional Laugh.[19]

---

[17] *Letters from the Grand Tour*, ed. S. Klima (1975), 420. In his edn. of Pope's *Works* (i, p. xxxvi), Joseph Warton commended Spence's *Essay on Pope's Odyssey* (1726–7), and added that his opinion of its merit was shared by 'three persons, from whose judgment there can be no appeal, Dr Akenside, Bishop Lowth, and Mr James Harris'. Harris subscribed to Spence's *Polymetis* (1747), a comparison of Latin literature with its surviving artefacts. [18] *Series of Letters*, i. 162.
[19] Harris Papers, vol. 18. part 4: 'Mr Harris's Travels through Flanders and Holland in 1732'.

In common with most of his compatriots, Harris was struck by the cleanliness of the Dutch towns and the pusillanimity of its criminals ('so fearfull at Execution, that they are forced to be dragged up ye Gallows by men'). His observations are rarely judgemental: the Dutch 'beg little and bow little', their manners are 'rough', and Dutch women of all ages are conspicuous for their 'Neatness'. From Amsterdam he travelled the 21 miles to Utrecht by canal, crossing the Rhine at Nijmegen on 21 August. At Cleves (Kleve) he was struck by the King of Prussia's park, calling it 'Romantic', and then moved south to Wesel, arriving in Düsseldorf on 27 August, where his chief interest lay in the Elector's palace, 'which has the largest and finest Collection of Paintings I ever saw'. He sketched a plan of this famous gallery, noting the number of paintings by Rubens, Van Dyck, Rembrandt, Jourdain, Dürer, Tintoretto, Raphael, Correggio, and Titian: 'Good Pictures come so fast, that our Memorys became quite distracted.' Recrossing the Rhine by ferry, he recorded that Cologne, usually reckoned the ugliest town in Germany at that time, had as many churches as days in the year, and in one of them discovered 'The Angel Gabriel . . . in wax work in a Red-Silk Vestment, Golden Wings and a bright powdered Perruque in the Nuns Chapple.' A post-chaise took him quickly on to Bonn and Aix-la-Chapelle. At the imperial city of Limburg Harris was again alert to its 'most Romantic' situation: 'it stands on the Top of a Rock, with a Precipice on almost all Sides.'

Harris's southernmost detour took him to the most celebrated of the European watering-places, at Spa, much favoured by the English. Here were seventy of his fellow countrymen at its springs, even though the town itself was 'but a village'. Again, he found the countryside around Liège 'very rocky and unequal, but Romantic', and took a keen interest in the barrier town of Namur, where he noted down its detailed fortifications and its garrison of 3–4,000 men (11 September). Two days later a diligence transported his small party to Tournai and Lille, and then north-west along the river Lys to Menen, Courtrai, Ghent (then the largest city in the Low Countries), arriving in the infinitely more sophisticated Brussels on 21 September. Here he jotted down a rare political remark:

Brussells has excellent Bread, Butter, Wine, Cheese, Beer and
Water, and all very reasonable: yet the Inhabitants are poor, and
the Streets, as well as those of Gand [Ghent], crouded with Beggars.
So little do the Natural advantages of a plentifull Countrey tend to
enrich a People, where the valuable possession of Liberty is wanting.

After three days of respite in Brussels the party took a
detour by way of Vilvorde, north by diligence to Antwerp
(then in the Austrian Netherlands), judged by Harris as, like
most places, inferior to London, and then back to Ghent. The
horrors of bad roads and bone-shaking coaches were then
exchanged for a 30 mile trip by *trekschuit* along the waterway to
Bruges. The comfort of this horse-drawn passenger vessel was
to be celebrated by Fielding, doubtless from first-hand
experience, in *Tom Jones*, Bk. XIII, ch. 1, and Harris was no
less enthusiastic: it was 'the most commodious of the Kind,
that can be conceived, as well as the Cheapest. It is as large as
a little Ship, and built in that manner, with several Apartments under Deck, according to the different Ranks of the
Passengers.' Disembarking at Bruges, he characteristically
headed for the paintings, noting plenty by Van Dyck and
Rubens. On 30 September he journeyed south to Ostend, two
days later scrutinizing the French monarch's barracks for
25,000 men at Dunkirk, and on 3 October paid his respects to
the College of English Jesuits at St Omer. Here he was shown
the mitre of Thomas à Becket 'dipped in his blood, preserved
in a large gilt Cup . . . saved from England at the Dissolution
of the Abbeys'.

Two days later he arrived in Calais by chaise, embarked for
England at noon, and arrived at Dover after an exceptionally
lengthy crossing at two o'clock the following morning,
Tuesday, 7 October 1732. Typically, Harris reached for the
succinctness of a classical quotation in order to round off his
travels: 'longae finis chartaeque viaque' (the end of a long
story and a long journey: Horace, *Satires*, 1. 5. 104). Henry
Fielding, or his editor, was to use a version of the same
quotation to conclude *A Voyage to Lisbon*, posthumously
published in 1755.

Harris had broken new ground in more than the obvious
sense of seeing some of the Continent. He was the first Harris
to do that, and in setting off without contacts, reputation,

social prominence, letters of introduction, and with relatively modest means, he was a natural tourist. Certainly, his eye was drawn to the public, historical, and broadly cultural and aesthetic sights offered by the tour, but he took with him remarkably little of the contemporary Englishman's instinctive suspicion of foreigners and their bizarre ways. In fact, his imagination was quickly fired by disparate monuments to man's intellectual and material achievements, ranging from the university of Leyden and the printing house of Elseviér to the almost overwhelming supremacy of Dutch engineering. Their systems of dykes were 'some of the greatest Works of Art [*sic*] in Europe'. Similarly, his aesthetic sense was unprejudiced, responding with equal enthusiasm to landscape either 'natural' and 'Romantic' or formed by human concepts of symmetry and design. Naturally, he admired and appreciated good roads, and as an Englishman he took a keen interest in military architecture, as in his detailed description of the fortifications at Namur. But his greatest enthusiasm was for the fine arts, and for painting in particular. These are the most exuberant passages in an otherwise largely factual travel diary. The Elector's palace in Düsseldorf defeated him: 'It would be impossible to enumerate the Beauties of these Paintings so I shall say no more of them but they infinitely exceed all I ever saw.' Within 4 years of his return to Salisbury he had a draft of his first essay, 'A Discourse on Music, Painting, and Poetry' complete, and the opening essay in *Three Treatises* (completed by 1738) is the theoretical product of this empirical experience in European collections.

Returning to England four months after his twenty-third birthday, the eldest son of a widowed mother, Harris assumed the management of the family properties. This gave him the leisure to indulge his zeal for speculative learning, and he was soon to discover the sorrows as well as the joys of the autodidact. His aim, perhaps not yet fully conscious, and certainly unaffected by foreknowledge of the hostility it would arouse, was no less than the redirection of England's Lockian and materialist philosophy and culture. Crossing the English Channel was as nothing compared with this intellectual adventure, and although his immediate physical horizon was largely bounded by Salisbury, Oxford, and an annual visit or two to London, his mind was to range over vast tracts of literature and philosophy.

III

A few yards separate the houses in Sarum Close once occupied by the Harris family and the Cradocks. The widowed Mrs Cradock and her three daughters occupied a spacious house leased from the Vicars Choral on the south side of St Ann's Street, adjoining St Ann's Gate, from 1714 to 1733. The gatehouse itself connects one side of the street with the other, containing an upper room which the Harrises denoted 'the Chapel Room', and which they used for musical and dramatic entertainments. Its north wall is part of the Harris house, whose Queen Anne front stands at right angles to the Cradock home. Mrs Cradock's three daughters were Charlotte, Catherine, and Mary Penelope (whose obscure life ended on 28 October 1729, at 24). Outside the Close, in a house near St Martin's church, Henry Fielding's maternal grandmother Lady Sarah Davidge Gould lived from 1723 to her death in 1733. By decree of Chancery, Henry was ordered to spend his Eton holidays and vacations with Lady Gould rather than with his father and his Catholic stepmother. Presumably, he complied with this order, and therefore spent such periods there from 1719 to 1725. Most of the shorter poems in his *Miscellanies* (1743) allude to his early social life in Salisbury and specifically to his courtship of Charlotte Cradock as 'Celia'. An untitled poem includes the lines 'Sarum, for Beauties ever fam'd, | Whose Nymphs excel all Beauty's Flowers, | As thy high Steeple doth all Towers'.[20]

In 1734 Fielding was approaching the pinnacle of his financially precarious dramatic career in London. He was married to Charlotte Cradock on 28 November 1734 in Bath by Walter Robbins, a native of Salisbury. Malmesbury's 'Memoir' lists Harris's earliest friends ('and merry companions they were') as John Upton (1707–60), John Taylor (1704–66), Arthur Collier (1707–77: the son of the metaphysician), John Hoadly (1711–76: the youngest son of the Bishop of Salisbury), and Henry Fielding (1707–54).

[20] *Miscellanies by Henry Fielding*, p. 79.

This was by no means a complete list, even of Harris's literary friends; yet each was to remain part of Harris's inner literary circle. By comparison with these, Harris's friendships with Lyttelton, Richard Owen Cambridge, Samuel Johnson, Samuel Richardson, Boswell, and others, are more or less circumferential. Malmesbury's list also omits a group of scholarly friendships and collaborative relationships of much greater importance to Harris's literary interests, an axial group which was to include John Upton, William Young, Jonathan Toup, the fourth Earl of Shaftesbury, Joseph and Thomas Warton, Lord Kames, Lord Monboddo, Sarah Fielding, Elizabeth Carter, and others. The almost constant entertaining of visitors in the Close spread Harris's friendships very widely. Some were casual friendships, but others were virtually institutions in their own right. John Hoadly thus described to David Garrick the membership, nature, and purpose of such occasions: 'I have been at Salisbury for a few Days, enjoying some *Noctes Atticae* with my old friend of the Close, Mr Harris, and Joe Warton; reviving an old and most laudable Custom, of spending Xmass Week together, which his being a transitory Great Man in the Treasury had for a few Years interrupted.'[21]

For twelve years, from his return to England in 1732 until the publication of his first book in 1744, Harris devoted himself primarily to a rigorous and self-directed programme of re-education in classical literature and philosophy. This was his 'fair seed-time', a compensation for the misdirection of Oxford's syllabus and a preparation for his long-term aim to rescue, restore, and present anew to English readers the treasury of classical thought. Under the twin influences of materialistic philosophy and its own tendency to complacent insularity, Harris believed, England's intellectual soul was at stake. He was not alone in his ambition to restore what he took to be the amnesiac cultural memory of English readers, nor was his timing askew. Between 1744 and 1746 the poets Akenside, Joseph and Thomas Warton, and William Collins, among others, published works whose aim was to 'Revive the just designs of Greece' ('Ode on the Death of Thomson').

---

[21] John Hoadly to David Garrick, 3 Jan. 1773, Forster Collection, Victoria and Albert Museum, 48.F.38f.41.

Thus in the summer of 1734 we find Harris beavering away in Oxford's libraries, taking time off to advise John Upton on how to write a 'rational and philosophical' commentary on Stoic doctrines, particularly necessary now because 'few of our scholars have even a glimpse of the Stoical Philosophy. They set out with an opinion, that it is all contradictory nonsense.' A good commentator, Harris urges, should be a 'sort of master of ceremonies to his author, in introducing all strangers to his acquaintance'.[22] Harris's notion of scholarship evolves here from the Shaftesburian model of social *sprezzatura*. This is not surprising, since Harris's own copy of Epictetus' *Enchiridion*, which he here describes to Upton, was that once owned by the third earl himself, its 'margin . . . almost filled with references, from one page to another'. Six years later Harris is busy collating Upton's version of Epictetus (published in 1769) with that of Salmasius and 'an old MS collation, which I lately purchased, written in the margin of the old Venice edition of 1528 . . . [with] the commentary of Simplicius upon it'.

When Dr Charles Burney met Harris some time in 1747 at Wilbury House, he recalled 'the most amiable, learned, and worthy of men, who then lived the recluse life of a man of letters'.[23] Harris himself told his daughter Gertrude that 'in the younger part of his Life he used to get up at four or five o'clock even in Winter, which he preferred to early rising in Summer because of the extreme stillness that then prevail'd and when he said his mind was freer [and most] clear, and when he could study more uninterruptedly'.[24] One Christmas he was studying before three o'clock in the morning: he rewarded the astonished town waits, who came to serenade him, with brandy. Harris studied hard, both for himself and others, but he was no recluse.

On Wednesday, 14 January 1736, he took up his commission

---

[22] J. Wooll (ed.), *Biographical Memoirs of the Late Revd. Joseph Warton, D.D.* (1806), 207–8.

[23] 'Fragmentary Memoirs No 44', Berg MSS Collection. Wilbury House, the seat of Fulke Greville, but now owned by the Benson family, was just over 5 miles from Salisbury. In the 1770s the Burneys and Harrises met at musical events, such as the benefit for Celestini at Carlisle House (see Fanny Burney's *Journal*, 8 June 1775): 'the Harris's . . . seem never so pleased as when engaged with the most eminent singers & players' (Berg Collection, fo. 36ᵛ). [24] Gertrude Harris, 'Memoir'.

as justice of the peace for the General Quarter Sessions held at New Sarum.[25] Then, as now, Wiltshire was part of the Western Circuit. The Lent assizes began in Winchester in the first week of March and generally proceeded to Salisbury at the end of that week, before proceeding on to Dorchester, Launceston, Exeter, and, by the end of the month, Taunton. The summer assizes also began in Winchester at the end of July, and then moved to Southampton, Salisbury, Dorchester, Launceston, Exeter, and, by the end of August, Wells. This was the route travelled by Henry Fielding from June 1740 until his appointment as JP for Westminster in 1748, and by Harris's younger brother Thomas until 1754, when he was appointed Master in Chancery.

The daily wages for a justice of the peace were 4*s*, an honorarium unchanged since 1388. A glance at the list of justices for Wiltshire for 7 July 1736 shows Harris, theoretically but not in reality, in the most distinguished honorific company. Formally, the Wiltshire justices included the Earl of Burlington, the Duke of Chandos, the Earl of Chesterfield, Viscount Cobham, the Duke of Devonshire, Lord Chief Justice Hardwicke, Lord Hervey, the Earl of Pembroke, John, Duke of Rutland, Charles Townshend, the Prince of Wales, Sir Joseph Jekyll, Sir Robert Walpole, 'Speaker' Onslow, Horace Walpole, and four members of the Wyndham family, including Thomas, the Chancellor of Ireland. In reality, however, the justices for the 1736 Lent assizes were less distinguished men: James Reynolds, chief baron of the Exchequer, and Sir Francis Page, a justice of the King's Bench. In the summer, they were Sir Lawrence Carter and Sir William Thompson, both barons of the Exchequer. But the routine work of the quarter sessions and assizes fell upon even less distinguished, though nevertheless respectable, members of the county *haute bourgeoisie* such as Harris. Their concerns lay with such ordinary things as traverses (including assault), felonies (such as the theft of a pickaxe), and recognizances (which included cases of the fathering of bastards). It is not

---

[25] Information from *Wiltshire Quarter Sessions and Assizes, 1736*, ed. J. P. M. Fowle (Devizes, 1955), vol. ii of *WANHS*: see pp. 127–9 and 21. Harris's first case was a claim of broken apprenticeship (pp. 66–7). Add. material drawn from PRO Chancery Lane, Assizes 24/24 Western Circuit 1735/6–1743–4 (Wiltshire).

clear how many such cases came before Harris himself, but in the Assize Book for the Western Circuit his name appears first on 4 March 1737, and thereafter twice a year, almost always in March and July or August, and on a total of forty-four occasions.

Harris sat in judgement on a dismal procession of thieves, highwaymen, and burglars, until his last appearance as a JP on 19 March 1761, when he was elected MP for Christchurch and replaced by Edward Young. For such felons who came before the Salisbury bench, the sentence was, at least for the justiciary, monotonously formulaic. Since the 'Act for further preventing Robbery, Burglary, and other Felonies, and for the more effectual prevention of Felonies' (1720), the Royal Mercy could be extended to those found guilty of such crimes as stealing a shirt, on condition that they be transported to the American colonies and plantations. The Order Book for Transportations is almost entirely a collection of printed pro forma orders with blanks left for the insertion of two names (i.e. the condemned and the justice condemning). Elizabeth Hilman, for example, stole a shirt and received 7 years' transportation on 5 March 1742. For burglary, a group of villains was transported for 14 years (31 July 1742); for stealing two fowls, the sentence was a public whipping; for assault, there was a fine of 1*s*. and six months' imprisonment. Harris and his fellow magistrates committed Thomas White to Fisherton gaol on 1 November 1752 for robbing John Symonds of 1*s*. on the king's highway, and on the following Friday James Skuse was put in the county gaol for 'privately picking the Pocket of John Goulding, of Downton, of seven Guineas'. As in the last two cases, the *Salisbury Journal* reported the sentences without comment, even when on 30 October 1752 it stated that Harris and Richard Willoughby had committed Edward Roots of Bishopton 'for an attempt to commit a Rape on the Body of a Girl under Ten Years of Age'.

Harris kept no personal record of these years in the service of the law, but perhaps felt some private satisfaction in ridding the county of undesirables. Five years after he himself had left the bench for the Commons, Harris's fellow magistrates were publicly commended for their efficiency by Sir John Fielding, although the real purpose of his notice was to point out how many wanted felons had escaped apprehension in Salisbury and gone to ground among the criminal population of London.

IV

During the whole period of his local magistracy, Harris was wrestling with questions rather more abstract than the punishment fit for shirt-stealers, and one of his refuges was the home in Wimborne St Giles, Dorset, of his first cousin and only relative on his mother's side, the fourth Earl of Shaftesbury. In the winter of 1738 came his first opportunity to register his intellectual allegiance to his role-model, the third earl. On 4 December he wrote to Thomas Birch, Fellow of the Royal Society and eventually its secretary, to say that 'the Papers of Lord Shaftesbury's Life have been finished for some time'. Birch had previously sent Harris a draft outline of a biography with a request to edit and supplement it with materials from the Shaftesbury papers at St Giles, for subsequent publication in the ninth volume (1739) of Birch's *General Dictionary, Historical and Critical* (10 vols., 1734–41), based on the model of Bayle. Two days later Harris wrote back to Birch again to say that 'your Papers shall be transmitted to you immediately. They will be delivered into your Hands on Monday morning next. You will find there has been made a large retrenchment of Quotations, which would not have been done without the concurring Sentiments of Lord Shaftesbury and Mr Martyn.'[26] The published version of the life of Shaftesbury contains 127 lines of text and 986 lines of notes printed in double columns, and although this

---

[26] BM Add. MSS 34079, Original Letters 1513–1839; letters to Birch 4308 and 4309. Around 1732 Benjamin Martyn (1699–1763) had been engaged by the fourth earl to write the *first* earl's biography, using the family papers and a preliminary sketch by Thomas Stringer, a resident of Salisbury and a friend of John Locke who died in 1702. Deemed unacceptable by the earl, Martyn's book was revised by Dr Gregory Sharpe, Master of the Temple, in 1766, and again by Andrew Kippis in 1771 (privately printed around 1790; see *DNB*). The British Library copy, 1 of only 3 surviving copies of Martyn's biography, lacks a title-page, and is a vindication of Shaftesbury, Locke, and LeClerc against the hostile views of L'Estrange, Burnet, Rapin, and Father Orleans. Sharpe sent Harris a signed presentation copy of his *Origin and Structure of the Greek Tongue* in 1767, and Kippis was Malmesbury's first choice of biographer for his father.

proportion of text and notes is not in itself unusual in the
Dictionary, only someone with direct and free access to the
unpublished papers could have supplied much of the information. Moreover, the note attached to the final sentence of the
biography, on Shaftesbury's favourite classical authors
(p. 186), had a special meaning for Harris himself:

Among those writing which he most admired, and carried always
with him, were the moral works of Xenophon, Horace, the
Commentaries and Enchiridion of Epictetus as published by Arrian,
and Marcus Antoninus. These Authors are now extant in his library,
filled throughout with marginal notes, references, and explanations,
all written in his own hand.

As well as copious extracts from Shaftesbury's published
works (*Inquiry Concerning Virtue*, the *Characteristicks*, *Letter
Concerning Enthusiasm*, *Letters written by a Noble Lord to a Young
Man at University*, *The Moralists*, etc.), there are also extracts
from Shaftesbury's unpublished papers, such as 'a little
treatise upon the subject [written] some years before the *Letter
Concerning Enthusiasm*' (p. 182), and two letters, one to the Earl
of Godolphin and the other to the Earl of Oxford (p. 183).
Like all good disciples, Harris's chief concern was to provide
accurate records for an understanding of his mentor. His
editorial principles were thus a clear and uncontroversial
explication of Shaftesbury's philosophical ideas, followed by a
defence of his reputation based on clear analysis of authentic
texts. Harris, accordingly, had cut out a quotation from Dr
Butler in Birch's original draft on the grounds that it not only
misrepresented Shaftesbury's intended meaning but would
also, if left, 'commence a Sort of Controversy with Dr Butler,
whom neither my self, nor I dare say the Editors of this Work
would desire to offend'.[27]

If Harris saw himself as the guardian and interpreter of

---

[27] The *Life* of Shaftesbury was signed by Birch himself (identified as 'T') in vol. ix
(1739), 179–86. It was the first extensive biography of the third earl. In a prefatory
note (i, p. iii), Birch claimed that his biographies had been condensed into a succinct
narrative from primary sources provided by others. As Osborn remarked, 'Initials
were given only to articles written by the editors themselves, so that no clues are
present to facilitate the identification of "other hands" ': J. M. Osborn, 'Thomas
Birch and the *General Dictionary*, 1734–41', *MP* 36 (1939), 25–46. Harris's reference is
to Joseph Butler (1692–1752), who had criticized Shaftesbury in *Fifteen Sermons* (1726)
and argued for the supremacy of conscience over self-interest.

Shaftesbury's reputation and moral philosophy, others were more likely to see him as a dupe, or as a misguided servant of moral degeneration. Acting on the first of these assumptions, on 6 November 1744, the printer Paul Vaillant wrote to Harris's brother George to use James's influence on the fourth earl to prevent a Glasgow edition of the *Characteristicks* from being printed (or, alternatively, to arrange for Vaillant and Nourse to 'come in for shares'). When Vaillant arranged for an extract of Harris's own *Three Treatises* to be published in the *Bibliothèque raisonnée*, vol. xxxii (1744), p. 479, translated by Mathieu Maty, it included this comment:

Le second de ces Traités est un Discours; les deux autres sont des Dialogues, à peu près dans le goût des Moralistes du comte de Shaftesbury . . . On trouve certainement dans ces Notes une érudition peu commune, qui ne pourra que plaire aux Amateurs de la Littérature.[28]

Acting on the second assumption, John Brown published a hostile review of Shaftesbury's ideas in his *Essays on the Characteristics* (1751) in which Harris's use of Shaftesbury's ideas in the same book is regarded as axiomatically misguided. There was much worse to come, but in one sense Harris had begun where Shaftesbury had left off, since the latter's unfinished 'Discourse on the Arts of Painting, Sculpture, &c.' was to provide the starting-point for Harris's first book. *Three Treatises* was not only dedicated to the fourth earl but also copiously illustrated from the third earl's own favourite Stoical reading. And as for their philosophical agreement, Harris's notes on Shaftesbury published in Birch's *Life* illustrate precisely values and ideas common to both men: the concern for individual freedom from an arbitrary system; the love of study; the loathing of religious enthusiasm and superstition; the importance of practical benevolence in ordinary living ('moral practice and love of mankind') as opposed to a merely theoretical interest in the theology of virtue ('dark speculations and Monkish philosophy'); and above all a defence of speculative freedom organic to both individual and social health ('that which raises them above the degree of brutes, is freedom of reason in the learned world,

---

[28] George Harris to James Harris, 18 Oct. 1744, Harris Papers, vol. 40, part 4.

and good government and liberty in the civil world. Tyranny
in one is ever accompanied or soon followed by tyranny in the
other': p. 184). Finally, Harris was confident enough to
summarize Shaftesbury's whole 'civic, social, and theistic'
philosophy in a single and combative proposition:

> that there is a Providence, which administers and consults for the
> whole, to the absolute exclusion of general evil and disorder, and
> that man is made by that Providence a political and social animal;
> whose constitution can only find its true and natural end in the
> pursuit and exercise of the moral and social virtues. (p. 184)

Harris thus crept into print for the first time, as ghost writer
of a moral programme which began with Shaftesbury but
which was to be developed as his own. Apart from the
unpublished biographical memoir of his friend Henry Fielding,
this was to be Harris's only attempt at the art of biography.
There could be no more important subjects for Harris's
personal loyalties, and Harris provided links between the
philosopher and the novelist. Fielding was to read *Three
Treatises* in manuscript, and to some readers his proximity to
Shaftesbury's moral philosophy was to be held against him.
Years later, *The County Magazine* was to publish an article on
the modern novel in which the consequences for Fielding's
reputation of such a connection were seen as uniformly
regrettable: 'His morality, in respect that it resolves virtue
into good affections and a sense of duty, is that of Lord
Shaftesbury vulgarised, and is a system of excellent use in
palliating the vices most injurious to society.'[29]

From 1731 to 1744 Harris moved between Salisbury,
Wimborne St Giles, Oxford, and London. On 19 April 1737
he was consulting his first cousin, the fourth Earl of
Shaftesbury, in London on the arguments soon to appear in
Three Treatises, 'a result of thinking on that common topic, so
often advanced, but so seldom examined, that all Arts are
united and associated together . . . They are all three
Imitations, and therefore . . . they all share Imitation as a
common Genus.'[30] In fact, Harris was to abandon this

---

[29] *The County Magazine*, 15: 1 (Mar. 1787), 235, 'Novels and Romances, with
Characters and Anecdotes of Writers' (anon.).

[30] Shaftesbury Papers, PRO Chancery Lane 30/24/28 part 2/26.

commonplace principle of mimesis altogether in the first and most theoretical essay on Art, but it is not clear why he did so. Apart from his fascination for Handel, the fourth earl is not known for his aesthetic interests. Indeed, apart from his musical interests, and his interest in the Free British White Herring Fishery, nothing at all significant about him has come down to us. Yet he clearly meant much to Harris, and not least because he was Harris's nearest and only maternal relative. The account of him in Malmesbury's 'Memoir' is therefore particularly valuable, even though it relates a story of failure:

I cannot mention this Ld. Shaftesbury without wishing to rescue his memory from a very unjust imputation fixd on it, by those who were slightly acquainted with him, of his being weak even to simplicity. His understanding was naturally good, his Ideas correct, his observation just, and no one related with greater truth and precision than he did what he had heard or seen. His defects were all in his outward appearance. What the French call his *formes* were against him—but he was an only child—Heir to a large fortune and title and from his infancy left solely in [the] charge of a fond mother, extremely anxious, to preserve so precious a charge; but by no means endowed with sufficient powers of mind to give it just and due value. She never suffered him to leave her—married him before he was 14 to a [blank] just turn'd of 13—and at 40 still treated him as a child. He never had an opportunity of comparing his own situation with that of others of equal rank with himself, and he had no incitement to emulation; accustomed to have his amusements and his occupations (if they deserved that name) carved out and prepared for him. His mind naturally gentle & diffident was never called upon to exert itself, it never acquired its energies, and Ld Shaftesbury passed his life with infinitely less utility to others and less credit and advantage to himself than by nature he was designed to do.[31]

Harris wrote to Shaftesbury to congratulate him on being the patron of Handel and encouraged his study of Cicero, but his

[31] Malmesbury, 'Memoir', pp. 11–14. The 4th earl (1711–71) was appointed lord lieutenant and custos rotulorum of the county of Dorset in 1734, FRS in 1754, recorder of Shaftesbury in 1756, and high steward of Dorchester in 1757. His 1st wife (d. 1758) was Lady Susannah Noel, daughter of the third Earl of Gainsborough, and his 2nd (d. 1805), Mary Bouverie, sister of the Earl of Radnor. See further Sir Egerton Brydges, *Collins's Peerage of England*, 9 vols. (1812), iii. 587–9.

chief value to Harris was that he and the countess were indefatigable supporters of Italian opera. On 5 May 1737 Harris replied to Shaftesbury's report of Handel's ill-health (a combination of rheumatism and partial paralysis): 'I can assure your lordship it gave me no small Concern—when the Fate of Harmony depends upon a single Life.'[32] Effectively, then, Harris was obliged to work out his own aesthetic theories in the relative isolation of Salisbury.

Shaftesbury was no intellectual match for him, and even less of a stimulus. But when his own scholarly interests and deep family loyalties came together, as when Pope published his fourth book of the *Dunciad* in 1743, Harris needed no assistance from an acolyte. Reacting indignantly to Pope's list of 'Aristotle's Friends', the modern Oxford Dunces who turn their backs on Locke, Harris lectured his cousin on the other side of the argument:

Gellius is a Dunce, for having read & common placed all the best Authors, who had written before Him. He is peculiarly valuable to Dunces at present, for containing the fragments of more than 200 Authors, whose complete Works are all lost. Suidas's crime was the writing an excellent Historical Dictionary, & Kuster's Crime was the publishing Him in a fair & correct Edition. Clarke, Bentley, & Hobbs, yr. Lordp well knows. Wasser perhaps you may not have heard so much of. He was a most ingenious Clergyman, ye best Scholar in England, next Bentley . . . For tho' this Dunciad-writer in his note on Verso 216 seems to make it the Character of a Dunce to be the *accurate* Editor of a mean Author, I think it infinitely more so, to be ye *in-accurate* and blundering Editor of a good Author.

Pope's own major blunder, Harris goes on, was to misread *The Moralists* completely, and thus reveal his ignorance of the strategies of classical rhetoric, as well as his cowardly plagiarism from Shaftesbury in the *Essay on Man*. Line 488 of the *Dunciad* ('Theocles in raptured vision saw') specifically attacks the third earl's *The Moralists*, and Harris's rejoinder is sharply pointed:

The Author of the Charact[eristicks] is introduced for having furnished all that is Sublime or Consistent in the Essay on Man, the

---

[32] Shaftesbury Papers, 30/24/28/27.

rest being no more than a Chaos pickt out of twenty opposite Systems. This however was a fact, wch. the multitude were not to know. A mist therefore must be raised, & the Author abused. The Species of Abuse is indeed more extraordinary than ye motive to it. Ld. S[haftesbury] is made an Atheist in a Treatise, wch he wrote on purpose to prove Theism. But it is by a most infamous & immoral Quotation of him, where by ye omission of ye Word (Thou) ye whole Sense is perverted. As for the other Quotation, abt. ye lazy Academic method of Philosophizing (wch. He would insinuate as ye Character of Ld. S[haftesbury]) this again is a shameful Perversion, equally base with ye former. At best it can only be referred to a thorough Ignorance of the Nature & Laws of antient Dialogue, where ye *Ego* is often made ye lower Character, & opposers sentiments, quite averse to the Author's.[33]

Harris admired Pope's technical ability, his wit, and his imagination; but as a thinker ('for Sentiments'), he held him in very low esteem. This was a clash between the ancients and the moderns in miniature. In defending high culture Pope is seen to be misreading precisely those writers whose labours seek to preserve it. In satirizing accurate textual scholars, Pope's poor judgement implicates itself; and in lampooning Shaftesbury, Pope exposed a painful ignorance of the culture he ostensibly defends. The poet himself was just as much a modern dunce in this respect as 'piddling Tibbalds'. Such an attack was perhaps predictable from a man whose immediate friends—scholars, editors, critics, and translators—were the understrappers for a threatened classical culture which was in danger of losing its skilled reading audience.

Three Treatises (discussed in greater detail in Chapter 3) finally appeared in May 1744. Its allusiveness is not obvious, but its shape and themes are nevertheless connected everywhere with Harris's life to date. The first treatise on Art is a dialogue in the manner of Shaftesbury and purports to be the record of a peripatetic conversation between the author and an unidentified 'Friend from a distant Country', returning to Salisbury from a visit to the Earl of Pembroke's treasure house of classical and modern art at Wilton (3 miles from the Close along the northern bank of the Nadder). In fact, there are two speakers and a silent editor. The narrator is the naïve partner

---

[33] Ibid. 30/24/28/38.

progressively drawn into a debate on aesthetics and eventually
embarrassed by his companion's rhapsodic obsession with a
theory. The 'Friend' becomes the interrogator who forces his
partner to answer questions for which he has a prepared reply.
The footnotes to Aristotle, Quintilian, Horace, Plato,
Ammonius, and so on are, as it were, the deep structure of the
treatise, demonstrating the classical roots and the radical
basis of the conversation in English. Their non-dramatic
effect, of course, is to demonstrate silently the indebtedness of
this 'English' conversation to a distinguished Latin and Greek
tradition. Eventually, the 'Friend' admits his obsession and
launches into a reading of a prepared essay. The footnotes
continue to provide an ironic commentary, since his 'original'
ideas are at every point tracked down to classical precedents.
On their return to Salisbury, the astonished citizens gawp in
amazement at the two intellectuals, and they are forced to
desist.

The 'Friend' is not named, although the dialogue is
dedicated to Harris's friend and college contemporary Floyer
Sydenham, the editor of Plato's dialogues. But Sydenham was
not the only friend to be taken on such philosophical rambles,
and the portrait is likely to be a composite one. Henry
Fielding, not on the face of it a likely candidate for an abstract
discussion on aesthetic theory, was nevertheless such a
companion, prone to indulge a favourite thesis. The opening
words of his very first letter to Harris, 8 November 1741, are:
'You may have forgot, that in one of our Evening Walks,
whilst I was busy explaining some Conceits of mine (for
probably they are no other) concerning the clear Distinction
between Love and Lust, our Conversation was interrupted by
several fair Objects of both those Passions.'[34] Another, and
perhaps more likely model, was Fielding's patron and Harris's
exact contemporary, George Lyttelton. He certainly stimulated
in Harris precisely such thoughts as we find in the dialogue on
Art. Harris recorded one such occasion in the proof copy of his
essay 'Upon the Rise and Progress of Criticism' (1752), along
with a manuscript dedication to Lyttelton which never
appeared in print. Here Harris reminds Lyttelton of their

---

[34] Fielding to Harris, 8 Nov. 1741, Harris Papers, vol. 40, part 4.

conversation on an unspecified evening uncannily reminiscent
of the dialogue:

> that striking one in particular, when I heard you with such
> attention, as we were walking together in the groves of Hagley,
> during the calm silence of a starry night. Yr. Lordship remembers
> the time, & knows wt I relate to be no poetical reverie. The scene
> was actual nature exquisite in its kind; the subject founded not in
> fiction, but in truth, and such a one, as might well become a wise &
> good man, the nature of whence those Beautys were derived.
>
> Neither the language, nor the sentiments of yr. conversation that
> evening are attempted here where the whole is nothing more, than a
> short but simple, and (as I have aimed at least to render it) an
> orderly narrative, with certain notes subjoined, explanatory &
> historical.
>
> 'Tis in these notes principally (tho' it has not been omitted in
> other parts of my writings) that I have endeavoured to mention with
> honour certain friends & litterary geniuses, some still living, &
> others now no more.[35]

Harris usually made no secret of his key friendships and
intellectual models. Thus, chapter 6 of the second treatise (on
music, painting, and poetry) contains his tribute to Handel, a
genius 'by far the sublimest and most universal now known
. . . without an Equal or a Second'.[36] Not surprisingly,
Shaftesbury's *Characteristicks* is cited three times in the second

---

[35] BM Add. MSS 18728, retitled 'Essay on Criticism', and intended to carry a
dedication to 'the right honourable George Lord Lyttleton'; corrected and enlarged in
1775 and eventually printed in *Philological Inquiries*, part 1, chs. 1–7. Harris was a
welcome visitor at Hagley in 1765. The following Sept. Lyttleton thanked the Harris
family for visiting him, remarking that 'I feel the honour of your Friendship a great
addition to the Happiness of my remaining Life. A Sympathy of Tastes and of
Sentiments is a natural Bond of Union between two Minds and I think the better of
my own from finding in it so much agreement with your's. Nothing shall be wanting
on my part to cultivate and improve that resemblance' (Harris Papers, Letter Book,
vol. 31, part i, p. 55). See also Wooll, *Biographical Memoirs*, p. 324, for Harris's letter to
Joseph Warton about his 6-day 'tour to Hagley' and the 'mental repast' shared with
Lyttleton in Sept. 1767. Harris and Warton surveyed the ruins of Clarendon Palace
for Lyttleton's *History of King Henry II* (1767–71) on 28 Dec. 1769 (BM Add. MSS
18729, fos. 1–7), and by 1771 their relationship was both familiar and regular. Harris
wrote from Salisbury to Gertrude: 'Ld Lyttleton came hither Saturday and went
yesterday . . . as usual . . . very entertaining and instructive in his Conversation'
(Harris Papers, PRO Kew 30/43/2/, Mr and Mrs Harris to their Daughters, 1767–
81).

[36] *Three Treatises* (1744; 2nd edn., 1765), 99. All quotations are from this revised
and corrected edn.

and third treatises, and the dialogue on Happiness directs the reader to the edition of Arrian's *Epictetus* by 'his learned and ingenious Friend, Mr Upton' (p. 286). Equally, in his published essay *Upon the Rise and Progress of Criticism*, Harris inserts Shaftesbury, Warton, Upton, Robert Lowth, Johnson, Elizabeth Carter, Maurice Ashley, and Reynolds into his list of outstanding examples of modern criticism, alongside Roscommon, Buckingham, Pope, Addison, and Mrs Elizabeth Montague. Reynolds, for example, had sent Harris presentation copies of several of the *Discourses on Art*, and they were included in Harris's literary survey because 'they who would write ably upon any liberal Art must write philosophically, and . . . as the Principles of these arts are all congenial and naturally converge, so when traced to their common source, they terminate all in *the first Philosophy*'.[37]

Celebrating his friendships in print was to remain a feature of Harris's public manner, just as *omitting* mention of his own name in the work he undertook for others became a hallmark of his friendship. In 1780, for example, Harris commemorated three of his friends in the category of textual editors (alongside the Scaligers and Bentley): Jonathan Toup, for his editions of Suidas and Longinus; Dr John Taylor, for his edition of Demosthenes; and John Upton, chiefly for his edition of Arrian's Epictetus. Of the latter two, Harris notes, 'These two valuable men were friends of my youth; the companions of my social as well as literary hours. I admired them for their erudition; I loved them for their virtue: they are now no more.'[38]

Upton, Toup, and Harris formed a busy cell. Each searched out the other two for expert advice, requesting and often receiving copies of particular texts, exchanging opinions on variant readings, and so on. Since publication was the eventual goal, this collaborative process included the compilation of subscriber's lists, collecting their money, advice on reviewing, reciprocal exchange of presentation copies of the published books, occasional meetings, and firm intellectual loyalties. Harris was generous in sharing the advantages of a good personal library and time for research in Oxford and

[37] BM Add. MSS 18728.
[38] *Philological Inquiries*, p. 397.

Cambridge. On 1 July 1739 Harris sent Upton a letter in Latin enclosing variant readings of Epictetus' *Manual*, and also enclosed Simplicius' edition of Epictetus printed at Venice in 1520, hinting that it had been 'stolen' from a Jesuit college. On Upton's behalf Harris had collated editions of 1535, 1640, and 1711, and went on to annotate the proofs of Upton's own edition of 1739–40. The results of all this were presented to Upton as 'a pledge of friendship'.[39] The financial cost to Harris was never mentioned, but was considerable. The bookseller John Nourse informed Harris on 5 November 1751 that on a recent trip through Flanders and France he had failed to acquire a copy of Simplicius, but had tracked one down in Payne's London bookshop at an extravagant price. Harris bought it.[40]

For much of his own work Harris needed not only books but manuscripts, and from a period no later than 1753 (and probably from a much earlier date) until 1765 he employed research assistants to make meticulous transcriptions of classical texts deposited in the Bodleian, British Museum, and Corpus Christi College libraries. One of these amanuenses was Edward Fawconer, rector of Upwey, and the surviving texts copied out by him include Proclus' commentaries on Plato's *Parmenides* and *Republic*, the works of Pletho, Bessarion, Philoponus on the *Arithmetic* of Nichomacus Gersinus, and Asclepius. Each is inscribed with the phrase 'in usum Jacobi Harissii Sarisberiensis'. On 10 November 1752, for example,

---

[39] A number of Harris's classical texts are in the Baillieu Library, University of Melbourne. For further discussion and a full transcript of Harris's letter to Upton, see C. T. Probyn, 'James Harris to Parson Adams in Germany: Some Light on Fielding's Salisbury Set', *PQ* 64: 1 (Winter 1985), 130–9. The Hurn Court library sale (Christie's, 9–10 Mar., 30–1 Mar., 27–8 Apr., 1950) records other MS transcriptions done for Harris: in 1746 he received Olympiodorus' *Hermiae*, and works by Demosthenes, Aeschines, and others (item 160); in 1753, 9 vols. of Greek commentaries (Proclus and the *Anthology* of Cephalus; item 582), the principal commentaries on Plato from Corpus Christi College, Oxford (item 158), and Greek MSS of Proclus and Cephalus in 9 vols. (item 582). The sheer size of the collection deterred detailed description. In Nov. 1989 Pickering and Chatto offered Proclus's commentary on the first book of Euclid's *Elements* for sale. This, the first printing of the earliest work on the philosophy of mathematics (Basle, 1533), was extensively annotated for Harris by Fawconer in 1751, using the Barocci MS in the Bodleian library.

[40] Harris Papers, Letter Book, vol. 40, part 4. Nourse bought the book from Payne for 3 gns.

Fawconer sent Harris 'a Sketch of Asclepius Trallianus' and offered to transcribe the whole manuscript. But before doing so he reminded Harris of his desire for a Fellowship at Merton. Harris's reaction was to solicit the interest of Sir Thomas Robinson, MP for Christchurch and a man whose influence Harris himself was to need for his own election as the Christchurch parliamentary representative in 1761. Robinson replied to Harris on 9 January with an assurance that he would do all he could for Fawconer's election.[41] Thus the pursuit of 'pure', curiosity-led research evidently depended on a well-understood system of personal favours, not to say polite blackmail, political allegiance, and personal ambition. Patronage is too bland a term for this complex interrelationship of human obligations.

Even before the publication of his first book, Harris was used by his ambitious friends to provide a seal of approval for their literary endeavours. On 29 December 1741 William Young (1702–57), Fielding's prototype for Parson Adams, was asking for Harris's advice on his hopefully money-spinning proposal for an anthology of 'Modern Greek and Latin Poetry Collected from near 400 Authors of almost every nation in Europe with an account of the Lives & Characters of the most eminent of them & Notes Historical Geographical & Critical'. The publication date was set for 1 August 1742, but Young had been advised by his 'very good Friends Dr Collier and Mr Fielding' to alter his proposals. He had done so, but wished to delay printing until Harris had been given the chance to revise and correct them.[42] On 22 March 1746, from Richards Coffee House, Temple Gate, John Upton put before his mentor an ambitious project which clearly assumed a particular relationship between his friend and Handel:

when last I came from Handel's Oratorio I was so charmed, that to work I went, and from Milton's Paradise Lost, drew out a plan of a new oratorio; I finished the two first acts, and wrote them out; and sent Handel an account of wt I had prepared for him; But he has given me as yet no answer; and I have not leisure or desire to go so far as his house. I told him in my lettr, if he approved of the design, the plan should be sent to you. I have religiously observed Milton's

---

[41] Thomas Robinson to Harris, 9 Jan. 1752, from Merton College: Harris Papers, vol. 40, part 4.   [42] Ibid., vol. 31. part 1, letter 75.

words, and tho' I have here and there varied the measure, yet I have strictly kept close to his words; the second act ends with a chorus of guardian angels, immediately after the Hymn, Yr. friends here are all well; Dr [Pentecost] Barber, Collier (who is going to the Bath), Sydenham, I believe is so; he keeps like the eastern Monarchs at a Distance.[43]

Like the earlier oratorio constructed from *Paradise Lost* by Mary Delany, Upton's was, presumably, politely declined; neither are heard about subsequently. But a few years later, Harris was helping Upton in a project better suited to his talents, his edition of Spenser's *Faerie Queene*. This time there was a fee in kind to be paid. Upton had been puzzling over 'that crabbed stanza' in *The Faerie Queene* (Bk. II, canto ix. stanza 22, on the House of Alma), and Harris effortlessly provided a numerological explanation of its use of harmonic and arithmetical means drawn from Plato's *Timaeus*. Recalling a 'silly and superficial' review he had received of *Three Treatises*, he then asks Upton to review *Hermes* for the foreign journals of literature, which Vaillant could get translated. To sweeten the request, Harris then adds: 'Lord Shaftesbury and myself subscribe royally to Spenser; Chancellor Hoadly, as a Plebeian. By this I am indebted to you in the sum of four guineas.'[44]

Upton's subsequent correspondence with Harris informs him of the progress of what he calls 'our acquaintance' among the Salisbury set, Sarah Fielding, Jane Collier, Dr Arthur Collier, and Henry Fielding. All four had made, or were about to make, a literary career among the shining lights of literary

---

[43] Ibid., letter 81.

[44] SRO DD/TP 11 (part 1). Harris here explained that the 'principles of all harmony [are found] in two tetrachords; one being 6.8, that is, from the key-note to the fourth; the other 9.12 from the fifth to the octave; while in the mean time 8.9, gives the proportion of one complete or whole tone'. Upton's published note included Harris's explanation from the *Timaeus*, but omits the musical reference (see *The Works of Edmund Spenser*, ed. E. Greenlaw *et al.*, Baltimore, 1933, ii. 478–80). On 25 Feb. 1752 Upton reported back to Harris; 'Can there be anything so ridiculous as the fellows who write our Monthly review of Books? Vaillant offered them what I had drawn up concerning Hermes; upon which a Scotchman looking over it, said *We have a good opinion of the book, and will do more to its honour: leave it to us Mr Vaillant*. I could not help telling Vaillant, that my friend despised the praise of Fools.—The Frenchman is more modest and is translating what is sent him' (Harris Papers, vol. 40, part 4). In fact, Harris was to appreciate the North Britons' warmer response to his philosophical works. See further ch. 5 n. 11 below.

London. The first and second had received very substantial help from Harris, and the last mentioned was his closest friend. Thus, on 12 February 1753, Upton wrote:

Mrs Fielding has published a 3d volume of David Simple; the world think it a meer 3d volume and not a new story, and thus the book stops with the booksellers. She should, I told her, have changed the title; for Novelty is the charm of the present age, Jenny Collier has almost printed her art of Teizing, Richardson has near finished another novel on his old plan of letter writing; wch makes the story long, tho natural; but he is an original. He showed me his Clarissa translated into French, into High-Dutch and Low-Dutch, with some Latin letters from the translators, letting him know how virtuous his designs were . . . our friend the Dr [Collier] has written a large book, being a scheme to provide for clergyman's widows; ye Dr is very earnest . . . I much like Fielding's scheme for providing for the Middlesex poor, but I am afraid [it] will meet with the fate of other schemes; I have heard cold water and cold reflexions cast on it by those who I wish would patronize it.[45]

Other members of Harris's circle were glad of his expertise. Thomas Warton's poems and his *Life of Ralph Bathurst* were proof-read by Harris in July 1761, and in July 1763 Joseph Warton composed a Latin poem, a Hymn to Peace, for Harris's son to perform in the Sheldonian Theatre in Oxford.[46]

[45] Harris Papers, vol. 31, part I, letter 82. Gertrude Harris's 'Memoir' suggests that James Harris 'had written a great part' of [Jane Collier's] *Essay on the Art of Ingeniously Tormenting; With Rules for the Exercises of that Pleasant Art* (1753). This is certainly possible, but both Jane Collier ('the lamb') and Harris were equally well placed to insert the many references to the works of Henry and Sarah Fielding which this work contains. Harris certainly helped Jane Collier with *The Cry: A New Dramatic Fable* (Mar. 1754). On 18 Mar. 1753, from the lodgings she shared with Sarah Fielding in Beaufort Buildings, Bath, Jane wrote to Harris: 'My Book waits on you in print to pay its thanks for the trouble you took with it in manuscript . . . My acknowledgements are due to you for your Dedication, the thought of which, is very proper to the thing, but I was advised to make neither serious or ironical address to any One as the book is in an oblique manner address'd to the Princess of Wales by the Compliment intended for her in the 4th Chap: of the second Part. My being the Author is now one of those profound Secrets that is known only to all the People that I know' (Harris Papers, vol. 40, part 4). Upton further refers to Richardson's *History of Charles Grandison* (1753–4), and Fielding's *Proposal for Making an Effectual Provision for the Poor* (1753). Collier's piece has not been traced, but a series of articles in Fielding's *Jacobite's Journal*, 21, 29–32 (23 Apr.–9 July 1748) is devoted to this scheme.

[46] Birch and Francis Wise also assisted Warton in his *Life . . . of Ralph Bathurst* (1761). James Harris's son sent his father a copy of Warton's 'Hymnus ad Pacem' on 24 July 1763. Not available to Wooll in a correct copy (see p. 294), it is found in Harris Papers, vol. 11, part 26: see app. V below.

The least known of Harris's circle is the one for whom the record is most complete. Jonathan Toup (1713–85) was described by Warburton as 'the first Greek scholar in Europe'. He was Upton's student at Oxford, and his extant correspondence with Harris covers a thirty-year period from 1747 to 1778. On 14 August 1747 Harris gives him an opinion on a passage of Longinus and some advice on Greek syntax.[47] Their correspondence began with a discussion of the second edition of Upton's *Observations on Shakespeare* (1746), which had included a 50-page attack on Warburton's shaky scholarship, and of Upton's recent edition of *Alcibiades*. Harris fully endorsed Upton's attack on the 'rash daring absurdities of Warburton', and informed Toup that Upton was wrong in attributing a Greek commentary on it to Olympiodorus (it was by Proclus). Harris also asked Toup to return Harris's two editions of Longinus by way of his friend Archdeacon Hele of Exeter. By lending him books from his own library, sending him news of forthcoming publications, and ploughing through Greek dictionaries on his behalf, Harris sustained this brilliant classical scholar in his remote Cornish parish.

On 30 June 1763 Harris asked his own bookseller John Nourse to send Toup a new edition of Herodotus as a token of friendship. Toup replied (13 October), that 'Such a Testimony from so real a Scholar, cannot but be most agreeable to one who has always been a Lover of Letters, and literary Men'.[48] In the meantime (2 July) Harris had also recommended Toup, 'an ingenious clergyman of Cornwall', to Joseph Warton, and when Warton produced his own edition of Theocritus (2 vols., 1770), it included many notes supplied by Toup.[49] One of them, note 37 on Idyll XIV (on sodomy), caused a scandal, and Dr Lowth (appointed bishop of Oxford in 1766) was so offended that he required Oxford's vice-chancellor to have the press delete the page on which it appeared. Given his close friendship with Warton, his patronage of Toup, and his Shaftesburian sense of free speech,

---

[47] BM Add. MSS 32565, fo. 6, J. Mitford Notebooks, vol. 7 (Harris to Toup, 2 July 1747). Harris promised to send him the emendations to Suidas on 14 Aug., and the former's hostility to Warburton was no doubt determined also by the latter's attack on Shaftesbury in his *Divine Legation of Moses*.

[48] Ibid., fo. 16.   [49] Wooll, *Biographical Memoirs*, p. 295.

Harris's reaction was predictable outrage. When the controversy had blown over Harris wrote to Toup, on 20 October
1770: 'The matter is now over and forgot . . . your Conduct as
a Critic and Man of Learning stands unimpeach'd and justly
arranges you among the first scholars of the time, whether at
home or abroad.' Both men kept in touch, and Harris took
particular pleasure in congratulating Toup when he was
appointed prebend of Exeter in 1775 and chaplain of Lichfield
at the hands of Bishop Hurd in the following year. Toup
presented Harris with his edition of Longinus in 1778, and
Harris's natural delight was tempered by the same uneasiness
which had accompanied Pope's misreading of Shaftesbury's
Moralists 34 years before: 'you will I hope by yr succeeding
Publications keep that taste for antient & classical literature
among us, which none can support but such acute and able
Masters as yourself . . . although I fear the number is too
few'.[50]

Harris's public reputation in the mid-1740s rested squarely
on his theoretical essay on the sister arts, *Three Treatises*, and
the first poetic testimonies to it came from his inner circle of
friends. The earliest was probably the undated and unfinished
poem inscribed to Joseph Warton and addressed to Harris by
William Collins. This fragment referred to the relationship
outlined in the second treatise between poetry and painting,
and marks Harris's work as a continuation of Shaftesbury's
aesthetic theories. Collins had carefully considered the form,
content, and provenance of the treatise:

> And thou, the gentlest patron, born to grace
> And add new brightness even to Ashley's race,
> Intent like him in Plato's polished style
> To fix fair Science in our careless Isle;
> Whether through Wilton's pictured halls you stray,
> Or o'er some speaking marble waste the day,
> Or weigh each sound, its various powers to learn,
>   Come, Son of Harmony, [Sweet Philosopher deleted]
>   O hither turn!
> Led by thy hand, Philosophy will deign
> To own me, meanest of her votive train.[51]

[50] BM Add. MSS 32565, pp. 42, 63.
[51] R. Lonsdale (ed.), *The Poems of Thomas Gray, William Collins, Oliver Goldsmith*
(1969), 528–9.

The second poem, unsigned and undated, is entitled 'To James Harris Esqre. with a Pastoral set to Musick by Dr Greene'. It includes these Newtonian lines:

> Friendship is Musick—and we find
> No sweeter Harris than of your Mind.
>
> .   .   .   .   .   .
>
> But You, great Master of the Lyre,
> To nobler Harmony aspire;
> You, by Analogy divine,
> Can *sounds* to Moral truths refine;
> This universal Fabrick scan,
> And read the Harmony of Man;
> Can see the Parts in Concert roll,
> And join to form one beauteous Whole.[52]

The 'Pastoral' has not survived among the Harris papers, but it was almost certainly *Phoebe: A Pastoral set to Music by Dr Greene* (1748), with a libretto by Harris's friend John Hoadly. It is likely that the poem was also by Hoadly, who was to publish eleven epigrams from Martial to Harris, including two extravagantly complimentary addresses to Harris's critical acumen in 1758. Dr Maurice Greene held every major musical appointment in the land at this time. He had been Master of the King's Musick since 1735, organist and composer in the Chapel Royal since 1727, and the purely honorary Professor of Music at Cambridge since 1730. His grandfather, the Revd Thomas Greene (1648–1720), had been a canon of Salisbury.

There was another and more specific occasion for Hoadly's poem. Harris had probably assumed the leading role in managing the annual Salisbury musical festival at this time (see Chapter 7 below). Certainly, Hoadly's high-toned address to the new maestro would have been particularly timely, for Harris, 'great Master of the Lyre', had turned from the theory of music and poetry to the practical propagation of Handel's music in his home town. The Shaftesburys kept Harris in constant touch with Handel's career in London, and his brother William, who had met Handel in London in August

---

[52] BM Add. MSS 37683, P. A. Taylor Papers, vol. 2, p. 108 (i.e. Revd Henry Taylor of Winchester). Information about Maurice Greene below is from Stanley Sadie (ed.), *The New Grove Dictionary of Music and Musicians*, 20 vols. (1980), vii. 684–7, xi. 192–3.

1745, assured his sister-in-law that she could expect a visit in Salisbury from Handel very soon.[53] William heard Handel's new oratorio occasioned by the Jacobite rebellion, *Judas Maccabaeus*, in Handel's own house the following February. When the Salisbury festival began, it was inaugurated by and continued to feature Handel's music for the next 40 years.

Upton, Toup, Young, Collier, Sydenham, and Hoadly provided Harris with a vital literary support system and a highly developed scholarly network. But there was one intimate relationship which had a much deeper meaning and which has so far only been touched upon. Harris's friendship with Henry Fielding not only combined philosophical compatibility with literary co-operation, but also extended to long-term financial relationships and a congeniality of temperaments somewhat surprising in two men of such different interests. They come together when Harris was writing *Three Treatises*, and it is to this first book by Harris that we now turn.

---

[53] *Series of Letters*, i. 3. William reports to Elizabeth Harris: 'I met Mr Handel a few days since in the street, and stopped him and put him in mind who I was, upon which I am sure it would have diverted you to have seen his antic motions. He seemed highly pleased, and was full of inquiry after you and the Councillor [i.e. her husband James, not, as has been previously assumed, Thomas, who did not become a Master in Chancery until 1754]. I told him I was very confident that you expected a visit from him this summer. He talked much of his precarious state of health, yet he looks well enough. I believe you will have him with you ere long.' The traditional belief that Handel not only recuperated in Harris's Salisbury home in the summer of 1746 but also performed 'in James Harris's band' is very probably based in fact, but unfortunately lacks corroborating evidence. The assertion was first made in W. Coxe, *Anecdotes of George Frederick Handel, and John Christopher Smith* (London and Salisbury, 1799), then in Malmesbury's article on the Harris family in *The Ancestor* (1902), then became a fact in R. A. Streatfield, *Handel* (1910), 109, 186, and was subsequently repeated in e.g. P. M. Young, *Handel* (1947). William Harris was a prebendary of Sarum, rector of Eccliffe, and chaplain and secretary to the Bishop of Durham, who lived in Grosvenor Square.

3

The Sister Arts

Where Nature foils us we prevail by Art.

MS note by Harris, *Three Treatises*

He who has knowledge of Nature and Art,
Will never lack Religion;
Whoever lacks a knowledge of these,
He must not lack Religion.

Goethe, *Zahme Xenien*[1]

HARRIS's first book was the product of a dozen years of private study, but it is characteristic of the man that two-thirds of its learning was to be cast in the sociable mode of a dialogue which takes place during a walk in the country. Having collected his knowledge in relative solitude, Harris was now to learn how to apply it by mixing with mankind. *Three Treatises* announced the method and aims common to each of his subsequent publications. Renovation of critical and aesthetic ideas taken from the classical period for a contemporary purpose was to show not just the value of ancient sources but the permanence of certain ideas about art, language, and human nature. The same assumption was to underpin the method of *Hermes, Philosophical Arrangements*, and his last posthumously published work, *Philological Inquiries*. In his first three books Harris almost completely effaced contemporary references and, more importantly, the insights which were both new and original to himself. Such patrician modesty courted and received misinterpretation from those who thought that Harris's aim was merely to show off his learning.

Nothing could be further from the truth. Behind Harris's eclectic method lies a conviction that the collective cultural memory of mid-eighteenth-century England had become parochial, insular, obsessed with its own present, and anti-intellectual. Soliciting support from Monboddo in 1775, in

---

[1] E. Gombrich, 'Nature and Art as Needs of the Mind', *The Fourth Leverhulme Memorial Lecture* (Liverpool, 1981), 15.

words reminiscent of Norris's *Theory of the Ideal or Intelligible World*, he deplored the fact that scholars in England had 'transferred the whole of Philosophy from the head to the hands; that is to say, from Syllogism and Theory, to Air-Pumps and the Electric Apparatus'.[2] Harris set himself the task of correcting mechanistic and empirical prejudices by a method of literary analogies designed to restore the past, rejuvenate metaphysics, and show that man's noblest faculty was his imagination. Ironically, Harris's own method was that of a scientifically rigorous examination of sources and an unswerving devotion to logical rigour. He was determined that neither the ascendancy of Locke (or, in Monboddo's words, the 'Philomaths . . . Empyrics . . . Newtonians'), nor the decline of the Greek language should be allowed to prevent access to valuable truth and the 'science of Universals'.[3]

What Harris said about the third Earl of Shaftesbury's 3-volume *Characteristicks* is applicable not only to his own *Three Treatises*, but also true of Harris's own series of books:

it is to be observed that the several pieces, which compose them, are not only perfectly finished in themselves, according to the nature and genius of that species of writings to which each belongs, but that all together they form a complete whole, whose parts have a certain order and relation to each other, and which cannot be inverted without the whole's being injured.[4]

But again, Harris's modesty has obscured the difference between mentor and pupil: Harris's own work is remarkable for its methodical and analytical clarity, for what Lowth and Monboddo, among others, termed his 'dividing method'. As a whole work, Harris's first volume moves from the initial idea of Art as a necessary and defining cause of an activity in man, through its various expressive media (music, painting, poetry), finally to the ultimate union of all aesthetic and ethical

---

[2] W. Knight (ed.), *Lord Monboddo and Some of his Contemporaries* (1900), 91 (11 Feb. 1775, referring to *Philosophical Arrangements*). Four months later, in reply to Sir John Pringle's observations on the 3rd vol. of his *Origin and Progress of Language* (21 June 1776), Monboddo similarly characterized modern natural philosophy as 'chiefly conversant with particular facts, and seldom rises above the air-pump or alambic' (ibid. 92).   [3] Ibid. 108 (23 May 1779).

[4] [Birch's] 'Life of Shaftesbury', *General Dictionary*, ix. 183. This lengthy footnote arises from an account of Shaftesbury's MS corrections. I am assuming that Harris supplied both the information and the footnote.

activities in the pursuit of the (Aristotelian) notion of happiness.

I

Harris's initial problem was one of presentation: how to convey complex and to some extent unfashionable ideas in an amenable and attractive form. In choosing the device of an imaginary dialogue, he not only allied himself with Shaftesbury, the greatest recent exponent of that form, but also implicated himself in the decreasing understanding of its rhetorical strategies. Even so, he borrowed intelligently, adopting not only the formal model but also a body of aesthetic, moral, and philosophical concepts which already existed in this conversational literary mode. In particular, he had learnt (if his readers had not) a precise sense of the strategies involved in using the dialogue form for the first and third of the treatises, and commentators without such an awareness have usually confused and misinterpreted Harris's own point of view. This misunderstanding continues to be perpetrated simply by reprinting only the 'enthusiastic' apostrophe to Art at the end of the first dialogue, from which Harris dissociated himself.[5] His letter to Birch in 1738 about the 'Life of Shaftesbury' points out the foolishness of attempts to refute the argument of 'the Sceptic Character' in Shaftesbury's *The Moralists*, 'whom the Author all along employs not to deliver his real Sentiments, but to oppose them as delivered by Theocles the Dogmatist'.[6] In Shaftesbury's own words, from chapter 2 of the fifth *Miscellaneous Reflections*, *The Moralists* was conceived as a drama,

and carries with it not only those features of the pieces anciently called mimes; but it attempts to unite the several personages and

---

[5] See e.g. Robertson, *Sarum Close*, pp. 239–40, where we are told of the 'self-conscious Pomposity . . . utmost artifice . . . affectation . . . high-flown rhetoric . . . the style, growing ever more ornate', in the 'Dialogue Concerning Art'. In such remarks one observes the real damage done to Harris's reputation by Johnson's remarks about him being a prig (cited by Robertson with unconvincing regret). On the dialogue form, see E. Merril, *The Dialogue in English Literature* (New York, 1911), and E. R. Purpus, 'The "plain, easy, and familiar way": The dialogue in English literature 1660–1725', *ELH* 17 (1950), 47–58. Neither study discusses Harris's contributions to this genre, and it remains true, as Purpus remarked in 1950, that there is 'no satisfactory account of the growth and development of the dialogue' (p. 47).

[6] Letter to Thomas Birch, 6 Dec. 1738, BM Add. MSS 4309.

characters in one action or story, within a determinate compass of time, regularly divided and drawn into different and proportioned scenes; and this, too, with variety of styles; the simple, comic, rhetorical, and even the poetic or sublime, such as is the aptest to run into enthusiasm and extravagance.

A footnote adds that the characters are 'neither wholly feigned . . . nor wholly true . . . 'Tis a sceptic recites, and the hero of the piece passes for an enthusiast'.[7]

Thus the 'I' in Harris's first treatise, 'Dialogue Concerning Art', is hesitant, somewhat obtuse, a man who agrees without understanding, an easy victim, unable to follow the argument unless it is spelt out in simple terms, whereas his tutorial colleague ('he', or 'my Friend') is sublimely confident, somewhat patronizing, and subject to enthusiastic fits. The latter's apostrophe to Art—'Thou distinguishing Attribute and Honour of Human Kind'—fits into a recognizable sub-genre within the dialogue form. Shaftesbury's Theocles deliberately indulges himself in a 20-page rhapsodic apo-strophe to Art, Nature, and to the 'designing active principle' in the universe which he himself describes as a 'fit', a 'sensible kind of madness' associated with the transported poet (ii. 97–118). To this feigned enthusiasm Philocles replies, 'you might well expect the fate of Icarus for your high-soaring'. An identical purpose and fate awaits Johnson's Imlac when he rhapsodically lists the poetic desiderata in chapter 10 of *Rasselas*, whereas Yorick's hymn to Sensibility in *A Sentimental Journey* repeats the apostrophic rhapsody but substitutes a self-correcting irony for the scepticism of an absent interlocutor.

The dialogue form evidently attracted Harris, and 'moral painting by way of dialogue' (to use Shaftesbury's own phrase), became something of a minor industry for Harris (and for Lyttelton after him) between 1744 and 1752. Apart from the first and third of the *Three Treatises*, there was 'Much Ado: A Dialogue', written in October 1744 and later given to Sarah Fielding for inclusion in *David Simple*, along with 'Fashion: A Dialogue' (1746); 'Knowledge of the World, or Good Company, A Dialogue' appeared separately in 1752,

---

[7] J. M. Robertson (ed.), *Characteristics of Men, Manners, Opinions, Times, etc.*, 2 vols. (Gloucester, Mass., 1900; repr. 1963). ii. 333–4. All references are to this edition and are given in the text.

with a note on the final page that it was 'Printed in the year MDCCLII, from the MSS of J. H. of S. in the County of W'; and number 30 of Fielding's *Covent-Garden Journal*, 14 April 1752, is 'A Dialogue at Tunbridge-Wells between a Philosopher and a Fine Lady. After the Manner of Plato'. This was signed 'J', and although there is no proof of Harris's authorship, it is a unique signature in the *Journal*, and repeats the situation and style of bantering ironies at the expense of the female mind which Sarah Fielding had gratefully received from Harris and published as 'Fashion'.

None of these whimsical dialogues was publicly owned by Harris, but the original manuscripts of some (and other) unpublished dialogues were retained among the Harris papers. 'Knowledge of the World', however, was hardly a whimsical exercise.[8] It begins by discussing one of the favourite themes common to both Shaftesbury and Harris—the relationship between the man of liberal education and learning on the one hand and the ignorance of the man of fashion on the other. It also contains the clearest evidence of Harris's conscious manipulation of the dramatic strategies in the dialogue form. Moreover, from the apparently commonplace and urbane chatter develops a serious outline of many of the ideas in Harris's first two books, a number of which are indebted to his reading of Shaftesbury.

In its 48 pages Harris writes his own version of Shaftesbury's dictum of knowing oneself (the 'familiar method of soliloquy' which forms the basis of Shaftesbury's *Advice to an Author*), the fundamental importance of sociability, the question whether 'good Company is . . . formed by people of fashion, and of birth' (p. 33), or whether 'there will be some bad Company more good than some good Company' (p. 35). The conversational victim of the argument remarks at one point: 'You imagined, I dare say, I should have surrender'd by this time; have acknowledged my errors; have recognized your wisdom; have acted with due decorum *the under hero of a modern dialogue*, that thing of wood, set up for nothing else, than for another to shew his skill, by tipping of him down. But this, you may be

---

[8] This dialogue, together with Harris's 'Upon the Rise and Progress of Criticism' (1752), which reappears in modified form as part I, chs. 1–7 of *Philological Inquiries* (1781), has been repr. in facsimile by Garland Publishing Inc., New York, 1971.

satisfied, will never happen on my part' (p. 41). Towards the end of the dialogue the real point is made, i.e. that the careless discourse of modern conversation, and its reduction of social knowledge to the fashionable and trivial commonplace of 'knowing the world' of polite manners, has obliterated ethical and intellectual duties of the first importance. Harris's interlocutor outlines nothing less than a philosophy of the mind's obligations as vast in its scope as Imlac's sketch of the poet's obligations and skills in *Rasselas*. To know mankind and human nature is thus 'to know the several powers of human action and perception . . . their concurring with reason . . . the various affections, whether selfish or social . . . the transition from what is social and rational into vitious habits . . . the slow and critical process of raising up better':

applied to a man's self, 'tis called the virtue of *prudence*; to a family, it assumes the name of *oeconomy*; when seen in the propriety of our common intercourse with others, 'tis recognized by the name of *civility* and *address*; when extended to the leading of states and empires, 'tis the *rhetoric* and *policy of the genuine statesman*; in a word, 'tis a Knowledge which differs in this from all others, that by possessing it we become not only wiser but better. (pp. 43–4)

Behind the trite phrase 'knowing the world' there is the imperative to look beyond the surface of things in order to 'gain a glimpse of that active Intelligence, the repository of all final causes, and the first mover of all efficient . . . In short, 'tis the union of these two sciences, (call the one wisdom, the other moral virtue,) which completes the just exemplar of perfect humanity; that consummate idea, which but to resemble and approach is the highest proficiency of the best men' (p. 45). The partner in this dialogue is crushed into silent acquiescence by such high-toned social reproof. Whereas Shaftesbury had stated that 'in dialogue . . . the author is annihilated, and the reader, being no way applied to, stands for nobody' (*Advice to an Author*, p. 132), there seems little doubt that Harris's own convictions provide the energy for the closing pages of this dialogue.

Shaftesbury's sometimes infuriatingly oblique 'way of chat', and his apparently 'random miscellaneous air' are simultaneously an attempt to imitate and reform the fashionable 'modern' style from within. The classical form of the Platonic dialogue teaches by leading questions, not by rigorous methodical argument; it conceals its art and purpose behind an apparently casual meeting and a conversational discourse which might conceivably take place between any liberally educated men, each of whom belongs to an exclusive club of amateurs. Harris undoubtedly found hints and suggestive lines of investigation in Shaftesbury's 3 volumes for almost every topic he was to write about: the necessary relationship between the arts of poetry, music, painting, and architecture; ideas and definitions of happiness expressed as a concept of moral rectitude and psychological harmony; intellectual freedom from censorship and freedom to speculate; unsociability expressed as a perversion of man's natural benevolence in a providential universe of design, in which 'everything is governed, ordered, or regulated for the best, by a designing principle of mind, necessarily good and permanent' (*Inquiry*, p. 240). All these ideas, together with Shaftesbury's horror of religious enthusiasm and his argument for an *instinctive* human joy in ideas of natural, moral, and psychological harmony, had been restated in Maurice Ashley's *Cyropaedia*, addressed to Harris's mother in 1728. The doctrines of benevolent sociability, an anti-Hobbesian optimism, and a conviction that metaphysics has no connection with politics, had become, as it were, Harris family property long before *Three Treatises* appeared in 1744.

Although it was Shaftesbury who said that 'the most ingenious way of becoming foolish is by a system' (i. 189), there is perhaps no other contemporary figure whose moral, ethical, psychological, and philosophical ideas combine so neatly and homogeneously. Scattered throughout Shafesbury's miscellaneous work could also be found an argument for the supreme gift of language, for the cultural value of classical literature and philosophy, and in *Advice to an Author* (i. 158, 166), an image of Aristotle as 'the prince of critics' and 'the grand master'. Here was the intellectual grammar for Harris's own discourse, the *langue* inside which his own *parole* could be expressed, and when he published his only major poem, *Concord*, in 1751, it was to place on public record a wholesale commitment to the third earl's metaphysic of benevolent social affections, cosmic design, and a concept of the mind as an active, creative principle of latent energy:

> Ere yet creation was, ere sun and moon
> And stars bedeck'd the splendid vault of heaven,
> Was God; and God was Mind; and Mind was Beauty,
> And Truth, and Form, and Order: For all these
> In Mind's profound recess, and union pure
> Together dwelt, involv'd, inexplicate.
> Then matter (if then matter was) devoid,
> Formless, indefinite, and passive lay;
> Mysterious Being, in one instant found,
> Nor any thing, nor nothing; but at once
> Both all and none; none by privation, all
> By vast capacity, and pregnant power.
> This passive nature th'active Almighty Mind
> Deeming fit subject for his art, at once
> Expell'd privation, and pour'd forth himself;
> Himself pour'd forth thro' all the mighty mass
> Of matter, now first bounded. Then was beauty
> And truth, and form, and order, all evolv'd,
> Was open'd all, that lay enwrapp'd and hid
> In the great mind of Godhead. Forth it went,
> Forth went the pure quintessence far and wide
> Thro' the vast whole; nor did its force not feel
> The last of minim atoms.[9]
>
> (ll. 42–64)

Unless he was to be subsumed under the influence of Shaftesbury, Harris had to find his own expressive and creative resources. The disciple had to assume leadership, and the first measure of difference between the two men was inevitably to centre upon style and method. It is, for example, impossible to separate what Shaftesbury says from the manner in which he says it: the style is the man. But in Harris's work, certainly after *Three Treatises*, there seems no organic

[9] *Concord* has not been reprinted since its first appearance in the 12th and final vol. of F. Fawkes and W. Woty's *The Poetical Calendar*, pp. 53–9. Dedicated to the Earl of Radnor, it is 168 ll. long, and identified as Harris's at the beginning. It appeared in the same vol. as some of Christopher Smart's poems, and Johnson's character sketch of William Collins. Among the subscribers to Fawkes's *Original Poems and Translations* (1761) were Benjamin Collins, the Salisbury printer, Garrick, Goldsmith, Hawkesworth, Hogarth, Johnson, Sir William Robinson, Reynolds, Richardson, Rousseau, Smollett, Sterne, Henry Thrale, the poet laureate William Whitehead, and Edward Young. See Appendix I below.

connection at all between the man and the matter, between the local magistrate and the universal grammarian, the Lord of the Admiralty and the logician, the MP and the philologist. Shaftesbury infuses his work with the persona of an aristocratic amateur: revealing one's philosophical rigour is a greater social sin than appearing a mere dilettante. Harris unashamedly attempts to prove a thesis by intensive and specialized scholarship: the style remains a transparent medium in the service of a strictly analytical purpose. Shaftesbury's greatest fear is overtaxing the reader's intellectual concentration, but it is often his periodic style that stands in our way. Harris warns the reader at the outset that he is in for a testing experience, although it must be said that *his* stylistic vice is a superfluity of connectives.[10] Shaftesbury avoids the topical and the specialized, but Harris thrusts his sources in front of us, sometimes citing from unpublished manuscripts in his own possession; the former builds into his essays episodes of rest and recreation, the latter's 'dividing' and steadily accumulative method of argument permits few if any asides and excursions. Shaftesbury pretends that we can all be his equals, Harris is always the sympathetic mentor keen to enlarge his reader's awareness by sharing his own discoveries. For the third earl, the envisaged audience is like Palemon himself, an airy man of fashion, or like Philocles, a man of quality fashionably in love and equally fashionably a 'half-thinker'. Both are to be seduced into thinking about morals and philosophy ('wide of common conversation, and by long custom so appropriated to the school, the university chair or pulpit', ii. 335), by a series of deceptions practised by Shaftesbury himself through his persona Theocles, who because he is a gentleman-philosopher is also the enemy of 'the gravity of strict argument' and of the boredom it commonly induces.

Harris, by contrast, practises no seductive rhetorical

[10] An anonymous reviewer of W. Mitford's *Inquiry into the Principles of Harmony in Language, and of the Mechanisms of Verse* (1774; 2nd edn., 1804), remarked that its 'truly Attic' style was 'made up with the smallest possible proportion of substantives and verbs; but conjunctions and particles of all sorts are sifted about every where: as if sand were more essential to mortar than lime, and mortar to an edifice than brick and stone. It recalls the writings of Shaftesbury and Harris. With them, connectives outnumber the things to be joined: like Xerxes, they provide more chains than captives.' See *The Critical Review*, 3rd ser. 3 (1804), 263.

deceits, yet is just as much aware as Shaftesbury of the
contemporary resistance to serious philosophic works. An
unpublished manuscript poem from his daughter's day-book
entitled 'The Author to his Books while in Manuscript',
dated November 1780, the month of his death, includes these
lines:

> Dear scrapes and scrolls you long I know
> In Nourse's shop to make a show
> Half, in stiff paste-board bound, and half
> Like Beaux in glittering gold and Calf.
> Hateful, and harsh you think your doom,
> To live immured in Chapel room,
> On dusty shelf your time to spend
> Redde once a quarter to a friend.
> Fly then, if thus to fly you lack
> But mark, you never can come back:
> Alas I fear it wont amuse ye
> Should surly critics dare abuse ye,
> Or should the gentle reader tire
> And in a gentle nap expire
> Yet still I augur your renown
> May live and thrive awhile in Town,
> Nay more as round the land you stray,
> To Windsor, you may find your way,
> And there, if you're in luck be seen
> By Britain's all accomplish'd Queen.
> The world will like you while in fashion
> But newer books will mar their passion.[11]

Such diffidence is understandable in relation to his last and
posthumously published book on the rise of criticism and the
literature of the Middle Ages, *Philological Inquiries*. But
neither *Three Treatises* nor *Hermes* suffered the fate of super-
session. In a series of eleven epigrams from Martial dedicated
to James Harris by his friend John Hoadly published in
Dodsley's *Collection of Poems* in 1758, Harris's taste and
scholarship have become the standard by which everyone else
is to be judged:

---

[11] Malmesbury Papers: *MSS Verses*, an unpaginated, beautifully illustrated vol. in
Gertrude Harris's hand, contains several poems relating to her father's works, and
some exquisite watercolour sketches.

Would'st thou, by Attic taste approv'd,
By all be read, by all be lov'd,
To learned Harris' curious eye,
By me advis'd, dear Muse, apply:
In him the perfect judge you'll find,
In him the candid friend, and kind.
If he repeats, if he approves,
If he the laughing muscle moves,
Thou nor the critic's sneer shall'st mind,
Nor be to pies or trunks consign'd.
If he condemns, away you fly,
And mount in paper kites the sky,
Or dead 'mongst Grub-street's records lye.[12]

II

Three Treatises is Harris's most urbane and accessible book. English contemporaries, particularly Johnson and Mrs Thrale, reacted chiefly to the last of the treatises, 'Concerning Happiness: A Dialogue', believing that Harris was simply continuing his uncle's deist heterodoxy. The first two treatises proposed an aesthetic theory which was both new and original.[13]

Like Shaftesbury before him, and Akenside and Hartley after him, Harris defines Art in terms of human needs and

---

[12] R. Dodsley, *A Collection of Poems*, 6 vols. (1758), v. 285–8 (see p. 288). The series closes with this poetical compliment to Harris: 'Is there, enrich'd with Virtue's honest store, | Deep vers'd in Latian and Athenian lore? | Is there, who right maintains and truth pursues, | Nor knows a wish that heaven can refuse? | Is there, who can on his great self depend? | Now let me die, but Harris is this friend.'

[13] See J. Malek, 'Art as Mind Shaped by Medium: The Significance of James Harris's "A Discourse on Music, Painting, and Poetry" in Eighteenth-Century Aesthetics', *TSLL* 12: 1 (Spring 1970), 231–9, for an argument that Harris's treatise 'necessitates consideration of a complex of questions of enduring critical worth' in the history of British aesthetics. The single and most valuable discussion, however, is 'Harris and the Dialectic of Books', in R. Marsh, *Four Dialectical Theories of Poetry: An Aspect of English Neoclassical Criticism* (Chicago, 1965), 129–70. G. McKenzie's *Critical Responsiveness: A Study of the Psychological Current in Later Eighteenth-Century Criticism* (Berkeley and Los Angeles, 1949), mentions Harris's critical theories only in passing, and in one paragraph (p. 56) he misleadingly speaks of Harris's 'undiscerning Aristotelianism' and seems to think that Harris's crucial distinction between the different media of the arts is less important because the idea is also found in Lessing's *Laocoon*, published 22 years later! Even so, Harris is grudgingly admitted to a critical tradition which includes Hume, Hartley, Kames, Beattie, George Campbell, and Hugh Blair.

psychological processes: it is 'an habitual power in man of becoming a certain cause'. Art begins from a need to compensate our sense of imperfection. As soon as we perceive the 'absence of joys, elegancies and amusements from our constitution, as left by nature', we are 'induced . . . to seek them in . . . arts of elegance and entertainment'. Art thus provides a substitute for an absent good, and is superior to our 'natural and uninstructed faculties' because it is learnt, conscious, and never inheres in any single individual. All productions of Art are either Energies or Works: thus a musician will exhibit his temporal art while performing it (the Art therefore being an Energy coeval with the artist), whereas the sculptor or novelist produces a permanent Work, which remains after the artist has departed. Harris thus divides faculties, powers, and capacities from completed action, and generates what is essentially a dialectical and dynamic theory of Art-as-process and Art-as-completion. Schematically, his system appears like this:

ENERGY	WORK
contingent	necessary
change, motion, transience	immutability, abstraction
particular	general and intellectual essence
operation	completion

Thus the human power which causes the effect of music produces a melody which is a process of transitional and *successive* sounds existing over a period of time; whereas the power which produces the art of architecture has as its effect a finished building whose parts are coexistent and are permanently 'fixed'. The former is recognized during performance as a Motion or Energy, the latter as a thing having been performed, and therefore a Work, or 'thing done'. The potential significance of such a distinction when transferred to *language* should be obvious. Harris seems to be only a whisker away from a distinction between *surface* and *deep* structure. In fact, such distinctions will be made about grammar in *Hermes* and about the 'latent force' of logic in literary discourse in *Philosophical Arrangements*.

In *Three Treatises* Harris is primarily concerned with the

psychological causes of Art, and the value of his aesthetic distinction here is that instead of recycling the old notion of all Art as a mirror of Nature (*ut pictura poesis*), he moves aside the mimetic function of Art in order to discriminate between the respective media of each of the arts, something which none of his predecessors had managed to do—either in Dryden's *Parallel betwixt Poetry and Painting* (1695), or in Charles Lamotte's *An Essay upon Painting and Poetry* (1730)—for Art is a conscious act of volition quite separate from either instinct or divine power. Art is a uniquely human activity—indeed, it is the highest activity of human life—arising both from our perception of imperfection and instability and from our need to reconcile the imperfect world of experience with an imagined world of perfection. The highest forms of Art thus embody man's innate notions of perfection and reflect his deepest needs. The *active* principle of mind works on the material of Nature perceived through the senses to produce something greater than both. Without Art human life is mere existence, and at this point Harris's speaker addresses Art not just as 'The Ornament of Mind', but 'Mind itself . . . most perfect Mind . . . of such Thou art the Form'. A gloss on Harris's meaning here, and at the same time a measure of his influence on Monboddo's thinking, can be seen clearly in the latter's remarks in 1776 that 'the capital difference' between ancient and modern physics is that 'the Ancients not only held Mind to be the first cause of all things, but the immediate cause of the chief operations in Nature. I use the word Mind in a large sense, so as to comprehend not only Intelligence, but Vitality, and whatever principle there is in Nature that produces Motion.'

In 'Dialogue on Art' the word imitation is not used at all. It occurs only in the notes in order to distinguish the more sublime intellectual arts (logic, rhetoric, moral virtue, etc.) from the less sublime 'plastic' and performing arts of sculpture, painting, and dramatic poetry, to which Harris then turns in the second treatise.

Harris found the distinction between Energy and Work at the beginning of the first book of Aristotle's *Nichomachean Ethics*.[14] In detail, Aristotle distinguished between activities

[14] *Nichomachean Ethics* 1. 8. 1098–9.

(*energeia*), products (*ergon*), actions, arts, and sciences (*episteme*), and ends (*telos*). Harris followed Aristotle's formulation of the highest realizable good, as happiness (*eudaimonia*) in his final treatise 'Concerning Happiness', and Fielding similarly adopted the Aristotelian distinction between the mere possession (*hexis*) of virtue and its enactment in a state of character and its active expression in conduct (*energeia*). Aristotle remarked that 'most men . . . take refuge in theories, and suppose that by philosophizing they will be improved—like a sick man who listens attentively to his physician but disobeys his orders. Bare philosophy will no more produce health in the soul than a course in medical theory will produce health in the body.'[15] Fielding's letter to Harris on the death of his wife Charlotte states that Harris had first taught Fielding precisely this lesson about the value of philosophy as an activity of the mind.

Harris's whole epistemology is built on this dynamic dialectic. His authority for transferring the distinction from ethics to Art and for rendering Art as a potent force in culture and as a necessary satisfaction of human needs, was Quintilian's *Institutes*. In France and Germany Harris himself was credited with the theory. Batteux translated the first two treatises into French in his *Principes de littérature* (1755), a year before the German translation of the complete *Three Treatises* by J. G. Müchler. Herder appropriated the theory of *Energie* and *Werk* for his *Erstes Waldchen*, but with no specific reference made to Harris.[16]

III

The second treatise, 'Concerning Music, Painting, and Poetry', turns from a discussion of the psychological causes of Art to a schematic and comparative analysis of particular art-forms, all of which are dependent, to a greater or lesser extent, on a mimetic theory. Harris argues that painting is superior to music in that it can raise ideas and depict a more universal reality; but both are inferior to the 'Master-knowledge'

---

[15] *Nichomachean Ethics* 1, 8, 1098–9.
[16] G. ten Hoor, 'James Harris and the Influence of his Aesthetic Theories in Germany', Ph. D. thesis (Ann Arbor, Mich., 1929).

provided by poetry. The three media are arranged as follows:

Art	Organ	Media	Limitations
Painting	Eye to mind	Colour and figure: (natural)	Visible objects only; *punctum temporis*; static
Music	Ear to mind	Sound and motion: (natural)	No visual dimension
Poetry	Ear to mind by 'compact'	Sound and motion: (artificial)	None

Poetry, of course, is restricted to words (sound and motion), but because words convey meaning 'by compact', poetry expresses everything conceivable. Whereas music and painting employ 'natural' media, poetry depends on an 'artificial' medium which has no necessary relation to and should not be judged by the criterion of pictorial accuracy. The essential superiority enjoyed by poetry is that it is of all the arts the one that can 'lay open the internal Constitution of Man, and give us an Insight into Characters, Manners, Passions, and Sentiments' (p. 84). It is poetry alone that can 'raise *no other* [Idea] than what every Mind is furnished with before' (pp. 77–8), since 'the Sentiments in real Life are only known by men's Discourse' (p. 89). Poetry's subject therefore appeals to our higher moral sense, 'an express Consciousness of something similar within; of something homogeneous in the Recesses of our own Minds; in that, which constitutes to each of us his true and real Self' (p. 89). No English critic before Harris had gone as far as this in relating the philosophy of mind to the study of language. The next step to be taken was into a philosophy of language known as 'universal grammar', and Harris again anticipates his next book in the penultimate chapter of this discourse, where he suggests a relationship between words and moral nature (p. 91):

Not only therefore Language is an *adequate* Medium of Imitation, but in Sentiments it is the *only* Medium; and in Manners and Passions there is no other, which can exhibit them to us after that clear, precise, and definite Way, as they in Nature stand allotted to the various sorts of Men, and are so found to constitute the several Characters of each.

Harris has essentially finished at this point, and although he concludes that of all the sister arts it is poetry that is superior, the last chapter attempts to rescue music from its subordinate status by reopening its special alliance with poetry. The first dialogue, on the psychology of art as a human creation, mentions no single artefact; the second examines no particular painting (Raphael, Titian, and Salvator Rosa are mentioned in notes only), no poems except one line from *Lycidas* and two references to *Paradise Lost*, and the only musician mentioned by name is Handel. But, for this last chapter, Handel is crucial, since Harris's final position on the hierarchy of the arts is that the union of poetry and music in opera or oratorio is 'a Force irresistible, and penetrates into the deepest Recesses of the Soul' (pp. 99–100). The explanation for this— apart from Harris's particular awe of Handel—is that poetry compensates for the temporary and transient effect of music, literally concentrating the mind on the otherwise vague 'raising of Affections'. Except in quite trivial ways, music is neither intrinsically mimetic nor intellectually stimulating, but it is the most effective means for eliciting 'affections to which Ideas may correspond' (p. 99). Handel is the master of such an alliance between the mind and the soul.

Charles Avison, generally regarded as the first important critic of music in England, was a careful reader of *Three Treatises*, and naturally acknowledged Harris's formative influence on his own thinking.[17] He adopted Harris's sceptical view of music's mimetic powers and agreed that the alliance of music and poetry was the most affective symbiosis between the sister arts. Handel is cited by both writers as the proof, and Avison does little more than echo Harris's theory:

Music as an imitative Art has *very confined Powers*, and because, when it is an Ally to Poetry (which it ought always to be when it exerts its mimetic Faculty) it obtains its End by *raising correspondent Affections* in the Soul with those which ought to result from the Genius of the Poem.[18]

---

[17] H. M. Schueller, 'Literature and Music as Sister Arts: An Aspect of Aesthetic Theory in Eighteenth-Century Britain', *PQ* 26 (July 1947), 195. Harris's treatise is the earliest cited here.

[18] C. Avison, *An Essay on Musical Expression* (2nd edn., 1753), 60. Avison quotes Harris on the mimetic limitations of music and adds: 'This has been already shown, by a judicious Writer, with that Precision and Accuracy which distinguishes his

The pleasure of music thus arises not from a narrowly mimetic purpose but from its unique ability to stimulate what Avison called 'a peculiar internal Sense . . . of a much more refined Nature than the external Senses'. Musical sounds have as their 'peculiar and essential Property, to divest the Soul of every unquiet Passion, to pour in upon the Mind, a silent and serene Joy, beyond the Power of Words to express' (p. 2). This is Harris's own view and also, as we shall see in Chapter 7, that of Thomas Naish in his sermon to the Salisbury Society of Musick in 1726. Avison goes no further than Harris when he says that the purpose of music is to 'fix the Heart in a rational, benevolent, and happy Tranquillity . . . to raise the *sociable and happy Passions*, and to subdue the contrary ones (p. 4). The source for this is undoubtedly Shaftesbury's *Letter Concerning Design*, and in the second edition of his *Essay*, Avison included John Jortins's *Letter to the Author*, in which the ethical implications of this musical theory are made even plainer: 'There is no *Harmony* so charming as that of a well-ordered Life, *moving in concert* with the sacred Laws of Virtue' (p. 41). Harris's poem *Concord* is a celebration in verse of precisely this idea.

Avison's most eloquent plea is for a national reform of music, starting with cathedral organists, who ought to be 'our *Maestri di Capella*' (p. 97). Their models should be the best Italians: Palestina, Carissimi, Marcello, Geminiani, Giardini, Pergolese, Bononcini, Stradella, Steffani, Corelli, Scarlatti, and the as yet little-known operas of the Frenchman Rameau.

Writings. To his excellent Treatise I shall, therefore, refer my Reader, and content myself, in this Place, with adding two or three practical Observations by way of corollary to his Theory' (pp. 60–1). Avison's Advertisement draws further parallels between the syntax of music and grammar ('*Passages* in Music are also like *Sentences* or *Paragraphs* in Writing'), and between painting and music ('*Design, Colouring* and *Expression*' are to the former as '*Melody, Harmony*, and *Expression* are to the latter) (p. 21). Further references to the *Essay* are given in the text. In his *Critical Reflections on Poetry, Painting and Music* (1719–33), Du Bos traced this idea back to Cicero: '*nothing is more naturally agreeable to our minds than numbers and sounds, for by these our passions are excited and inflamed, and by these also we are soothed and taught to languish.* [*De Oratore*, p. 3] By this means the pleasure of the ear is communicated to the heart.' See T. Nugent's trans., 3 vols. (1748), i. 362 (item 821 in the Hurn Court library sale). Writing of the 'intimate relationship' between Handel's orchestral composition and the dramatic context, Winton Dean remarks: 'The orchestra is never a mere background or a cradle for the voice; it supplies the atmosphere in which the music breathes and through which it draws its life-blood; *Handel and the Opera Seria: The Ernest Bloch Lectures* (1970), 185.

Lulli is to France as Scarlatti was to Rome and Handel is to England. Harris was not of course a cathedral organist, but Avison's list could find no better practical exemplification than the 2 volumes of his arrangements published by Joseph Corfe, who was the cathedral organist at Salisbury, around 1800.[19] The arrangements included the works of Pergolesi, Scolari, Jomelli, Vinci, Perez, Martini, Beretti, Geminiani, and Cocchi. Unpublished during Harris's lifetime, these adaptations were regularly performed in Salisbury. Thus, Harris's essay on the theory of music (or *hexis*) had the practical concomitant (*ergon* or *energeia*) of sustaining a brilliant provincial repertoire of sacred music (*episteme*). His conviction that Handel's sacred oratorios synthesized the mind and the soul would eventually lead to that Wagnerian sense of opera as a *Gesamtkunstwerk*.

In a volume that implicitly offers the materials for a revaluation both of Harris's musical theories and of his interpretation of Aristotle in particular, Elder Olson remarks that 'Harris has had far less than his due from the generality of latter-day critics. There is . . . more sound and original thought manifested in this little work than in the combined works of most of Harris's detractors: and this treatise deserves to be remembered as an important investigation of the problem of how imitation is possible in the arts concerned.'[20] We should remember that before Henry Pye's translation of the *Poetics* in 1788, English-only readers had access to Aristotle's work solely through Rymer's translation of Rapin's *Reflections on Aristotle's Treatise of Poesie* (1674), the English version of Mme Dacier (1705), and a poor anonymous translation of 1775. Harris's expertise in Greek made *Three Treatises* a major critical and textual event, not least for

---

[19] See Robertson, *Sarum Close*, pp. 232, 244–7; and ch. 7 below. Joseph Corfe (1740–1820), cathedral organist from 1792–1804, dedicated his 2 vols. of Harris's arrangements to the Earl of Malmesbury and records his gratitude for Harris's patronage 'at an early age . . . and during a long course of years' (*Sacred Music*, i. 2). Joseph's son succeeded as organist until his death in 1825.

[20] E. Olson (ed.), *Aristotle's 'Poetics' and English Literature: A Collection of Critical Essays* (Chicago, 1965), p. xxii. Olson includes H. J. Pye, *A Commentary Illustrating the Poetics of Aristotle* (1792), and T. Twining, *Aristotle's Treatise on Poetry*, ed. D. Twining, 2 vols. (2nd edn., 1812), i. 3–65. Harris's essay is the first in this collection. Thomas Twining was Dr Burney's acknowledged classical consultant for his *General History of Music*, 4 vols. (1776–89) i, p. xix.

transmitting the concept of *energy*. Certainly, the excellence of Thomas Twining's later translation of the *Poetics* in 1789 was in some ways indebted to Harris's definition of imitation and his sceptical view of what Johnson was to call 'representative metre'. Twining also cited *Hermes* and *Philological Inquiries* with much approval, and silently appropriated the term first applied to Fielding's novels by Harris, 'comic epopee'.[21]

Harris's specific rescue of music from its subordinate position as a merely decorative scaffold for words to a potent ally of poetry was, in the long run, of even greater moment. Unlike Du Bos, Voltaire, Rousseau, Diderot, Grimm, D'Alembert, Marmontel, Arnaud, La Harpe, Beaumarchais, Algarotti, Avison, and Beattie, all of whom stressed the imitative function of music, Harris determined that music is 'a power which consists not in imitations . . . but in the raising affections to which ideas may correspond' (pp. 40–1). At least one commentator has noticed the implications of this for eighteenth-century opera. Gluck's reform in *Orfeo* (1762) and in the preface to *Alceste* (1769) was to move away from the 'abuses' of Italian opera in the direction pointed by Harris's treatise: 'I sought to reduce music to its true function, that of supporting the poetry, in order to strengthen the expression of the sentiments and the interest of the situations, without interrupting the action or disfiguring it with superfluous ornament.' As this critic points out about Harris's treatise, 'we seem to be listening to Gluck himself'.[22] Finally, and as a measure of Harris's distance from prevailing orthodoxies, we might recall the remark cited in the *Encyclopédie* from Algarotti's *An Essay on Opera* (1762): 'All music, that paints nothing, is only noise.'[23]

IV

'Concerning Happiness, A Dialogue' takes place over two days and is dedicated to Harris's Platonic friend Floyer

---

[21] Olson, *Aristotle's 'Poetics'* p. 21 (Twining's trans.).

[22] E. Newman, *Gluck and the Opera: A Study in Musical History* (1964), 238. Newman analyses Harris's essay on pp. 257–61, noting that Harris 'seems to be nearer the modern aesthetic of music when he deprecates too strong an insistence on the merely imitative function of the art' (p. 259).

[23] F. Algarotti, *An Essay on Opera* (1762), Engl. trans. dedicated to William Pitt (Glasgow, 1768), 47.

Sydenham. The outer form is like an occasional letter,
enclosing the inner form of a written conversation between a
Platonist and a Stoic, both elements framed by an extensive
commentary pursued in footnotes. Harris himself appears
nowhere as a character (although the 'under-character' is said
to have musical as well as literary and philosophical interests),
but the real authorial presence is, literally, marginal. The
annotations are not appendages but integral, serving a
number of important purposes. They function educationally,
like a dictionary of classical philosophy, defining terms, ideas,
movements, telling us who wrote what and under whose
influence. Their main purpose, however, is to illustrate the
timelessness of the question, 'What is the sovereign good?',
and the reply to this question foreshadows, if not directly
influences, the ethical speculations in the work of two of
Harris's close friends, particularly in Johnson's *Rasselas* and
more generally in Fielding's *Tom Jones*.

Harris is very specific about the composition of this final
dialogue ('Finished Dec. 15, A. D. 1741' appears on the title-
page). It also involved some of his friends. We know from
their correspondence that the second treatise had been
discussed in private letters between Harris and Shaftesbury,
but the third was actually read in manuscript by Fielding (as he
explains in his *Essay on Conversation*, 1742),[24] and when it
appeared in print it contained a generous public acknowledgement to John Upton's edition of Arrian's Epictetus (1738).
Aware that its dramatic form and its parade of Stoical views
would not prevent personal charges of impiety, Harris was
perhaps nervous about the reception of this dialogue. Much of
his own thinking on ethics was indeed contained in the
dialogue, but it was never conceived as an argument for
orthodox Christian piety. A horror of religious extremism and
a belief in reasonableness hardly constitute the creed of an
infidel, but in the event this was the piece which was to dog his
reputation with scepticism. Harris's position is nowhere better
illustrated than in the epigraph at the beginning of this
chapter, but a comparison with a German Romantic was to do
him no good in the 1740s.

To Harris the 'True and Rational Life' is that 'where the

---

[24] Miller (ed.), *Miscellanies by Henry Fielding, Esq.*, i. 122.

Value of all things is justly measured by those Relations, which they bear to the Natural Frame, and real Constitution of Mankind; in fewer Words, a Life of Virtue appears to be the Life according to Nature' (p. 174). In chapter 22 of Johnson's *Rasselas* (1759) the prince visits an academy and listens to a charismatic philosopher who states that 'to live according to nature, is to act always with due regard to the fitness arising from the relations and qualities of causes and effects; to concur with the great and unchangeable scheme of universal felicity; to co-operate with the general disposition and tendency of the present system of things'.[25] As narrator, Johnson enters with a censor's irony at this point to remark that here was 'one of the sages whom he should understand less as he heard him longer'. The possibility that Harris and his *Dialogue* were both in Johnson's sights when he was attacking philosophic theories of happiness is supported by many other details in addition to this apparent reproof. Harris and Johnson both use imagery of vegetation and seasonal change to depict the folly of man's psychological need to make perpetual that which is only transient: we 'seek the Vegetables of Spring in the Rigours of Winter, and Winter's Ice, during the Heats of Summer' (p. 118); 'No man can taste the fruits of autumn while he is delighting his scent with the flowers of the spring' (*Rasselas*, p. 72). Both regard the effort as a futile, but also as an ineradicable motive in the search for human happiness. The ideal country imagined by Harris ('an Elysian Temperature of Sunshine and Shade', p. 132) is, like the Happy Valley in *Rasselas*, a creation of the mind which the mind itself will destroy by satiety and tedium. In both works the life of contemplation is ruined by the sublunary fact of fogs which dim the prospects and the cares of life which molest us. Harris points out that even the best men are subject to casual accidents ('Pests may afflict their Bodies; Inundations o'erwhelm their Property; or, what is worse than Inundations, either Tyrants, Pirates, Heroes, or Banditti', p. 177), a caveat which Pekuah and Nekayah are forced to acknowledge when Pekuah is abducted by the Arabs! Johnson's 'wise and happy man', the stoic philosopher of 'rational fortitude' in chapter

---

[25] *The History of Rasselas Prince of Abissinia*, ed. J. P. Hardy (Oxford, 1968), 56. All references are to this edn. and are given in the text.

18, echoes the preface of *Hermes* when he compares Reason to the sun, constant, uniform, and lasting, but he loses his fortitude on the death of his daughter. Harris's 'stargazer' and Johnson's astronomer are both used as emblems of intellectual hubris, the one a child in wisdom, the other a tragic example of the 'uncertain continuance of reason'. Most damaging of all to Harris's careful theory of happiness is Johnson's fundamental assertion that life is simply not characterized by free choice. Harris had put it this way:

in the Moral Art of Life, the very Conduct is the End . . . of all Arts is this the only one perpetually complete in every Instant, because it needs not, like other arts, Time to arrive at that Perfection, at which in every instant 'tis arrived already. Hence, by Duration it is not rendered either more or less perfect; Completion, like Truth, admitting of no Degrees, and being in no sense capable of either *Intension or Remission*. And hence too by necessary Connection (which is a greater paradox than all) even that Happiness, or Sovereign Good, the End of this Moral Art, is itself, too, in every instant, Consummate and Complete; is neither heightened or diminished by the Quantity of its Duration, but is the same to its Enjoyers, for a Moment or a Century. (pp. 189–90)

Imlac had already rejected the notion that fulfilled happiness rests upon 'incontestable reasons of preference', since such a life of constant acts of choosing would be spent not in living but in 'inquiry and deliberating'. Of more pointed reference to the passage from Harris is that even the naïve Nekayah is allowed to see through such empty optimism: 'There are a thousand familiar disputes which reason can never decide; questions that elude investigation, and make logick ridiculous; cases where something must be done, and where little can be said' (p. 71). The specific mention of logic being made ridiculous may refer to Harris's particular argument for happiness: the italicized words *intension* and *remission* carry distinct logical connotations, and are cited in Johnson's *Dictionary* accordingly,[26] although the general point of Nekayah's speech is that no matter what degree of rational precision and verbal nicety is used, life always moves too quickly and is too complicated to permit the luxury of

---

[26] Johnson cites Locke's usage: 'The difference of intention and remission of the mind in thinking, every one has experimented in himself.'

detached ratiocination. 'Very few . . . live by choice,' Imlac remarks, since every man 'is placed in his present condition by causes which acted without his foresight, and with which he did not always willingly co-operate.'

Harris's proposition that happiness is a matter of pursuit rather than achievement is illustrated by the example of the sportsman who admits of his companions that 'Completion of their Endeavours was so far from giving them Joy, that instantly at that Period all their Joy was at an end' (p. 194). Johnson's travellers, and his readers, learn that there is not only no joy in arriving, but no arrival either, only a fresh beginning: 'The conclusion in which nothing is concluded', as the final chapter is entitled. Paradoxically, this is not form subverting the expectations aroused by content, but exactly the reverse. If there is no achievable state of human felicity in the sublunary world, and no choice of life in a world dominated by contingency and accident, then there is no destination except destiny itself. The end of *Rasselas* is not, therefore, ambiguous, but it is insufficient. If the choice of life is not a real choice, then neither is Nekayah's sudden transference to 'the choice of eternity'. What is the evidence to guarantee any greater possibility of success in something we can know nothing about? In the secular terms of Johnson's moral fable there can be no answer to this, apart from retreat to the convent of St Anthony in Pekuah's case. Though we are doubtless to infer from the penultimate chapter that a providential God awaits the worthy pilgrim after death, human ignorance and insufficiency make the 'choice of eternity' the only alternative available: a non-choice in fact. Imlac, who for so much of *Rasselas* provides a pseudo-Johnsonian bedrock of disillusioned sense, is strangely silent. His willingness to be 'driven' along the stream of life without directing [his] course to any particular port' is a humiliating end to his magnificent and ambitious vision of the poet as legislator of mankind in chapter 10. Imlac, ultimately, fades: he becomes a drifter and a dilettante because Johnson cannot give to Art (and poetry in particular) the function and status of revealed religion. The end of the quest is therefore necessarily gloomy and necessarily incomplete without the spiritual sanctions Johnson wishes us to infer.

But Harris could, and did, bestow on Art the capacity to perceive a transcendental happiness. Harris's interlocutor relates the story of Theophilus (perhaps Harris's amalgam of Shaftesbury's Theocles and Philocles in *The Moralists*), when he is asked 'whence it should happen, that in a Discourse of such a nature, you should say so little of Religion, of Providence, and a Deity' (p. 222). The question has perplexed readers of Johnson's *Rasselas* for the same reason. Johnson's Imlac has nothing to say, but Harris's Theophilus was also a poet and is positively brimming over with answers. He it is who bears the responsibility in the *Dialogue* for relating his scheme of rational, social happiness to the demands of a providential God. Through Theophilus, Harris offers an essentially Platonic vision of good, aspects of which are observable in the 'multitude of mixed imperfect characters', but the totality of which no individual character can combine in himself. Theophilus's subsequent speech is characterized by what Harris's narrator calls 'a large Portion of that rapturous, anti-prosaic Stile, in which those Ladies [the Muses] usually choose to express themselves' (p. 224), a direct anticipation of the Prince's reaction to Imlac's description of poetry in chapter 10 of *Rasselas* ('Imlac now felt the enthusiastic fit . . . Enough! thou hast convinced me that no human being can ever be a poet'). Theophilus, 'warmed to a degree of rapture', proceeds:

The more diligent our Search, the more accurate our Scrutiny, the more only are we convinced, that our Labours can never finish; that Subjects inexhaustible remain behind, still unexplored.

Hence the mind truly wise, quitting the Study of Particulars, as knowing their Multitude to be infinite and incomprehensible, turns its intellectual Eye to what is general and comprehensive, and through Generals learns to see, and recognize whatever exists. (pp. 226–7)

For Augustan aesthetics this direction becomes a desideratum, advanced as a fundamental prerequisite in Reynolds's *Discourse Four*, in Pope's *Essay on Criticism*, in Johnson's *Life of Cowley* and *Preface to Shakespeare*. But by common consent its *locus classicus* is Imlac's speech in *Rasselas*, beginning: 'The business of a poet . . . is to examine, not the individual, but the species;

to remark general properties and large appearances.' Imlac's poet rises to 'general and transcendent truths', and is 'the interpreter of nature, the legislator of mankind'. Theophilus maintains that such a mind enjoys a 'transcendent faculty', which 'becomes a canon, a corrector, and a standard universal'. Imlac's poet must 'preside over the thoughts and manners of future generations; as a being superior to time and place'; Harris's Theophilus regards the mind's proper aspiration 'to regard the universe itself as our true and genuine country, not that little casual spot where we first drew vital air'. As a preparation for this awesome task Imlac stores his mind with 'all the appearances of nature . . . all that is awfully vast or elegantly little'; Theophilus's dictum is no less comprehensive: 'the whole train of moral virtues . . . my own stock, my own neighbourhood, my own nation . . . the whole race of mankind, as dispersed through the earth'. Unlike Imlac, Theophilus is allowed to finish his speech, and he concludes with an appeal for Stoic acquiescence in a providential universe which has little overt parallel in *Rasselas*, but which does anticipate Johnson's earlier verse essay on the theme of happiness, *The Vanity of Human Wishes*. If 'Wisdom . . . makes the Happiness she does not find', as the last line of Johnson's poem assures us, then Theophilus's doctrine that we should acquiesce in 'every obstruction, as ultimately referable to thy providence' produces the same conclusion: that is, that we can thereby become 'happy with that transcendent Happiness, of which no one can deprive us; and blest with that Divine Liberty, which no Tyrant can annoy' (p. 235).

Yet there is one fundamental difference separating Johnson's use of Imlac and Harris's use of Theophilus. Though they correspond to each other in the differentiae of argument, aesthetic and moral theory, manner, and place (both being identified with natural or, rather, poetical landscape as the proper context for such raptures), Imlac talks exclusively about the mind of the poet and Theophilus, though a poet, extrapolates his argument into a concept of 'the perfect moral character'. Johnson cuts off Imlac's 'enthusiastic fit' as a species of obsession, thereby implicitly denying the extension of his poetical argument into Harris's kind of universal moral animism. Harris's references to 'that animating Wisdom, which pervades and rules the Whole—that Law irresistible, immutable, supreme, which leads the Willing, and compels the Averse, to co-operate in their Station to the general Welfare—that Magic Divine' (p. 233), would have made no sense to Johnson because it provided no practical substitute for the mysteries of Christianity. To Johnson man's greatest burden is *self*-consciousness, the pain and paradox of his existence. The prince's diagnosis, 'that I want nothing . . . or that I know not what I want, is the cause of my complaint', is followed by an appeal, 'give me something to desire'. Within the secular terms of Johnson's fable there can be no satisfaction of such needs: the 'choice of life' is a mirage because it is the nature of life to be driven along by contingency, not ratiocinative choices. The best one can hope for is a moment of apparent stability poised on the edge of flux. Harris's leap from Stoic philosophy into a metaphysical piety is a trick of the moonlight. Indeed the seeds of Johnson's critical treatment of Imlac's visionary poetics may also lie in Harris's sceptical narrator's reaction to Theophilus' rhetoric: 'Then the Business of the Day gently obliterated all' that had seemed so convincing when surrounded by 'the night's beauty and stillness, with the romantic scene where we were walking . . . and left me by Night as little of a Philosopher, as I had ever been before' (p. 236). In the same way, Imlac's poetic theory begins among 'the crags of the rock and the pinnacles of the palace . . . the mazes of the rivulet . . . the changes of the summer clouds' (*Rasselas*, p. 62). Unlike the Platonism which Harris is forced into adopting, Johnson's rigorous separation of secular from transcendental, and of life from Art, prevents Imlac's visionary poetic from imposing a neat and schematic order on the transient and the imperfect.

Thus the relationship between Harris's *Dialogue* on happiness and Johnson's *Rasselas* is complex. The function of poetry and the mind in each is hyperbolically presented, a core of typically Augustan desiderata carried too far; the choice of life which is a central theme in each work issues out into metaphysics in the one, and is denied as a possibility in the other. Johnson up-ends (to use Wasserman's word)[27] our

---

[27] 'Johnson's *Rasselas*: Implicit Contexts', *JEGP* 84: 1 (1975), 1–25.

narrative expectations just as he had up-ended every other category of stable, durable, and worthy human attempt to deny earthly transience in the course of *Rasselas*. Inevitably, the question of Johnson's indebtedness to Harris arises as a consequence of the parallels in theme, content, and attitude. The evidence is mostly internal and has already been presented above. Johnson was certainly aware of the intellectual relationship between Harris and Shaftesbury and he infers from this that Harris's obsession for classical philosophy had eroded his Christian orthodoxy.

Harris provides a type of the well-meaning busy philosophic intelligence, particularly imbued with a faith in logic and ancient philosophy, which Johnson frequently criticized as an example of intellectual hubris. The space between Harris and Johnson was very small. Harris was a regular contributing member of what Boswell called Johnson's 'good company', a friend, though an exponent of linguistic and logical views unfashionable at the time, not least with Johnson himself. In the history of ideas as currently presented Harris is less prominent than Shaftesbury, but as far as Johnson's intellectual life is concerned Harris's presence represented an actual social embodiment of the classical humanist tradition and of its Aristotelian bases upon which rested Johnson's own literary and moral outlook. As such, Harris was an existent, if not a 'soft', target for Johnson's critical revisions to that tradition, although tact dictated that he could not be named as a specific target. At least this much may be said for their proximity. But on fundamental issues and in philosophic outlook there was an intellectual chasm between the two men. On one occasion Johnson was asked by Boswell why he, Boswell, found it difficult to get on with his own father when he found it easy to get on with Johnson, who was approximately of the same age-group. Johnson's reply characterized Boswell's father as a man who took 'all his notions . . . from the old world', and himself as 'a man of the world. I live in the world, and I take, in some degree, the colour of the world as it moves along'. In the history of ideas it was a generation gap. Harris tried hard to revivify classical philosophy as a homogeneous and universal system of conduct for 'men of action and business; men of the world'. But the world had changed. Johnson loved to have his wisdom operate on the business of living, and Harris, along with his friend Monboddo, it seemed, inhabited the mental world of the Aristotelian past. After Locke had ushered in a new world of rational empiricism it was no longer a place for a neo-Aristotelian conquering hero. Indeed Shaftesbury himself bemoaned the fact that philosophy itself was 'no longer active in the world . . . if some few maintain their acquaintance, and come now and then to her recesses, 'tis as the disciple of quality came to his lord and master, secretly and by night'.[28] Johnson's most damaging remark on Harris's expertise as the 'eminent Grecian' was the cutting reply: 'His friends give him out as such, but I know not who of his friends are able to judge of it.' Johnson's remark was one of the meanest he ever made. It was also disingenuous.

[28] From *The Moralists*. See Robertson's edn. of Shaftesbury's *Characteristics of Men, Manners, Opinions, Times*, ii. 4–5.

4

A Particular Friendship
Henry and Sarah Fielding, 1740–1769

Let my Lamp at midnight hour
Be seen in som high lonely Towr,
Where I may oft outwatch the *Bear*,
With thrice great *Hermes*, or unsphear
The spirit of *Plato* to unfold
What Worlds, or what vast Regions hold
The immortal mind that hath forsook
Her mansion in this fleshly nook.

Milton, *Il Penseroso*

HENRY FIELDING's itinerary as a lawyer on the Western Circuit assizes from the summer of 1740 onwards made Salisbury both a necessary and a congenial place to visit. His metropolitan social zest was put at the service of his more retiring friend James Harris, and several letters show Fielding mischievously surprising Harris with the extent of his knowledge of Salisbury news, both real and fictional. From Bath, 24 September 1742, he relays some newspaper report about Edward Goldwyre (1707–74), the Salisbury surgeon who had been Harris's next-door neighbour, as well as Charlotte Cradock's before her marriage to Fielding. Evidently enjoying a reputation as a beau and a dancer, Goldwyre was

lately married to one Miss Harris a young Lady of &c—in Return for wch yo may if you please acquaint me that one Henry Fielding Esqr. commonly known by the Name of *Beau Fielding* did to the utter Confusion of all his Brother Beaus open the first Ball at Bath with a Minuet. As strange as this News may appear, it is as true as that I am with the most perfect Sincerity and most ardent Esteem . . . Yr. most obliged affectionate humble servant.[1]

---

[1] Harris Papers, vol. 40, part 4, 'Letters chiefly on Literary Subjects'. All the following extracts from Henry Fielding's letters and others are from this source and are identified hereafter by date only. Edward Goldwyre (1707–74) married the eldest daughter of William Harris of the Close on 8 September 1742: see *The Gentleman's Magazine* (12 Sept. 1742), 499, and *Salisbury Journal* for 18 Aug. 1746. He witnessed the codicil to the second James Harris's will (24 Aug. 1731) and subscribed to Fielding's

I

As the tone of this letter makes plain, Harris's most significant correspondence in the 1740s was also his most intimate. His relationship with Fielding is best described in Fielding's own words. Only two of Harris's letters to Fielding have survived, but the commonality of minds between the two men was always stronger than the apparent dissimilarity of their interests. The letters to Harris are, of course, the only significant group of Fielding letters to survive, largely because Fielding found letter-writing a chore. But as subsequent discussion will show, he was nevertheless an expert in the art of the informal letter.

Their friendship obviously pre-dated Fielding's initiation of an epistolary relationship, which actually began on 8 September 1741. Annoyed by the dull company available in the otherwise pleasant Bath, and in the immediate company of Serjeant Burnet and his ex-schoolfellow Charles Hanbury Williams, Fielding tells Harris of his secret delight in hearing Burnet praise his Salisbury friend. The letter opens with a recollection of an earlier event suggestively prophetic both of the opening paragraph of the 'Dialogue Concerning Art' (which was being written at this time), and also of a theme common to *Joseph Andrews* and *Tom Jones*:

You may have forgot that in one of our Evening Walks, whilst I was busy in explaining some Conceits of mine (for probably they are no other) concerning the clear Distinction between Love and Lust, our Conversation was interrupted by several fair Objects of both those Passions. Had I then proceeded, I should have told yo, as perhaps one Instance of their distinct Existence in my Mind, that nothing was ever more irksome to me than those Letters which I had formerly from the latter Motive written to Women, nor any thing more agreeable and delightful than I have always found this Method of conversing with the absent beloved Object . . .

For my own Part, I solemnly declare, I can never give Man or Woman with whom I have no Business (which the Satisfaction of Lust may well be called) a more certain Token of violent Affection

Miscellanies (1743). His father, William (1666–1748), also a surgeon, was Charlotte and Harris's neighbour. J. Paul de Castro suggests that 'the famous Surgeon' who mended the heroine's nose in *Amelia* was Edward. See *Amelia*, ed. M. C. Battestin (Oxford, 1983), 68.

than by writing to them, an Exercise which, notwithstanding I have in my time printed a few Pages, I so much detest, that I believe it is not in the Power of three Persons to expose my epistolary Correspondence. Receive this therefore as the Fruit of an Affection which hath been long growing, and hath taken the deepest Root whence it originally sprung, & how it hath been cultivated I shall not intimate: for as Flattery sufficiently distinguishes the two letters I have mentioned above from each other, this must bear all the Marks of the purer Passion, since Narcissus would never have inspired me with the other. I intend yo therefore no Compliment in this nor any future Letter, unless my Friendship be one which, as it bears no gawdy Outside, I shall endeavour to reconcile you to by its Qualities.

Fielding also offered his services to Harris's brother Thomas, even though Fielding himself was living in temporary accommodation and had no spare room. Harris replied on 17 September, and the exchange began, the gregarious and mobile Fielding promising gossip from Bath and political news from London for the relatively reclusive bachelor in news-starved Salisbury, hard at work on aesthetic theory.

In fact, Fielding was also to sound out Harris on some of his most important concerns as a moralist and writer of fiction. On 29 September, from the Bath bookshop of Samuel Richardson's brother-in-law, James Leake, Fielding broached the subject of comic theory. He was prompted to this by a disdainful smile directed at him by a beautiful woman in the Pump Room, and the anecdote shows Fielding elevating social trivia to a level appropriate for a philosopher's attention: 'I began to philosophize on this Occasion . . . I soon resolved with Mr Hobbs, that it was the Effect of Pride and she desired no more than to acquaint me with her Contempt. I carried this still farther and am in doubt whether that Laughter which entitles to the general Character of Good Humour be not rather a Sign of an evil than a good Mind.' After fleshing out his argument with Horace, Juvenal, and Homer, Fielding cites Democritus ('the first and greatest of all the Philosophers'), and reminded Harris of his favourite Montaigne's remark that 'there is more room for Laughter to exert itself as we are not so full of misery as inanity'. Of his own state of mind, there is this defensive remark: 'my Learning like my Life is less deficient in good Matter than Method, for want of which I have the small Satisfaction of knowing both to be somewhat better than they are allowed by those who have no great Right to call either in question.' This lengthy letter, so clearly written to the moment, is dramatically curtailed by the news of the death of Hanbury Williams's mother: 'I am in a true sense not in the vulgar Phrase concerned for him, who is capable of being a Son and a Friend, which latter Word when in the strictest Sense yo apply to me you will do Justice and Honour both.' One of the chief characteristics of the ensuing correspondence was Fielding's clear assumption that Harris would always be ready to discuss questions of a theoretical kind.

On 27 March 1742 the London news from Fielding was political. The Committee of Secrecy established to investigate the conduct of Sir Robert Walpole had met, and Fielding enclosed a list of its members, including Harris's cousin Edward Hooper. It seemed that the Committee would bring down a judgment favourable to the Opposition, and although Fielding assumes that this will please Harris, Fielding himself was too engrossed by his wife's illness to worry overmuch about Walpole (this was a 'very dangerous Illness in wch she was given over', he later reported, 6 September 1743). Fielding's sense of harassment persisted through the autumn, and from Bath on 24 September he wrote to tell Harris of his plan to spend the next Monday afternoon or evening with him either at Salisbury or Amesbury, adding that 'there is no Man living in whose Esteem I so eagerly desire a very high Place'.

Fielding's additional anxiety lay in the urgent need to find enough material to make up the first volume of his *Miscellanies*. Harris was recruited to help in the two matters of bulk and poetry, and he therefore sent some 'Verses' by Fielding preserved by Lord Hardwicke, the Lord Chancellor and the dedicatee of Harris's second book, *Hermes* (1751). Harris believed the verses were worth publishing, but Fielding thought his friends had overvalued them, Hardwicke by 'long preserving' them, Harris by 'commenting on what was originally writ with the Haste & Inaccuracy of a common Letter, & wch I shall be sorry if any Scarcity of Matter under the *Poetical* Article should oblige me to publish'. There are insufficient clues for identifying these poems, but perhaps one of them was 'A Description of U[pton] G[rey] . . . 1728', where Fielding was living in the 1720s.

A more substantial matter was Fielding's version of 'The First Olynthiac of Demosthenes'. Henry Knight Miller's uneasiness about the unevenness of this prose piece turns out to have been well founded. Miller remarks that this

mysterious little exercise in Translation . . . rendered by Fielding almost without any mark of Henry Fielding to be found in it [is] the almost literal rendering, one feels, of a student, not the mature translation of a man in his full powers . . . Unless one is to suppose that he revised with greater skill and knowledge an earlier academic exercise, the curious mingling of competence and naivete in his production of the Greek text seems almost inexplicable.[2]

The truth is that Fielding himself was embarrassed by its poor quality and had sent it to Harris, the Greek expert, for revision. On 27 February 1743, from Brompton, he asked Harris to 'correct' it and also to 'give the Argument of it in as few Words as yo please. The Hurry I am in at present & not having a Demosthenes by me, brings yo this Trouble wch with yr many other Favours will be always gratefully acknowledged.' Fielding did not in fact acknowledge Harris's editorial hand when the piece was published. It would have been characteristic of Harris to wish his name kept out of it, and in any case there was little or no time for such niceties. Fielding asked Harris to send the results directly to Andrew Millar for printing. It was a rushed job, since only eight days later *The Daily Post* announced delivery of the *Miscellanies* to its subscribers. From the Flask at Brompton, near Knightsbridge, Fielding thanked Harris privately for his assistance: 'I am infinitely obliged to yo for the Trouble yo have given yrself on my incorrect and low Translation of the first Olynthiac.' He also acknowledged the receipt of money from Harris (14 March).

Within three weeks the Olynthiac was in print, almost certainly the last item in the volume to be revised. It silently incorporated both Harris's revisions and his head-note. The task of classical consultant was, after all, a role which Harris

[2] H. K. Miller, *Essays on Fielding's Miscellanies: A Commentary on Volume One* (Princeton, 1961), 364, 361, 362.

was always ready to perform unobtrusively for those who wished to spread the gospel of classical literature, and Fielding had characterised his own command of Greek rather modestly ('*Tuscan* and *French* are in my Head; | *Latin* I write, and *Greek* I—read').[3] Although it is impossible to allocate Harris's precise share in this collaborative translation, its combination of naïvety and competence invites the explanation that Fielding's original work was partially but not completely corrected by his friend, and with a predictable result.

Of more general interest in Fielding's letter is what he says about his own poetical ineptitude in relation to the 'superior' art of the prose writer. Undoubtedly, Fielding knew that he was teasing his friend by such rhetoric, since his wording and general argument refer to Harris's 'Discourse on Music, Painting, and Poetry', where the poet is placed at the top of the artistic creation. Fielding acknowledged that he had read 'Concerning Happiness, A Dialogue' in manuscript, but the following remarks suggest that he had read the entire volume:

> my Talents (if I have any) lie not in Versification. My Muse is a true born Briton & disdains the slavish Fetters of Rhyme. And must not yo or the greatest admirer of Numbers allow that, even in the Hands of the best Versifier, a noble Sentiment or a noble Expression at least, may be sometimes lost by the unfortunate length or shortness of a Word? For the Poet must imitate the Recruiting Officer who rejects Strength & Symmetry for a long ill-made Fellow of 6 Foot. What beautiful Expressions are to be found in the antient Prose-Writers wch would not have yielded to Numbers . . .

> I apprehend it will be readily admitted that every Part of Stile except the Sublime may be reached by Prose: what if even this may not only be atchieved by the Prose-Writer; but it should be found that the Dignity & Majority of Prose would be superior to that of Verse. Will yo pardon me if I think Paradise Lost is writ in Prose . . .

> If yo are . . . offended at my Observations, yo may revenge your self on me by imitating Ld Shaftesbury, who says of a Philosopher & a Poet who decryed Courage, that they were both Cowards, so may yo conclude that I am no Poet.

Fielding's self-deprecatory remarks on poetry allude directly to Harris's work-in-progress, and he goes on gracefully to turn

---

[3] 'To the Right Honourable Sir Robert Walpole' (1730), ll. 55–6, in Miller (ed.), *Miscellanies by Henry Fielding*, p. 58.

Harris's argument for the supremacy of the poet's imagination against Harris himself. After a lengthy quotation from Bacon's *Advancement of Learning* as an example of sublime prose, he adds that 'I might have spared this long Quotation by referring yo to one Part of a Manuscript in yr own Possession, in wch the Sublime is as greatly exemplified as in any wch I have ever seen'. Fielding was remembering the closing pages of Harris's second treatise where he writes of the combination of poetry and music which 'to the Muses's Friends . . . is a force *irresistible*, and penetrates into the deepest Recesses of the Soul' (pp. 99–100). Harris had not, of course, adjudicated between the rival claims of poetry and prose. He had not discussed prose at all, and this was Fielding's point. But as to the pre-eminence of poetry, there was no question that 'Music, when alone, can only raise Affections, which soon languish and decay, if not maintained and fed by the nutritive Images of Poetry. Yet it must be remembered, in this Union, that Poetry ever have the Precedence; its Utility, as well as Dignity, being by far the more considerable' (p. 102).

In an undated letter, probably written in September or October of 1743, from Bath to Harris's country house at Great Durnford, Fielding acknowledged receipt of a very different production from Harris's pen. Entitled 'The History of the Life and Actions of Nobody', this was a 15 page *jeu d'esprit* dedicated to Fielding as 'the Author of the Life of Joseph Andrews' (see Appendix II below). It had arisen from one of their conversations, specifically on the depredations of Time. Harris's dedicatory epistle offers the piece to Fielding as his patron and as a token of friendship. More particularly, this mock biography of Nobody, descended from 'several collateral and honourable Families of ye Privatives, Negatives, Non-Entities, Nullities', and described in Lucretius, Aristotle, Homer, the Stoics, and by memorials in every churchyard thereafter, is seen by Harris as a sequel to Fielding's own exercise in the paradoxical encomium, 'An Essay on Nothing', which is the third prose work in his *Miscellanies*. Fielding took some time to reply to Harris's 'excellent History of Nobody', but eventually acknowledged the flattering dedication and apologised for the delayed reply. He blames the effects of damp, which, as it rises in the mind, is called Dullness. After a page and a half of outrageous Scriblerian puns, mock-Greek etymologies on the subject of Nobody, and references to the rheumatic, dullness-inducing climate of Salisbury, he turns to the 'Essay' itself:

The Vein of Humour is rich and close; and yo luxuriously throw away more Learning in yr Merriment than would set up a voluminous modern Author in the gravest Treatise. Others may be surprised to see such Excellence in the Ridiculous flowing from the same P[er]s[o]n from whose Genius and Knowledge the World will shortly derive Treasure of so very different a Kind, and of so much greater Value: but I am not of this Number: for to me Wit and Philosophy have always learned to bear a closer Alliance than they are allowed by those who have little Acquaintance with either . . . you can have had Assistance from no one at Sarum unless from a Namesake of yr Hero who is there famous for Wit and Learning.

During this November the sociable Fielding tried to seduce Harris away from his books and the final stages of writing *Three Treatises*. He first tried a delightful legal spoof. On 14 November he warned Harris of an 'approaching danger . . . a Process against yo in the Court of the Snash Bench at Bath to wch I advise yo not to fail to appear in Person the very next Court Day'. Fielding included a 'true Copy' in his own hand of a warrant for Harris's arrest issued by authority of Richard, 'great Snash' (i.e. Richard 'Beau' Nash, the master of ceremonies at Bath), and addressed to 'our right forward and well assured' Edward Young (Snash's deputy in Bath), and Francis Swanton (Snash's deputy in Salisbury). They were empowered to:

take James Harris Esqr if he should be found &c and him safely keep so that he have his Body before us on the morrow of our Great Tuesday's Ball whensoever we should then be in Bath to answer to a Plea of Gallantry of Dame Margaret Brown Wife of Robert Brown Bart, wherefore he the Fan of the said Margaret in his Presence dropt did not take up . . . it is sufficiently testified to us in our Ball Room that the sd. James Harris doth hide himself and run about in yt. Assembly.

Evidently, Fielding was thinking that it was about time for Harris to take dalliance with the opposite sex a little more seriously. There are reasons for thinking that Fielding was at least instrumental in finding a suitable partner for his elusive friend.

Ten days later, Fielding tried again, this time urging Harris to complete *Three Treatises* as his guest at Twerton:

> As yr Notes are the only impediment to yr Journey hither, I hope that Objection will be removed by my assuring yo that from the public Shops here and from some of my Acquaintance I can furnish yo with any Books wch shall be necessary to yo for that purpose . . . I need not tell yo the Pleasure I should have in sitting at the same Fire with yo while we were pursuing different Studies. I therefore repeat my Desire to yo that yo will fill a vacant Room in my House at Tiverton, and I sincerely assure yo there is nothing in wch the Interest of my Family is not immediately concerned wch would give me half the Happiness . . . N.B. My Small Beer is excellent.

The idea of Harris and Fielding engaging in serious peripatetic conversations of a philosophical sort, and then working on their very different literary works around the same fire while drinking home brew, is tantalizing.

In October 1744 Fielding was again oppressed by his wife's illness. If she were to stand any chance of recovery she would need to be moved from London to Bath, and on 10 October Fielding took up Harris's previous offer of a loan, made when they last met at Salisbury. Charlotte's state of sickness worsened, however, and she died in the middle of November. This was the bleakest period of Fielding's life, the ultimate suffering in a private life not characterized by prolonged periods of happiness.[4] Characteristically, Fielding managed to sustain a social guise of equanimity, but in private he revealed his true desolation of spirit to his 'dearest Friend'. This is his most revealing letter (24 November), and one which must have given almost as much pain to its recipient as to its author:

> Nothing could afford me greater Consolation at this Time than so kind a Letter from the Man whom I esteem most of any psn in this

---

[4] In the preface to his *Miscellanies* (ed. Miller, p. xvi) Fielding wrote: 'I was last Winter laid up in the Gout, with a favourite Child dying in one Bed, and my Wife in a Condition very little better, on another, attended with other Circumstances, which served as very proper Decorations to such a Scene.' The Harris family lost an infant daughter, Elizabeth, on 13 Apr. 1749, and 3 years later (9 Dec. 1752) a son, John Thomas, died before his 2nd birthday.

World, and in whose Friendship I propose some of the greatest enjoyments wch my future Life will be capable of. I have some late Obligations to you as extraordinary in the Manner as the Measure, for wch when I can find Words I shall return yo proper thanks, in the mean time as yo can not be ignorant of them I shall leave yo to conceive my Sense of them. I have however one Obligation to you wch as you may not probably suspect, it is proper to mention. You, my Friend, first taught me that Philosophy was not a bare Name, not a fruitless vain Pursuit of Something as chimerical as the Grand secret of the Alchemist, or rather resembling those specious Fantoms the common Honour and Friendship of the World, wch however gaudy and beautiful they appear vanish entirely if we endeavour to apply them to Use. In a Word, you first awakened an Idea in me that true Philosophy consisted in Habit only, without cultivating wch in our own Minds, we should no more become Philosophers by reading the Pages of Plato and Xenophon, than great Poets by perusing the Works of Homer and Virgil.

When I had recd this Hint from yo Fortune presented me with sufficient occasion to exert it, by entring me in the School of Distress, through all the Classes of wch, I have, I think proceeded with greater Diligence and Celerity than Boys generally use in going through the Schools of Eton or Westminster. Why should not I be as pleased with being arrived at the Head of this my School, and with equal Satisfacõn look back on the several Whippings Head-achs and Heart-achs wch I have suffered in my Progress, since I can now be lashed no more.

This is in good earnest the present Situation of my Mind, wch you will find at our next meeting neither soured nor deprest; neither snarling like a Cynic nor blubbering like a Woman. Indeed I spent as pleasant a day with yr Friends Collyer and Barker yesterday as I have known some Months. Nay I was so little a mourner that I believe many a good Woman, had she been present, would have denied the Possibility of my having ever been a good Husband.

You, my dear Friend, who are a Philosopher in earnest . . . will not laugh when I tell yo, I have often asked myself, How would Socrates have acted on this Occasion?

It is unnecessary to emphasize what Fielding himself says so eloquently here. As a confirmation of Harris's theme of the human value of philosophy and of its practical application at such moments, as well as a clear statement of Harris's seminal influence on Fielding's own intellectual character, this remarkable letter is clear enough. Moreover, as Fielding could not have been unaware, he was writing to a man whose Stoicism was no mere theoretical posture. Harris's own experience of grief was only a little less than Fielding's own. On 8 January 1741 his half-sister Catherine Knatchbull died in childbirth. Harris had been very close to Catherine, and Malmesbury was to recall his parents speaking of her as 'a woman of wit and humour. She had a great affection for my father.' In January of 1744 Harris's mother, Lady Elizabeth Harris, died, leaving her son in absolute charge of the family. Harris was still a bachelor, of course, and if he was unable to match the special depth of Fielding's grief for his wife, it seems that Fielding himself would soon turn his attention to his friend's anomalous status.

We should also notice that if Fielding's statement at the beginning of his second paragraph is accurate, then Harris's philosophical influence on Fielding began back in the mid-1730s, immediately prior to Fielding's marriage. Fielding's 'School of Distress' may have begun with the death of an infant daughter, Penelope, buried 6 February 1740. In November 1740 his father Edmund Fielding was imprisoned for debt in the Fleet, and died within its jurisdiction in June 1741. Hounded by creditors from 1739 to 1743, he also lost his first child, Charlotte, on 9 March 1742. His wife's death was, as he put it, the final lash, although litigation over his uncle George's will was also to drag on into 1745.

In his most autobiographical novel, *Amelia*, published six years later, it has been suggested that the Christian Stoicism of Booth's philosophical mentor Dr Harrison may be a compliment to Harris.[5] This novel, unlike the previous two, is set in a period earlier than its composition, i.e. around 1733, a few months after Harris returned from the Continent. By the spring of 1737 Harris's first book was in draft form. Its third treatise, on Happiness, read by Fielding in manuscript, argues the point about philosophy being properly a 'Habit' or cultivated inner disposition of mind independent of external fortune. Part 2 offers the example of Socrates, whose 'Happiness' (in the words of Plato's *Crito*) 'was derived not from *without*, but from *within* . . . He it is, who, when Wealth or Children either come or are taken away, will be seen to grieve,

---

[5] See Martin Battestin's edition of *Amelia*, pp. 75, 137.

nor to rejoice in excess, from the Trust and Confidence he has reposed in himself' (pp. 198–9). Equipped with what Harris calls a 'manlike Magnanimity' (p. 245), the ultimate consolation is to be found 'in our own Minds':

> Be not shocked at the apparent Greatness of the perfect Moral Character, when you compare it to the Weakness and Imperfection of your own. On the contrary, when these dark, these melancholy Thoughts assail you, immediately turn your Mind to the Consideration of Habit. Remember how easy its Energies to those, who possess it; and yet how impracticable to such, as possess it not. (pp. 237–8)

In *Amelia*, viii, p. x, it is the Philosopher who passes on to Booth and to the reader what Fielding had learnt from Harris: 'By Philosophy I do not mean the bare Knowledge of Right and Wrong; but an Energy, a Habit, as *Aristotle* calls it; and this I do firmly believe, with him and the Stoics, is superior to all the Attacks of Fortune.'[6] Whether Dr Harrison was partly modelled on James Harris is, in the end, of little importance. What really matters is that Fielding's letter to Harris was not only an expression of grief at Charlotte's death; it was also a tribute to a man whose influence had enabled him to survive it, and whose philosophical ideas were to shape some aspects of the novelist's moral art in a profoundly personal way.

II

On 15 January 1745 Fielding replied to an earlier enquiry from Harris: 'I will when I see you (I hope in a few days) tell yo every thing I think of the young Lady yo mention; but as to the Question you ask of her Morals & Temper, I assure you in my Opinion neither Socrates nor St. Paul could have objected to them.' Fielding does not identify the subject of Harris's enquiry, but the fact is that within six months of this letter Harris was to take a wife. There is at least a strong presumption that the socially inexperienced scholar had asked

---

[6] Ibid. 350–1. Hugh Amory remarks that Fielding's characteristic use of the term 'energy', rather than the Latin form 'activity' seems to have originated with Harris. Fielding was well able to discover this particular inflection for himself, of course, in the *Nichomachean Ethics*, but his letter to Harris does suggest that the discovery was at the very least prompted by Harris: see A. N. L. Munby (ed.), *Sale Catalogues of Eminent Persons* (1973), viii. 134–5.

his worldly friend to survey the matrimonial market on his behalf. If the John Clark and John Clark jun. who appear in the subscribers list in the *Miscellanies* (7 April) were the father and brother of Elizabeth Clarke, then Fielding already knew Harris's wife-to-be. There is no more likely explanation for Harris preserving Fielding's otherwise somewhat embarrassing character reference for an unknown woman among his private papers. In the previous November, Fielding told Harris: 'I had the Pleasure of hearing of you last Saturday Night by Mr Clark's Family whom I accidentally met at the Inn [at Twerton] just as they arrived at Bath. They were all in good Health but in some Distress for Lodging.'

In Salisbury itself the impending marriage of this most eligible 36-year-old bachelor was an open secret by mid-March. On 17 March Jane Collier, admittedly recovering from an ague and also uneasy at the imminent departure of her brother Charles for the Flanders campaign, sent this waspish account of Harris's pre-marital manœuvres to Mrs John Barker in London:

Mr. Harris goes to London tomorrow. He talks of making a shorter stay there than usual at this time of the year, and I don't doubt but he will come down according to his appointment for I believe he has a very strong inducement so to do. I make no sort of doubt but soon after his return the Lady at Close Gate will remove to Ann Gate; which if it should happen I will say (entre nous) that I shall set down our Friend in the list of those many wise men that have plainly proved that superior sense and ingenuity is of less use as to directing them in the plain and common affairs of life than a vulgar John-Trot understanding. I believe that in time a very absurd action (especially in regard to love and marriage) will be the most certain proof that can be given of a man's great learning and taste. Amongst all the clever men we know (we were reckoning the other day) we could find but very few exceptions to this rule, that the more sense a man had the more likely he was to make a foolish choice.[7]

The 'Lady at Close Gate' was Elizabeth Clarke, and Collier's letter also confirms that Harris's regular visits to London throughout the 1740s were based at the Cecil Street

[7] J. P. de Castro, 'A Presentation Inscription by Fielding', *N&Q* 178 (Jan.–June 1940), 338. For some further information on Jane Collier, Harris, and Dr Barker, see J. P. de Castro's 'Ursula Fielding and *Tom Jones*', *N&Q* 178 (Mar. 1940), 164–7.

home of Dr John Barker. Although James was to break with the family tradition of marrying upwards into the Church or aristocracy, he was to disprove Jane Collier's theory about an inverse ratio between general intelligence and marital wisdom. Henry Fielding was the shrewder judge of his friend's character and of his fiancée's suitability. Their marriage was performed on 8 July 1745 by Revd Nathaniel Goodwin in the parish church of St Mary's, Alderbury,[8] a plain roughcast building with a wooden turret about 3 miles south-east of Salisbury, and demolished in 1857. The choice of this somewhat obscure alternative to the cathedral was no doubt determined by the fact that the bride's family had settled in the parish of Alderbury after their arrival in the Salisbury area. A positive attraction of Alderbury was its close proximity to Longford Castle, whose owner, William Bouverie, second Viscount Folkestone (1725–76) was soon to represent Salisbury, from 1747–61. Bouverie's sister Mary was to marry James Harris's cousin, the fourth Earl of Shaftesbury, on 20 March 1759, and one might fairly suppose that some kind of friendship existed between the Harrises and the Bouverie family in 1745. Longford itself, possibly the model for Sidney's Castle of Amphialeus in *Arcadia*, could not rival the treasure house of Pembroke's Wilton, but Bouverie was the outstanding candidate for the title of local Maecenas of the visual arts. His dazzling collection of paintings included Holbein's portrait of Erasmus, and others by Velazquez, Matsys, Tintoretto, Murillo, Rubens, Poussin, Michelangelo, Correggio, Teniers, Van Dyck, Franz Hals, Gainsborough, and Reynolds. Perhaps the newly-weds were treated to a reception at Longford. In any event, five days later they had arrived at their cottage at Dibden, on Southampton Water. On 13 July Fielding wrote to them at the Star Inn to wish them well from Boswell Court, using his favourite name for a romantic lover from La Calprenède's *Cassandre*.[9] 'My Dear Oroondates', he wrote, was to be assured that his new wife was 'a Woman who having I am convinced fixed her affections on y° will always endeavour yr Happiness, and who being so much the Object

---

[8] WRO Alderbury Parish register, 1966/2. The tripartite marriage contract, WRO Box 1/13/7M54/221, is dated 1 June 1745.

[9] See M. Battestin, 'Pictures of Fielding', *Eighteenth-Century Studies*, 17: 1 (1983), 5.

3. Henry Fielding to James Harris, 13 July 1745, on Harris's marriage.

of yrs will always be able to accomplish it'. Again, Fielding seems to speak from personal acquaintance with Elizabeth, and he goes on to weave all three of them into a fantastic scenario: they are soon to expect a visit from 'a very large Counsellor at Law . . . a Sort of Monster', like a hungry giant from Romance who will require pacifying 'with some Bacon and Beans and a bottle of Southampton Port. I hope he will do you no harm.'

Elizabeth Clarke, thirteen years younger than James, was the daughter of John Clarke, MP, originally from Sandford near Bridgwater, Somerset, but more recently settled in the Close. Malmesbury's 'Memoir' states that John Clarke had been 'a very gay man in his youth', and of considerable means. But whereas Harris's father had risen by the South Sea, Elizabeth's father had sunk. The loss could not have been catastrophic, however, because he was able to hand over a dowry of £3,000, the same sum that Lady Elizabeth Ashley had brought Harris's father in 1707, when he signed the marriage contract on 1 June. On the death of Clarke's son, who seems to have been mentally retarded as a result of a childhood illness, his fortune passed to his wife and then, in 1779, to Elizabeth.[10] Elizabeth's mother, Catherine Clarke (1691–1773) was the daughter of John Bowles (1649–1700), MP for Shaftesbury from 1685 and a lawyer of the Middle Temple. She was the kind of grandparent on whom harassed parents look with particular favour, and to whom children with educationally ambitious parents look for relief. Harris's son was to record that 'The happiest days of my infancy and boyhood were passed under her roof where I was exempt from tasks and lessons and indulged in all the luxuries of Pope Joan and Gooseberies; and when a little older in the delights of a very trusty but very singular servant of the name of John Ware.'

III

From June 1745 to July 1746 the recently married Harris and the recently widowed Fielding were plunged into self-

---

[10] Malmesbury records that Elizabeth's father died in 1747.

defensive litigation occasioned primarily by Harris's friendship with Dr Arthur Collier, Jane's brother. Fielding's constant impecuniousness placed particular strains on his friendship with Harris, and the Collier business made it worse. But Fielding's desperate need for money simultaneously drew Harris further into literary collaboration.

Back in September 1739 Arthur Collier had borrowed £400 from a Salisbury acquaintance, Tristram Walton.[11] In spite of repeated demands for repayment, it was eventually recognized that Collier would not pay up voluntarily. On 14 June 1745 Walton's attorney, Alexander Powell, presented a plea of debt against Collier in the King's Bench, Westminster. An order for 'special bail' was obtained, and this involved Harris and Fielding because they had offered their 'lands and chattels' as surety for Collier's debts. Collier was offered the choice of settling the debts or surrendering himself to the Marshalsea prison. He did neither. Instead, he tried delaying tactics. A writ of error was entered on 6 November, this being a legal objection to the judgment made against him. His objection was overruled, and the court found in favour of Walton for the second time, awarding additional costs to Walton on 12 November. Collier's two sureties, both of whom were lawyers, of course, tried a delaying tactic, submitting another writ of error to the Exchequer Chamber on 19 November. Eleven days later, acting in the double capacity of surety and plaintant, Fielding advised Harris that any attempt to prosecute in order to recover the debt would undoubtedly ruin Collier. The case dragged on into the new year. In the middle of all this, on 19 February 1746, Harris was asked by Fielding himself for another loan. Even to the writer, it seemed to be asking for more good money to be thrown after bad. The whole letter bespeaks a general penury and the awkward syntax a painful sense of his own embarrassment at being forced to beg (see Fig. 4). Harris, we must assume, alleviated his friend's distress once more. Yet these months of stress and financial embarrassment were to demand their closest co-operation. On 9 April a son, James, the future first Earl of Malmesbury, was born to Mrs Harris. Fielding wrote to

[11] Information on the Collier episode is taken from J. P. de Castro, 'Fielding and the Collier Family', *N&Q* 12th ser. 2 (1916), 104–6.

My dear Friend

If you dare venture fifty Pounds more on the same bad Bottom to wit y.ᵉ have trusted a hundred, it will be of very particular Service to me at this Time.

The Moment my Reversion of the Annuities is Sold the whole shall be repaid you, and I think the Devil can not prolong that affair much more.

If this be either inconvenient or disagreeable to you, if you have not the sum at present in y.ʳ Command, or if you have any other use for it, or it will give y.ᵉ Pleasure to retain it tho y.ᵉ have really for it; in any of these Cases, I must insist on y.ʳ granting me the Favour to burn this Letter without answering it, or ever mentioning it to me at our Meeting.

Be assured there is no Sum of money in my Estimation equal to y.ʳ Friendship. Make therefore no ill Interpretation of my Action, and I promise y.ᵉ, I will be so candid in accounting for y.ʳ not complying, that it shall never make me with less Regard

Dear Sir
y.ʳ Sincere, affect. Friend
and oblig.ᵈ hble Servant

Henry Fielding

Boswell Court Feb 19 1746
My Complim.ᵗˢ wait on y.ʳ Lady.

4. Henry Fielding to James Harris, 19 February 1746.

congratulate the parents, but also to point out some less welcome prospects:

The writ of Error will be run out in the next Term at which time Judgment will be affirmed, after which you have 4 days to surrender the Principle and in Default of such Surrender will become together with my self chargeable for the Debt, as will likewise his own Brother, who is I think responsible enough and more entitled to pay than either you or I . . .

As I told both yourself and the Doctor, at the time of your so kindly becoming Bail for him, the Friendship was entirely yours, for from you the Money would be expected in Failure of the Principal.

By 13 May 1746 Harris and Fielding both knew that Collier would be unable to pay up. In Fielding's view, this left Captain Collier, Arthur's brother, and James Harris as the next step in the law's process of recovering the debt. Whereas Fielding's own personal financial distress is clear, Harris's embarrassment can only be imagined. As Fielding correctly surmised, the original judgment against Collier was upheld (4 June 1746). The writ of error was not only rejected; it was given as the reason for additional damages against Collier. Walton was to receive an additional 11 guineas 'by reason of the delay of execution of the said judgment on pretence of prosecuting the said Writ of Error'. Harris and Fielding's legal ploy had failed, and it seems that Fielding himself paid all or part of Collier's debt, though perhaps with the aid of a further loan from Harris. A month or so before his death Fielding referred bitterly to Margaret Collier and 'the execution taken out against me for £400 for which I became bail for her brother'.

Fielding was still in debt to Harris, and the former's letters become for a time less graceful and more peremptory as his health and his financial situation worsened. On 15 October 1748 Fielding still could not repay Harris. Extremely painful gout had forced him to hire an amanuensis; he was unable to sleep; and in any case Harris was less in need of the money than Fielding himself: 'I have not indeed the Curse of thinking you will be distressed by my Incapacity; but I am convinced you will be inconvenienced by it, and there is no Inconvenience I would not subject myself to to prevent it . . . I hope . . . you will believe that nothing less than what must have made Socrates unjust can make me less so to you.'

The new year correspondence included more pleasant topics. A gift from Harris of 'one of the finest chines that ever graced the Back of a Hog with four Legs, or the Table of one with two' was acknowledged on 16 January 1749, but ten months later Harris is asked to defer drawing on Fielding for three months, and then only for £25, because Fielding had spent his money on the purchase of a small estate at Twickenham, that most exclusive of literary suburbs. A month later Fielding is still hoping for better times and asking for his friend's forebearance.

There is no evidence to suggest that Fielding ever repaid Harris the money he had borrowed; nor is there any reason for thinking that Harris ever stopped supporting his friend with credit, bacon, and the completest understanding. Indeed, when Harris sat down to write a biographical memoir of Fielding nine years later, the tone and content of his writing suggests nothing less than the deepest admiration for the way Fielding's battles with adversity had elicited his finest qualities.

IV

The Collier business was not the happiest of Harris and Fielding's shared activities, but it had no damaging effect on their literary partnership. Indeed, as in the earlier example of the *Miscellanies*, Fielding turned once again to Harris for money-making assistance, and this time with an unexpectedly pleasing result. In the same letter which informed Harris that Fielding had 'taken Care of yr Affair with Dr Collyer' by the writ of error (5 November), Fielding enclosed the first number of his *True Patriot*, published that day. As we might expect of a new weekly journal appearing in the middle of a national crisis, the contents of Fielding's journal were almost entirely given over to the Jacobite rebellion. Space was found, however, for a lead article on Fashion, a warm tribute to the recently deceased Swift, and news of preferments, including the election of Thomas Smyth as mayor of Salisbury. *The True Patriot* ran until 17 June 1746, and each of the early issues is marked by journalistic parodies, mischievous remarks on property-marriages, and sensational reporting of various kinds. The private allusiveness of the journal is indicated by the fact that it is 'Abraham Adams' who signs the lead articles for Tuesday, 17 December, and 21 January 1746.

The lead article for number 10, 7 January, was written by James Harris, and Fielding wrote to Harris on 11 January about the editorial changes he had made and its successful reception:

I hope yo will pardon the following offences.
1. publishing Yr Lre a week sooner yn I promised.
2. the Alteracōn of Italian Judges wch I thought too flagrant an Instance to be given so early, & 3dly the 2 or 3 small Additions I have ventured to make.

The Town, I assure yo, hath recd yr Wit with the Applause it deserves; however yo may despise their Approbation, I know yo will be pleased with it when yo hear, that it hath had the consequences of serving me much by raising the Sale of the Paper. A News Writer, yo see, hath transcribed it, wch is the first Honour of that Kind conferred on me, and this merely on his own Mocōn witht. ye least Application from myself.

Harris's article is dated 14 December 1745 and purports to be an imaginary journal of 'an honest Tradesman, living in the busy part of the City', who gives an eye-witness account of the state of England after a Jacobite victory. It is Whig propaganda, a lament for a lost constitutional monarchy and free press, and, as Fielding remarks, it was topical and well-enough written to be immediately reprinted in *The General Advertiser* for Saturday, 11 January 1746.

The journal begins on 1 January with the public acclamation of the 'supposed Conqueror' by Highlanders and Friars, following a massacre in the previous week. The entries from then until 17 March chart the rise of totalitarianism through an apocalyptic reign of terror, 'Confusions, Uproars, Committments, Hangings, Burnings', retrospective legislation, sudden arrests and disappearances of prominent citizens, summary executions after mock trials, the restoration of land to the monasteries, the award of peerages for political services (without land ownership qualification), the burning of heretics, and the gradual replacement of English Protestants by French, Italian, Irish, and Scots Catholics. Fielding had deleted 'Italian' in the first day's entry:

Jan. 2. A Proclamation issued for a *free Parliament (according to the Declaration)* to meet the 20th Instant. The twelve Judges removed, and twelve new ones appointed, some of whom had scarce ever been in *Westminster-Hall* before.

Jan. 3. Queen *Anne*'s Statue in *St Paul's* Church Yard taken away, and a large Crucifix erected in its Room.

Jan. 4, 5, 6. The cash, Transfer Books, &c, removed to the Tower, from the Bank, South Sea, and India Houses, which ('tis reported) are to be turned into Convents.

Jan. 10. Three Anabaptists committed to *Newgate*, for pulling down the Crucifix in Paul's Church Yard.

Jan. 12. Being the first Sunday after Epiphany, Father *Mac-Dagger*, the Royal Confessor, preach'd at *St James's*—sworn afterwards of the Privy Council—arrived afterwards the *French* Ambassador with a numerous Retinue.

Jan. 20. The *free* Parliament opened—the Speech and Address filled with Sentiments of *civil and religious Liberty.*—An Act of Grace proposed from the Crown, to pardon all Treasons committed under Pretext of any Office, *civil or military*, before *the first Declaration's being promulgated, which was in the Isle of Mull, about 19 Months ago.* The Judges consulted, whether all Persons throughout *Great Britain* were intended to be *bound* by this Promulgation, as being privy to it. 'Twas held they were, because *Ignorantia legis non excusat.*

Jan. 22. Three Members, to wit, Mr. D —— n, Mr. P —— t, and Mr. L —— n, were seized in their Houses, and sent to the Tower, by a Warrant from a Secretary of State. The same Day I heard another Great Man was dismissed from his Place, but his Name I could neither learn nor guess.[12].

The tribal memories of 1688 are reactivated: an Italian Jesuit is appointed Privy Seal; the Crown suspends all laws; Father Mac-Dagger becomes president of Magdalen College, Oxford; the bishoprics of Winchester and Ely are given to the General of the Order of Jesuits (resident in Italy); and an office is opened in Drury Lane for selling pardons and indulgences.

---

[12] For an annotated text, see H. Fielding, *The True Patriot and Related Writings*, ed. W. B. Coley (Oxford, 1987), 171–80. There is no doubt whatsoever that this article is Harris's work. The possibility and extent of Harris's *additional* contributions to Fielding's journal, however, remains mysterious.

Harris's journal sets out to articulate the near-hysteria of the time. In the context of war with France and invasion at home, present fears trigger anxieties in the national psyche. When it was all over, Fielding himself described the crisis in *The Jacobite's Journal* for 13 August 1748 as 'far more terrible to all the Lovers of Liberty and the Protestant Religion, than this Age had ever seen before, or is, I hope, in any Danger of seeing again'. Harris's imaginary journal thus projected the Englishman's loathing of ecclesiastical tyranny and his pride in established institutions, expressed in the usual xenophobic terms. Without the twin bastions of a constitutional monarchy and an independent judiciary, England could become another France or Italy.

Harris lacked neither an ideological interest in, nor first-hand information about, the progress of the rebellion. Apart from letters from Fielding himself on 5 October 1745 and 2 January 1746, the Earl of Shaftesbury had sent him a copy of the Bishop of Salisbury's letter (dated 9 September) addressed to the Dean, Chapter, rectors, vicars, and curates within the diocese. This called upon the recipients to 'act like ENGLISH-MEN, and like PROTESTANTS', and to mobilize their congregations against 'Indifference and Unconcernedness under such an Attempt upon our Constitution, [which] would be fatal Symptoms. Popery and Slavery will never have a fairer Opportunity of returning with Power, than when the People become unaffected with the Fears and Apprehensions of them.'[13] On 10 December William Harris wrote from London that 'The imaginary approach of the rebels last Friday evening caused a general consternation among us, and it is beyond the power of words to describe to you the hurry both Court and city were in.'[14]

The panic was as undeniable as the crisis itself, and it was not a time for subtlety. When Harris's article appeared in *The True Patriot* its marginal messages (added by the printer) screamed at the reader 'No Pretender', 'No Popery', 'No Slavery', 'No Pretender', 'No Wooden Shoes', 'No Arbitrary Power'. Yet within Harris's own article there is scepticism of such journalistic hyperbole. The imaginary journal ends with what looks like comic mayhem: '*March 17*. Fresh Rumours of a

---

[13] *Series of Letters*, i. 6. [14] Ibid. 21 (to Mrs Harris).

Plot—a Riot in the City—a Rising in the North—a Descent in the West—Confusions, Uproars, Commitments, Hangings, Burnings, &c. &c.—verbum non amplius addam.' Even so, on the very day that Harris's article came out his brother William writes again that 'I assure you nobody can be more subject to panics than the whole City and Court seem to be at present. Very late on Tuesday night last, long after my doors were locked up, an express arrived at the Duke of Newcastle's, with an account that the French were actually landed at Pevensey Bay, in Sussex.'[15] Throughout March 1747 the letters of his brothers Thomas and William are crammed with details of the aftermath, the latter sending an 'original' of Hogarth's unflattering portrait of Simon Fraser, Lord Lovat, at his trial, for which the political turncoat had gladly posed. This was Harris's only venture into topical journalism, and even though it was a considerable success, one may easily detect beneath its inventive and specific ironies a profound dislike for the whole business of distorting language for a populist motive.

V

In 1751 Fielding's last novel, *Amelia*, and Harris's *magnum opus*, *Hermes*, were published simultaneously. The former's most autobiographical and therefore most personal novel coincided with the most abstract and conceptually most difficult work Harris was ever to undertake. Fielding's 'ART of LIFE' accompanied Harris's essay on Universal Grammar. Harris's public reputation as 'Hermes' Harris was about to be made, and Fielding's career was nearing its untimely end. On 21 December 1751 the philosopher wrote to the novelist:

I hope you have received *Hermes*, for which in return I have the assurance to Desire that you would send me Amelia. Should you ask with indignation, what four Books, for one—I might answer, if I would, that put them in the Scales, and see how much heavier grammatical Speculation is than Wit, and Humour.

In the critics' estimation, the praise went to *Hermes* and the blame to *Amelia*. The former had no precedent and its

---

[15] Ibid. 23 (to Mrs Harris).

strenuously analytical discourse demanded and received respectful reading; the latter's serious comedy was not in the mould of *Joseph Andrews* or *Tom Jones* and went unrecognized or unread. One was a success and the other a failure. Two days later Fielding replied from Bow Street:

I do sincerely assure you, you would have recd my damned book (for so it is) by this very Coach, even without your having mentioned it. If you read it, you will do it more Honour than hath been done it by many here. Indeed I think I have been more abused in a Week than any other Author hath been; but tho' *our* favorite Authors have not taught me to write so as to avoid Censure, they have at least taught me to bear it with Patience.

I thank yo for yr. very valuable Present. (I mean that of yr. Book) I sincerely think it among the best Books in our Language; and I can with great Truth compliment you on yr. Success with all sensible Readers. Happily for you, you will not have to deal with many others, from your Subject, which hath no Invitation for those Critics who condemn whatever doth not make them laugh.

The inner circle of Harris's and Fielding's friends also considered the two books together. John Upton had been one of the first to receive a presentation copy of *Hermes*, and on 17 December he wrote to its author to say that it was 'the *only* book I have seen for these many days from our English presses . . . fit for a scholars reading & library. I cannot tell you how much I am pleased with it, nor how you are to be commended for yr order & perspicuity . . . I believe I shall never lay it out of my hands.' Eleven days later Upton unhappily confirmed the general opinion of *Amelia*:

Our friend's Amelia does not answer people's expectations in reading, or the bookseller in selling; they say 'tis deficient in characters; & see not a Parson Adams, a Square & Thwackum & Western in it. In short, the word condemnation, tho not Damnation, is given out. Miller expected to get thousands, & there chiefly the disappointment lies; for as to Fielding himself he laughs, & jokes, & eats well, as usual; & will continue so whilst rogues live in Covent Garden, & he signs warrants.

As *Hermes* was being rapturously received (on different levels of comprehension) by Joseph Highmore, Sarah Fielding, Jane and Arthur Collier, Charles Yorke, Robert Stillingfleet, John Hoadly, Samuel Richardson, William Hoare, Robert Lowth,

and others, Fielding's last major novel was faring badly.[16]
There is one exception to this general pattern, however, which
reached Harris, if not Fielding, at the end of the year. The
fourth Earl of Shaftesbury recognized the stark contrast
between the two books and confessed an intuitive fondness for
Fielding's novel and a defeat at the hands of *Hermes*:

To confess my weakness to you, the tender Scenes between Booth
and her in the 1st Vol:, when he leaves her to go to Gibraltar, so
affected my weak Nerves and Spirits, that I was forced to lay it
aside, and have just got courage to proceed, so that I have not yet
got through the Second Vol: I find it is not a greatly admir'd
performance but I think there are many fine sentiments in it and it is
a further confirmation to me of the Humanity and tender disposition
of the author . . . Though I do not presume to give my Opinion (who
know not the Art and rules of Grammar in Hermes), yet I hope it
will not be thought too presuming in me to say that I think the 5th
Chapter of the 3d. Book with the Notes one of the most charming
things I ever read.

Perhaps neither Fielding nor Harris would themselves have
been charmed by such a limp recommendation of their
respective books, but in the event Harris's remaining obligation to Fielding was a precise consideration of his intimate
friend's 'Humanity and tender disposition', although under
the unhappiest circumstances imaginable.

VI

Henry Fielding died on 8 October 1754 at Junqueira in
Portugal. Somehow, Harris managed to keep track of his
friend up to his last few days in England, and related the
following anecdote as an appendix to his biographical
memoir:

Two Friends made him a visit on his leaving England, when his
Constitution was so broken, that twas thought he could not survive a
week. To explain to them his Indifference as to a short protraction of
Life, he with his usual humour related them the following story. A
Man (sd. He) under condemnation at Newgate was just setting out
for Tyburn, when there arrived a Reprieve. His Friends who recd. ye

---

[16] Harris preserved letters from each of these recipients. The critical reception of
Hermes is discussed in ch. 5 below.

news with uncommon Joy, prest him instantly to be blooded; they were sure (they said) his Spirits on a change so unexpected must be agitated in the highest degree. Not in the least (replied the Hero) no agitation at all. If I am not hanged this Sessions, I know I shall ye next.

Harris's loyalty to his friendship was still to find its full expression. There were two connected duties, the first to Henry and the second to his sister Sarah. Sarah Fielding was soon to settle in Bath under the protection of Ralph Allen. In 1758 she was putting the finishing touches to her novel *The History of the Countess of Dellwyn* (1759). In this year Harris was also at Bath, assisting Sarah in her translation of Xenophon's *Memoir of Socrates* (1762). Before this task was completed Harris had drafted a 16-page memoir of her brother which was never to reach publication. Dated 5 February 1758, it is called 'An Essay on the Life and Genius of Henry Fielding, Esqr.' It is a remarkable expression of Harris's deep admiration for Fielding the man, his mind, his works, his company, his conversation, his industry, and his vehement passions. As one who knew more about him than almost anyone else, Harris was able to give a balanced view. He makes no secret of his friend's improvidence with money, his occasional alcoholic intemperance (which never 'improved' his ingenuity), nor of his love of pleasure. The latter had run Fielding into debts which Harris had been called upon to settle. But the overwhelming impression left by Harris's memoir is of a man whose best was elicited under adversity and whose enormous social appetite was served by a captivating conversational wit. The whole of this Memoir is reproduced in Appendix III below.

Harris preserved twenty letters from Sarah Fielding, one of which is jointly signed by Jane Collier, from the period 28 September 1751 to 15 March 1762. This covered Sarah's translation of Xenophon, a project which Harris did not initiate but one in which he eventually became deeply involved, the completion of his own memoir of Henry, and the appearance of Arthur Murphy's 'An Essay on the Life and Genius of Henry Fielding, Esq.' (the identical title Harris had chosen), prefixed to his edition of Fielding's works in 1762. Both Harris and Sarah found Murphy's essay deeply offensive.

Sarah and Jane Collier were on Harris's list for a presentation copy of *Hermes*. They both wrote to thank him for the book on 28 December 1751, from Beaufort Buildings, Bath, and for both women he was clearly perceived as a generous mentor requiring a somewhat stiff deference:

> We should indeed verify the truth of Mr Pope's observation that 'a little Learning is a dang'rous thing' should we vainly attempt to express our approbation of your work; and by such an attempt deservedly should incurr the censure cast on those Women who, having pick'd up a few scraps of Horace, immediately imagine themselves fraught with all knowledge . . .
>
> As little Children then Sir give us leave to consider ourselves, and as our kind Instructor accept our thanks for turning our studies from the barren Desarts of arbitrary words, into cultivated Plains where amidst the greatest variety we may in every part trace the footsteps of Reason, and where how much soever we wander, yet with such a guide we may still avoid confusion.

For having read and corrected the manuscript of her *The Cry: A New Dramatic Fable* (March 1754), a piece heretofore regarded as a collaborative enterprise between Jane Collier and Sarah Fielding, the former writes to thank Harris on 18 March 1753:

> My Book waits on you in print to pay its thanks for the trouble you took with it in manuscript . . . My acknowledgements are due to you for your Dedication, the thought of which is very proper to the thing, but I was advised to make neither serious or ironical address to any One as the book is in an oblique manner address'd to the Princess of Wales by the Compliments intended for her in the 4th Chap: of the second Part. My being the Author is now one of those profound Secrets that is known only to all the People that I know.[17]

Sarah's project to translate Xenophon began with an awkward silence from Harris. Desperate for his approval, and having already published proposals for its publication by subscription, she had communicated her anxiety to Hoadly at Bath. She then contacted Harris (25 December 1760): 'I sometime ago sent you a Translation, or rather a Schoolboy's Exercise wherein I aimed at putting into English Xenophon's Defence of Socrates; your not returning any Answer to that

---

[17] Jane Collier referred to Harris as 'a gentleman of no less erudition than judgment' (*The Cry*, i. 8).

Letter, I impute to your Politeness and Good-Nature, that would not suffer you to tell me how very ill it was done.' In fact, Harris had been preoccupied with winning the parliamentary seat of Christchurch. But he soon wrote back to explain the delay, and enclosed François Charpentier's French translation of Xenophon for Sarah's use. He also enclosed a list of potential subscribers which included the name of his own son. Sarah's reply reveals that although she had perused the French translation 'for half an hour', she was determined to go through the 'Greek and Latin only'. No doubt delighted by Sarah's scholarly determination, Harris subsequently gave her help with the style of the translation (Xenophon's essence being 'Simplicity'), with etymologies, and with particularly difficult concepts such as divination. He finally corrected her translation in his own hand but nevertheless insisted that she should make up her own mind about disputed passages. By 18 August 1761 she reported that 'I am now almost come to a conclusion of the work you have so kindly assisted'.[18]

Harris's memoir of her brother had been completed on 5 February 1758. As may be seen (Appendix III below), it is an attempt to put on record his own deep admiration for Fielding's character. But there was another motive. Harris assigned the memoir to Sarah Fielding in the hope that it could provide her with a financial claim on Henry's publisher, Andrew Millar. To some extent it was planned as a collaborative work, Sarah providing the notes. Some were written, but have not survived, and may not have been completed. The real problem lay with the recent death in 1757 of William Young, the authoritative source for a detailed

[18] *Xenophon's Memoir of Socrates with the Defence of Socrates Before his Judges. Translated from the Original Greek by Sarah Fielding* (Bath and London, 1762). The subscribers included Ralph Allen (3 copies), Dr Arthur Collier, Dr Patrick Delany (one of Swift's executors: 7 copies), Lady Gould, James Harris sen. and James Harris jun., Mrs Harris, Revd William Harris, Benjamin Hoadly, Sir Wyndham Knatchbull, Miss Knatchbull, the Earl and Countess of Shaftesbury, Dr John Taylor (now canon of St Paul's), and Wadham Wyndham. Sarah acknowledges Harris's translations, paraphrases, and commentary on pp. 2, 8, 10, 11, 24, 92; and 2 notes (p. 248 on the Socratic dialogue, and pp. 313–14 on Dialectic) are actually signed by Harris. Salisbury had also invested substantially in her earlier *Familiar Letters* (1747): James Harris, Mrs Harris, and Benjamin Hoadly each subscribed for a Royal Paper copy; Thomas Harris for 2 sets.

chronology of Fielding's life. On 30 July Sarah wrote to Harris:

I am greatly obliged to you for sending me your Essay . . . it came into my hand ye 28th . . . I will endeavour if possible to do something by way of Notes in the way you mention (by way of Dates our Friend Mr Young would have been a great help). If I should be so happy as to do any thing proper, I am sure I am very sincere, when I say I shall think the Honour infinitely on my Side, to have any thing I can do printed under the same cover with any Writing of yours.

But Sarah had misunderstood Harris's intentions. From the outset he determined to have nothing whatsoever to do with publishing the memoir. That was to be solely Sarah's task. Thomas Harris was deputed to deal with any corrections. On 1 July 1758 Sarah asked Harris to check the notes which she had (evidently) left with Harris on his last visit to Bath, adding that 'I would wish it to be printed early in the Winter'. She received the essay from Harris on 1 November with a letter outlining the method of dealing with its publication by Millar. Revelation of its authorship was expressly forbidden. Sarah replied immediately:

the Essay came last night to Bath . . . and I received it safe this Morning being Saturday. Give me leave to assure you Sir I have too much respect for whatever you write, to subject it to the Criticism of any Booksellers Agents in any other Manner than by mechanical Weight and Measure, and shall take great Care to give particular Charge about it, and desire the Correcting the press may be remitted to your Bro: in Lincoln's Inn Fields which will be an absolute security against any such Attempt. I expect Millar in Bath tomorrow, if he keeps the time appointed, and will then settle with him, but if he should not come will write to him next week. I ever intended to make Millar enable me to beg your acceptance of an Edition of my Brother's Works . . . What he will give for the Essay I am not apprized of, but think he must make some consideration for it, for even he had knowledge enough to be eager for the printing it, and as this perhaps will be the last Edition of my Brother's Works, at least in my time, I was desirous, if you had no Objection, to have it published . . .

Sarah badly underestimated Harris's desire to remain behind the scenes. Initially, she had felt that Harris was not unwilling to be the named author (4 October), and had mentioned the existence of the essay in a letter to Millar. On 6 September she reported the possibility of Millar publishing it as a prefix to *Tom Jones* or *Joseph Andrews*, and further asked if Harris knew of any additional unpublished writings by Henry: 'If you, and your Brother think there is nothing worth Printing to be collected, which I am afraid is the Case, I would not put any thing trifling and frippery to yours, nor depreciate him with such Accounts as I have often seen after the Subjects of them are dead.' By 9 April some progress had been made. Millar had taken the essay, with Harris's permission, with the intention of publishing it 'this Winter'. But it did not appear. Sarah thought Millar was waiting for a commercially more opportune moment, and her illness during the summer of 1760 had shifted the responsibility of supervision to Thomas Harris:

Had not Master Harris of Lincoln's Inn seen your Essay, because I gave Millar your strict Charge that he alone was to have the Management of it?

I was obliged to send you back the part in which I had the favour of your signing your Name, with your kind Assurance of Friendship because it is connected with the other Side. I am greatly oblig'd to you for giving me leave to credit my-self with your Authority.

But in March 1762 the real reasons for Millar's delay became horribly apparent. On 4 March Sarah wrote from Walcott, near Bath, in a state of acute anxiety, apparently unaware until now that Millar had advertised proposals for an edition of her brother's work, by Arthur Murphy, as far back as 19 May 1759:

I have been greatly perplexed and uneasy about an Affair wherein your great Kindness to me will seem to have been neglected, and yet in plain and simple terms, it is absolutely out of my power, to avoid the Appearance, although I know that in fact I am very far from having been guilty of any such design. Notwithstanding the reiterated request to obtain your leave for prefixing the Essay you so obligingly wrote, to my Brother's Works, yet they are going to be published without it, and another long Essay *weighing* heavy in every Sense will be published instead of it. Millar was here lately, and it was sent to him here, he shewed it to me, you never saw such a shocking Creature as it had made of my Brother, and not only of him but of his Father too. Millar himself desliked it, but peculiar Reasons (besides him having agreed to pay for it) too long to be explained unless I could have the pleasure of seeing you, made it necessary to print it. The Writer, a Mr Murphy, knew little or nothing of my Brother, yet, as Mr Millar informs me, rejoyces much to have hear'd the truth, and confesses he had his strange Information from Bow-Street [i.e. from Fielding's half-brother Sir John Fielding]; so that however dull it may be, it will be altered from giving such a wretched Picture of him. Millar says he has printed but a small Number now, and if I will write another Life of my Brother, for he doats on Length, and with your leave, add your Essay to it, he shall rejoyce to prefix it to another Edition. I have your Essay very safe again, and knew not till lately but it would have been published.

O that the Translation of Xenophon may not be thought very bad by you! which would be a great Gratification.

A letter from Harris (now lost) forbade the further circulation of his essay and expressed anger that the obscure and ill-informed Murphy had been permitted to batten on the fame of Fielding. Sarah agreed wholeheartedly and assured Harris (15 March) that 'The Opprobrium put into Murphy's Essay will be altered, so that it will only be tedious and dull.' In the meantime, however, Harris had expressed his approval of Sarah's Xenophon, and this was no small recompense to her.

Millar had evidently given Sarah pre-publication sight of Murphy's essay, since the latter is dated 25 March, ten days after her letter to Harris complaining about it. It is impossible to say whether her outrage led Millar or Murphy to changes in the printed essay. Certainly, Murphy's biographical knowledge of Fielding was slender, second hand, and ten-dentious. His picture of a dissolute youth, a spendthrift adulthood, a careless and commercially minded artistry, and a career summarized by 'disappointment, distress, vexation, infirmity, and study' was particularly damaging because it was half true. Murphy claimed that he had consulted 'the ablest and best of the author's friends', but the essay could never have passed Harris's scrutiny, even assuming the improbable chance of Harris being asked for an opinion. Obviously, Sarah herself was shocked when she saw it for the first time. For all its dullness, sly innuendo, and superfluous material on Genius, Pope, and Marivaux, Murphy's essay nevertheless became the publicly uncontested version of Fielding's life which was to determine his reputation for the next 150 years.

Harris was the one man who could have set the record straight: that he never did so suggests that his loathing of public journalism outweighed his sense of injustice. It also meant that, until now, his intimate friendship with Fielding would never become public property. Yet this somewhat unexpected friendship between L'Allegro and Il Penseroso was clearly sustained by deep roots, Harris savouring the social vivacity of the novelist, and Fielding turning to his philosophical friend both for money to sustain his style of life, and, more importantly, for philosophical consolation at times of great loss, as well as for literary assistance.

5

The Furniture of the Mind
Hermes, 1751

> The imagination (as a productive faculty of cognition) is very powerful in creating another nature, as it were, out of the material that actual nature gives it. We entertain ourselves with it when experience proves too commonplace, and by it we mould experience, always indeed in accordance with analogical laws, but yet also in accordance with principles which occupy a higher place in Reason . . .
>
> Immanuel Kant, The Critique of Judgement (1790)

THERE is a good joke about Harris's second and most important book. Sir John Hawkins recorded the following conversation between himself and Samuel Johnson:

> Harris's *Hermes* was mentioned. I said, 'I think the book is too abstruse; it is heavy.' 'It is; but a work of that kind must be heavy.' 'A rather dull man of my acquaintance asked me,' said I, 'to lend him some book to entertain him, and I offered him Harris's *Hermes*, and as I expected from the title, he took it for a novel; when he returned it, I asked him how he liked it, and, what he thought of it?' 'Why, to speak the truth,' says he, 'I was not much diverted; I think all these imitations of *Tristram Shandy* fall far short of the original!' This had its effect, and almost produced from Johnson a rhinocerous laugh.[1]

Both Harris and Sterne write about the relationship between words and things. 'Hermes' is Harris's title, but the other name for Hermes is Trismegistus, the name that Sterne's hero was denied. Hermes Trismegistus was the Greek name for the God Thoth, the supposed author of *Hermetica*, learned books written in Egypt by men of Greek speech. The learned Harris writes a book in English about universal linguistic relationships based on Greek writing. Sterne reduces all universal systems to subjectivity: relation becomes relativity, and the

---

[1] Cited in G. B. Hill (ed.), *Johnsonian Miscellanies*, 2 vols. (Oxford, 1947), i. 70–1, from J. Cradock's *Literary and Miscellaneous Memoirs*, 4 vols. (1828), i. 208.

basis of language in logic becomes a logical parody. As Socrates surmised in the *Cratylus*, 'the name Hermes has to do with speech, and signifies that he is the interpreter . . . or messenger . . . the God who invented language and speech'. Sterne defined his writing as conversation, and meaning as a collaborative function of author and reader. Harris, taking his cue from Platonic dialogue, also defined his writing as a co-operative function of a relationship between author and reader: 'What then is requisite, that [the hearer] may be said to understand?—That he should ascend to certain Ideas, treasured up within himself, correspondent and similar to those within the Speaker.' Harris had little toleration for the philosophy of John Locke, and one of his favourite books was Xenophon's *Cyropaedia*, on the education of the Persian ruler Cyrus. Sterne parodied Locke's theories on language through the effort of Walter's efforts to compile a *Tristrapaedia* for a misbegotten son who is imprisoned by language and cannot even rule his own text.

It was a good joke, but it was a malicious joke, designed to be dismissive of Harris's book. For Johnson, it was all the more welcome for that reason.

The publication of *Hermes* in 1751 was greeted by a flurry of respectful approval, not least, as we have seen, from Henry Fielding and John Upton. It was to associate Harris permanently with the subject of universal grammar, and henceforth he was to be known as 'Hermes' Harris, just as 'Dictionary' Johnson, 'Demosthenes' Taylor, and 'Corsica' Boswell. For those able to judge, *Hermes* seemed an important book on an important subject written by an expert. For those unwilling or unable to comprehend its significance, it seemed an exercise in obfuscation. It gained Harris a wide and lasting European reputation, but in his own country its reception was neatly divided between those pragmatists who saw language as a transparent and neutral medium and those who conceived language as an artificial construct just as much subject to rules and conventions as any other social activity. It divided the scholars of North Britain from their colleagues south of the Tweed, and it announced a challenge to those who thought that the Lockian orthodoxies had become axiomatic. In particular, it flushed out the deep hostility of Samuel Johnson.

Attacks on Harris the man were the screen for suspicions of his style, his *écriture*. Incomprehension dressed in ridicule was a not uncommon form of response. Johnson's own copy of *Hermes* contains not the slightest evidence that it was carefully read, but Macaulay's copy of *Hermes* contains plenty of evidence of uncomprehending fury. He pencilled in marginal comments such as 'A poor bad book', 'Greater trash I never read', 'Stuff', and 'What a fool you are!'[2] The vinegarish but linguistically adept Horne Tooke blasted the book: it was 'an improved compilation of almost all the errors which Grammarians have been accumulating from the days of Aristotle down to our present days of technical and learned affectation'.[3] The notion of Authority to Tooke was as a red rag to a bull. Having helped to organize the parliamentary resistance movement of John Wilkes, he saw in Harris another example of the tyranny of the Establishment, a potent enemy to be cut down. Tooke was right in assuming that *Hermes* was to be widely accepted. It appeared in a dozen editions between 1751 and 1841 and was translated into German in 1788, French in 1796 (by order of the Directory), and last reprinted in 1975.

The pro-Harris camp has been rather more distinguished. James Burnet, Lord Monboddo, the closest ally and a correspondent of Harris, remarked that *Hermes* was 'a work that will be read and admired as long as there is any taste for philosophy and fine writing in Britain'. He also described the intellectual battlefield on which they were both fighting:

---

[2] Johnson's own copy of *Hermes* is owned by Prof. Arthur Johnston, who kindly supplied this information; Macaulay's annotated copy is in the British Library (BM 1568 3310), where there is also a 29-page MS, 'Harris's Hermes Abridg'd', signed T. F. Winman (BM Add. MSS 41844B).

[3] J. H. Tooke, *A Letter to John Dunning, Esq.* (1778), 18. In *The Diversions of Purley* (1798), Tooke nevertheless acknowledged the authority of *Hermes*: 'Hermes . . . has been received with universal approbation both at home and abroad; and has been quoted as undeniable authority on the subject by the learned of all countries' (p. 120). The former pamphlet of 71 pages was presented to Harris by its author. For the king and queen's reaction to Tooke's attack on Harris, see *Series of Letters*, i. 388–9. Tooke had been imprisoned in the King's Bench prison by Lord Mansfield for his treasonable pro-American views. On his release, Tooke again savaged Harris in the *Diversions*. Vol. i appeared in 1786, and its 2nd edn. (1798) incorporated *A Letter* in chs. 6–8. Vol. ii appeared in 1805, the third (if it ever existed) was thrown on the fire. Tooke thought that Harris's co-worker Monboddo was 'incapable of writing a sentence in common English' (i. 286), and for Tooke's own account of his trial for seditious libel, see *Diversions* i. 74–80, and M. C. Yarborough, *John Horne Tooke* (New York, 1926), 146–78.

Mr Locke wrote at a time when the old philosophy, I mean the scholastic philosophy, was generally run down and despised, but no other come in its place. In that situation, being naturally an acute man, and not a bad writer, it was no wonder that his essay met with great applause. And I must allow, that I think it was difficult for any man, without the assistance of books, or of the conversation of men more learned than himself, to go further in the philosophy of mind than he has done. But now that Mr Harris has opened to us the treasures of Greek philosophy, to consider Mr Locke still as a standard-book of philosophy, would be . . . continuing to feed on *acorns* after corn was discovered . . . [Mr Harris] has applied his knowledge more to the study of Greek philosophy, than any man that has lived since that period.[4]

Coleridge thought that *Hermes* had been written with the precision of Aristotle and the elegance of Quintilian, and of the two eighteenth-century linguists discussed by Noam Chomsky in *Cartesian Linguistics* (1966), Harris is central and quoted at length.[5] Moreover, when Thurot's 1796 translation was reprinted in 1972, its editor claimed that *Hermes* remains one of the most important grammars of the classical age, meriting comparable status in the eighteenth century to that enjoyed by the Port-Royal Grammar in the seventeenth century.[6]

---

[4] J. Burnet, Lord Monboddo, *Of the Origin and Progress of Language*, 6 vols. (Edinburgh, 1773–92), i. 52.

[5] N. Chomsky, *Cartesian Linguistics: A Chapter in the History of Rationalist Thought* (New York and London, 1966), 15–16, 83–4, 88. Chomsky (p. 19) does not acknowledge the possibility that Humboldt's characterization of language as *energeia* rather than *ergon*, or work, was introduced into German philosophy by trans. of Harris's *Three Treatises*.

[6] A. Joly (ed.), *Hérmès ou recherches philosophiques sur la grammaire universelle*, trans. F. Thurot (Geneva, 1972), 9. Joly's lengthy introduction is the most detailed treatment of Harris's linguistic significance to date. But for a point by point analysis of Harris's contribution to grammatical theories, see I. Michael, *English Grammatical Categories and the Tradition to 1800*, part 2 (Cambridge, 1970). Since Harris's work is better known in Germany and France than in his own country, I have concentrated here on English reactions. But for a brief outline of Harris's important influence in France see A. Joly's introd. and notes, to: F. Thurot 'Discours préliminaire à "Hermes" ', *Tableau des progrès de la science grammaticale* (Bordeaux, 1970), 28–41. For the connections between Harris and Humboldt, see W. Lammers, *Wilhelm von Humboldts Weg zur Sprachforschung, 1751–1801* (1936); and for a brief discussion of Thurot's *Discours préliminaire* to his translation of *Hermes*, and of contemporary French reviews, see Joly's introduction to his 1972 reprint cited above (pp. 10–11). Among his immediate contemporaries in England, only Hume took up the comparison between Harris and his French antecedents, writing to the Abbé Le Blanc on 24 Oct. 1754: 'Mr. Harris,

I

Harris had begun writing *Hermes* on 20 August 1747. Apart from the dedication to Hardwicke, the preface, index, and a few of the notes, it was finished on 7 December 1750.[7] Even before its publication in 1751, however, he was already well enough known to be referred to as a philosopher of mind. Precisely what this phrase signified would become clearer only after *Hermes* had appeared, but by the last decade of the eighteenth century it was possible for Herder to praise Harris and Monboddo together for making possible, for the first time, a 'Philosophie des menschlichen Verständes', a philosophy of mind.[8]

In England *Hermes* was first reviewed in Fielding's *Covent-Garden Journal*, number 21, 14 March 1752. The author of this review is not known, but its favourable recommendation might suggest Fielding himself. This is just as implausible as it is to suggest that Harris himself wrote the review. Although public renown for his writings was of course not unwelcome, self-advertisement was anathema. When Fielding wrote to him on 11 January 1745 about his contribution to *The True Patriot*, he remarked how little concern Harris himself would have for the consequent applause of the Town. Most contemporary reviews of *Hermes* restrict themselves to a synoptic account of its aims and methods, but this review goes further. It explains the meaning of the book's title, emphasizes the work's 'Perspicuity and Order', and makes some more

about two Years ago, publish'd a Book, which he calls Hermes or Universal Grammar. Notwithstanding his Affectation of Greek, & His Mimickry of Aristotle, he is a good Writer; & this Performance, in my Opinion, equals or surpasses that of Abbé Girard, which has Merit', J. Y. T. Greig, *The Letters of David Hume*, 2 vols. (Oxford, 1932), i. 208. Hume refers to Gabriel Girard (1677–1748), grammarian and author of *Synonymes françois* (1718).

[7] Manuscript note on p. 427 in Harris's own copy of the 1st edn. (University College, London).

[8] B. Suphan (ed.), *Herders Sämtliche Werke* (Berlin, 1877–1913), xv. 183, cited in H. Aarslef, *The Study of Language in England, 1780–1860* (Princeton, 1967), 42. This was Herder's comment in his prize essay on the origin of language ('Abhandlung über den Ursprung der Sprache', 1772). Herder praised Harris in his introduction to E. A. Schmid's trans. of Monboddo's *Origin and Progress* (Riga, 1784–5). For a sketch of the similarities between Harris and Herder on the idea that rationality is the 'freedom from stimulus control', rather than a 'faculty with fixed properties', see Chomsky, *Cartesian Linguistics*, pp. 15–16.

detailed textual points which betoken close reading, an informed background, and a determination to commend. The final paragraph bestows high, if innocent, praise on Harris's championship of Aristotle against 'modern Philosophers': 'In a Word our Author is so much *Aristotle* both as to the *Matter* and *Form* of his Treatise, that he perpetually keeps him in his Eye.' In passing, we might note Fielding's own ironic defence of Aristotle in his modern glossary (*Covent-Garden Journal* for 14 January 1752): '*Nonsense*—Philosophy, especially the philosophical writings of the ancients, and more especially of Aristotle.'

The author of this review, not only the earliest but also the most specific and informed appreciation of *Hermes*, is not known for certain. But the likeliest candidate is John Upton. Back in October 1744 Harris's brother George had informed Harris that the printer Paul Vaillant had arranged for an extract of *Three Treatises* to be translated into French by Matthieu Maty. This was subsequently published in *Bibliothèque raisonée* (xxxi. 479, incorrectly identifying the author as 'S. H.'). The author of the original extract was not John Upton, and to avoid another such 'silly and superficial review' of *Hermes* Harris asked Upton to review his second book for translation and publication in French, again using Vaillant as the agent.[9] On 25 February 1751 Upton wrote back to Harris to complain that Vaillant's offer of Upton's review to 'the fellows who write our Monthly review of Books' had been turned down as insufficiently commendatory, adding that 'the Frenchman [Vaillant or Maty?] is more modest & is translating wt is sent him'. On 14 January 1752, the very day the *Covent-Garden Journal* review appeared, George Harris wrote again to his brother to say that Upton 'has made a short epitome of Hermes wch will be sent down for his [James's] inspection'. George was not referring to the review published by Fielding but very probably to an anonymous 3-page printed pamphlet entitled *A Short Account of the Four Parts of Speech according to Aristotle, as explained in Hermes, by James Harris Esquire, The First of the Moderns that seems to have understood his Meaning.*[10] Two months later, in volume 7 of Maty's influential

---

[9] Harris to Upton, 21 Dec. 1751: SRO DD/TP 11, part 1.
[10] Copy in BL.

Journal Britannique (March 1752, pp. 307–20), published at
The Hague, there is a review of *Hermes* signed 'J. V.', a unique
and so far unidentified signature, but beyond any doubt the review written by Upton in February 1751.[11] Upton's collaboration with Harris on an edition of Epictetus, his enthusiasm for
Aristotle, and his regular correspondence on scholarly matters
right up to his death on 2 December 1760, qualified him better
than anyone else among Harris's friends for such a task. Quite
clearly, Upton was only too ready to engage himself as a
propagandist for Harris's most important book, and nothing
would have been easier or more welcome than the chance of
reviewing it for Fielding.

Commendation apart, *Hermes* had to wait some time before
its full effect was taken up. Even today, two centuries later, the
most detailed discussions of *Hermes* are in German and
French.[12] The most important contemporary critical reaction
in England was not Johnson's but Horne Tooke's. The
former's bluff but influential hostility may have deflected
some readers, but the latter's *ad hominem* attack, together with
an informed specificity about linguistic definitions, had a
prolonged influence. Johnson's criticism informed the common
reader uncorrupted with literary prejudices that *Hermes* could
be safely ignored. Tooke influenced the grammarians until the
first third of the nineteenth century. He also converted
linguistic theory into power politics.

Seven years after the publication of *Hermes* Johnson
exploited the opportunity of a covert attack on Harris's type of
scholarly discourse. The occasion was an essay on style; the
victim, one of Harris's intellectual allies: and although the

---

[11] This unique signature is not identified in U. Janssens, *Matthieu Maty and the
Journal Britannique 1750–55* (Amsterdam, 1975). Maty wrote his articles in England in
French and sent them to his printer in The Hague. His father, Paul Maty, had
reviewed Petvin's *Letters Concerning Mind* in the *Journal*, 6 (Sept. 1751), 39–69.
Matthieu's path crossed Harris's at several points. He was successfully recommended
for the post of under-librarian of the British Museum by Philip Yorke in 1756, and
succeeded Thomas Birch as secretary of the Royal Society in 1765. Harris was
admitted to the British Museum Reading Room in 1763 and became one of its
Trustees on 22 Nov. 1765 (see A. Sherbo, 'Some Early Readers in the British
Museum', *Transactions of the Cambridge Bibliographical Society*, 6: 1 (1972), 57. Paul
Vaillant was to become the BM printer and bookseller.

[12] See Joly *Hérmès*, and O. Funke, *Englische Sprachphilosophie im späteren 18.
Jahrhundert* (Berne, 1934). Funke sees the linguistic–philosophical ideas in *Hermes* as 'a
constituent part of the English pre-Romantic movement' (p. 18).

book was not by Harris himself, it is inconceivable that Johnson could have been unaware both of Harris's role in its publication and of the implications for Harris himself. In *Idler* 36, for 23 December 1758, Johnson describes a style of writing 'by which the most evident truths are so obscured, that they can no longer be perceived, and the most familiar propositions so disguised that they cannot be known'. This new genus of obscurity has three species: 'This style may be called the *terrific*, for its chief intention is to terrify and amaze; it may be termed the *repulsive*, for its natural effect is to drive away the reader; or it may be distinguished, in plain English, by the denomination of the *bugbear style*, for it has more terror than danger, and will appear less formidable as it is more clearly approached.' Johnson's chosen illustration was [John Petvin's] dauntingly entitled *Letters Concerning Mind, To which is added A Sketch of Universal Arithmetic, Comprehending the Differential Calculus, and the Doctrine of Fluxions* (1750). For a man who had said that 'we are perpetually moralists, but we are geo-metricians only by chance' ('Life of Milton'), Petvin's book was a perfect opportunity to prove the obscurity and dubious utility of metaphysical speculation: it discusses Pythagorian number systems, 'Being, or Essence, Identity, and Diversity', the idea of a triangle, the harmony of Nature, the 'Universal' Mind', the beauty of Virtue and Happiness, and our ideas of Truth and God. In letters 5 and 11 praise is heaped upon the Stoics and their modern champion, the Earl of Shaftesbury. There was much here to antagonize Johnson, but he mentions none of these things. Instead, he attacks the style of the discourse as a whole, citing the beginning of letter 6:

The author begins by declaring, that *the sorts of things are things that now are, have been, and shall be, and the things that strictly* ARE. In this position, except the last clause, in which he uses something of the scholastick language, there is nothing but what every man has heard, and imagines himself to know. But who would not believe that some wonderful novelty is presented to his intellect, when he is afterwards told in the true *bugbear* style, that the *ares, in the former sense, are things that lie between the* have-beens *and* shall-bes. *The* have-beens *are things that are past; the shall-bes are things that are to come; and the things that* ARE, in the latter sense, *are things that have not been, nor shall be, nor stand in the midst of such as are before them, or shall be after them. The things that have been, and shall be, have respect to present, past, and future. Those likewise that now* ARE *have moreover place; that, for instance, which is here, that which is to the east, that which is to the west.*

All this, my dear reader, is very strange; but though it be strange, it is not new: survey these wonderful sentences again, and they will be found to contain nothing more than very plain truths, which, till this author arose, had always been delivered in plain language.[13]

The author of this ponderous prose was a vicar of Ilsington in Devon, but Johnson did not dignify such writing by identifying the author. One inhibition may have been due to the fact that Petvin had recently died; another is almost certainly due to the fact that it was James Harris who had put the book through the press and recommended it one year later in *Hermes*. As the preface states, Harris's editorship was a labour of friendship towards his dead friend. But it was also an act of intellectual loyalty, as Harris's own annotations in his copy of the first edition of *Hermes* make clear. In the first, there is this:

They were first written in Short-hand, and, being all transcribed from the original Characters, have been corrected by a Gentleman highly esteemed by the Deceased, and well skilled in all Parts of Polite Learning, and Science, but especially in that Chief Science which relates to Mind: They who are at all versed in this kind of Literature, will easily recognize, under this Character, the Author of a Book called Three Treatises.[14]

---

[13] *Idler and Adventure*, ed. W. J. Bate, Yale Edition of the Works of Samuel Johnson, ii (New Haven and London, 1963), 113–14. Johnson contrasts Petvin's argument with the style of R. Cumberland's *Philosophical Inquiry into the Laws of Nature* (1750). After John Petvin's death in 1745, his papers were collected and deciphered from the shorthand by his brother William Petvin, at Harris's request. Harris's brother George advised on the ordering of the various 'letters' and wrote the latter half of William's preface. A Mr Jack was paid to check the mathematical letters, and the whole volume was then submitted to James Harris for publication (Harris Papers, Letter Book, vol. 31, part 1, letters 83–90, 1746–65).

[14] *Letters Concerning Mind*, preface, p. iii. Like Harris, Petvin argues for the priority of Mind in all things: thus, an idea of a triangle precedes the perception of an individual triangle (p. 38). Aristotle is described as 'the greatest Genius, that lived in the politest Age the World ever saw' (pp. 37–8); Xenophon and Shaftesbury are linked as masters of the inductive method, whose beauty lies not in an exhibition of method but in its subtle persuasion of the reader. Petvin has little respect for Locke, and on p. 86 describes the Mind as the agent of happiness and as a reflection of the Platonic Good. Thus Petvin's book was an important confirmation and source of Harris's own cultural and intellectual loyalties. Petvin's later book, *Letter Concerning the Use and Method of History* (1753) similarly represented Harris's view of the human

In Harris's own revised and heavily annotated copy of the first edition of *Hermes* there are forty-four references to Petvin's book and a total of fifty-nine specific page references to analogous passages in *Letters Concerning Mind*. The second edition of *Hermes* does not mention Petvin's book, but a generous acknowledgement was inserted from Harris's notes in the 1801 and 1841 editions of his *Works* (p. 167).

Johnson attacked Harris by means of a book review of Petvin. The target in both was a style of discourse which seemed obfuscatory, saying plain truths in obscure ways. But before assenting to Johnson's ridicule, we should also notice how selective Johnson had been. In mocking Petvin's style, he had also suppressed some points about aesthetic theory for which Harris himself was to be acknowledged by Reynolds. It was thus left to others to see value in Petvin's ideas. The opening sentence of *Thoughts on Outline, Sculpture, and the System that Guided the Antient Philosophers* (1796) states:

If there be Beauty in Virtue, remarks the learned Mr Petvin, in his Letters Concerning Mind; 'the mind must have a feeling of it, whilst it has under view, no less than a feeling of harmony, when presented to the ear. It must be felt and understood *together, we must be in some measure what we behold*; and a man must be tolerably good before he can have any tolerable notion of goodness.'[15]

The author of that remark was George Cumberland, in a book illustrated by William Blake. Six years previously, the same idea had appeared in Joshua Reynolds's paraphrase of Harris's own version of empathic identification in *Discourse Four*: 'we must feign a relish, till we find a relish come; and feel, what began in fiction, terminates in reality.' Although this sentiment is virtually synonymous with the closing lines of Johnson's own *Vanity of Human Wishes*, it was not a critical principle which Johnson could comfortably apply to art. His nervousness when in the presence of devotional or love poetry, or witnessing the death of Cordelia at a performance of *King Lear*, necessitated for him a proscenium arch (either real or

value of classical history and the doctrines of philanthropy, benevolence, and inner harmony in Shaftesbury's writings (pp. 16–19). There is no evidence that Harris also published this book, but the circumstantial probability is high. Rivington printed both books: see n. 11 above.

[15] *Thoughts on Outline* (1796), p. i (opening sentence). Blake etched 8 of the plates.

metaphorical) between aesthetic enactment and the private self of the reader or spectator. To say that this principle of being what we perceive had its source in Shaftesbury was enough by itself to elicit a hostile reaction. In *Hermes* Johnson found even more to disapprove.

Between the first and second editions of *Hermes* Harris made a change to the title. Originally subtitled *A Philosophical Inquiry Concerning Language and Universal Grammar*, the second edition omitted the words 'Language and'. The reason for this change is not clear. Although *Hermes* does treat Universal Grammar as its ostensible subject, it is essentially a book about language; or rather, it treats Universal Grammar as part of an extremely wide philosophical framework of language as a system of symbols of general ideas. Universal Grammar is only one aspect of a metaphysical doctrine of universals. In Harris's own words, 'the same Reason has at all times prevailed; . . . there is one Truth, like one Sun, that has enlightened human Intelligence through every age, and saved it from the darkness both of Sophistry and Error'.[16] *Hermes* is an essentially eighteenth-century book, a product of the age's need to order the arts of discourse and to establish rational grounds for discriminating between them. It is a logical extension of what Harris had been attempting in *Three Treatises*, except that here he is concerned with the generation of spoken and written language, and with words as the symbols of human ideas. His general theory is that these ideas are themselves reflections, both particular and general, of the divine mind. Language is a defining attribute of man the sociable animal; it is the medium through which his unique mental capacity is expressed.

---

[16] *Hermes* (2nd edn., 1765), p. x. All quotations are from this rev. and corrected edn., and are hereafter given as page refs. in the text. Harris evidently regarded *Hermes* and his 2 subsequent books as corresponding to the ancient division of the medieval trivium, i.e. Grammar, Logic, and Rhetoric. A manuscript note in Harris's copy of the 1st edn. (University College, London, p. 7) states: 'The reason why the author has preferred Grammar, and postponed the other two, is, that Grammar has not been so fully handled in a philosophical way, whereas the Logic and Rhetoric of Aristotle have exhausted their several subjects, and left nothing to be added.' As Chomsky remarks (*Cartesian Linguistics*, p. 59), 'the universal grammarians of the seventeenth and eighteenth centuries have made a contribution of lasting value by the very fact that they posed so clearly the problem of changing the orientation of linguistics from "natural history" to "natural philosophy" and by stressing the importance of the search for universal principles and for rational explanation of linguistic fact'.

II

Hermes, then, is not just about grammar and the parts of speech; it is about the furniture of the mind, the actual syntax of thinking, and a link to what Harris was to call that 'Providential Circulation which never ceases for a Moment thro' every Part of the Universe'. Harris's God, like Fielding's, is a god of Order and Harmony, and the scholar's role is to establish the working evidence for such harmonies in the human arts of living, particularly in verbal discourse (liter-ature), moral behaviour (ethics), the visual arts (painting, sculpture, architecture), and in the more abstract arts of music and philosophy. At one point in *Hermes* Harris ventures an analogy between classifications in language and the system of botanical classification invented by Linnaeus: the point being that if order systems are indeed fundamental to the human world they may be found in any area if man is a part of an harmoniously ordered whole.

In the first two books of *Hermes* Harris postulates a connection between language and thought, and between words and things. The premises for Universal Grammar were disarmingly simple. If language is the image of thought and if thought is subject to the laws of Reason, then language itself must reveal and illustrate the laws of Reason. The anonymous author of the entry on Universal Grammar in the first edition of the *Encyclopaedia Britannica* (1771) defines the mechanical but not the metaphysical implications of Harris's theorem:

Grammar considered as a *Science*, views language in itself; neglecting particular modifications, or the analogy which *words* may bear to *each other*, it examines the analogy and relation between *words* and *things*; distinguishes between those particulars which are *essential* to language, and those which are only accidental; and thus furnishes a certain standard, by which different languages may be compared, and their several excellencies or defects pointed out. This is what is called PHILOSOPHIC or UNIVERSAL Grammar.[17]

---

[17] *Encyclopedia Britannica* (1771), 728; cited in *English Grammatical Categories*, p. 179. Cf. Chomsky's definition (*Cartesian Linguistics*, p. 59), and Aarslef, *The Study of Language in England*, p. 14.

The credit for enunciating such principles is usually given to Lancelot and Arnauld's Port-Royal *Grammar* of 1660, also known as *The Art of Speaking*, and the same authors' Port-Royal *Logic* of 1662, also known by the title *The Art of Thinking*. There is only one passing reference to the *Grammaire générale et raisonnée* in *Hermes* (p. 81). Although Harris owned a copy of this highly significant work, there is no reason to think that it directly shaped this thinking. Harris said that he had learnt everything he knew about Universal Grammar, not from French or English works, but from the sixteenth-century Spanish grammarian Sanctius, whose Latin grammar was first published in 1562, and whose fully developed linguistic theories appeared under the title of *Sanctii Minerva, seu de Causis Linguae Latinae* in Salamanca in 1587.[18]

Of the nine references Harris makes to Sanctius in his published works only one, on the notion of impersonal verbs, permits Sanctius to make a point not otherwise illustrated from the works of the ancients (Aristotle and Cicero), or more recent neo-Aristotelians such as Ammonius, Scaliger, and others. It would seem that Harris used the *Minerva* to steer by, but that his linguistic intention was to illustrate everything he needed to say from classical sources. His note on Sanctius in *Philological Inquiries* is, however, a clear acknowledgement of his indebtedness, and in *Minerva* he found what he was looking for, a rationalistic grammar which sought to uncover the logical (and Aristotelian) bases for language acts. The fact that *Minerva* was itself a source for the Port-Royal Latin grammar of 1664 was of no special significance to him. Unlike the linguistic materialism of Horne Tooke, who regarded words used in the art of reasoning as merely 'the machines used by the mind; as the lever, the wheel, or the screw, are but faint representations of their power and their utility', Harris's view of language is shaped by idealism. The crucial point is that both discourse (spoken and written) and thought

---

[18] Gertrude Harris, 'Memoir', states that her father 'had acquired all his rational ideas of Grammar from Sanctius's Minerva'. See M. Breva-Claramonte, *Sanctius' Theory of Language: A Contribution to the History of Renaissance Linguistics* (Amsterdam, 1983), for the most recent and extensive account. There is no mention of Harris in this book. A footnote in *Philological Inquiries* says that Minerva is the book 'to which the author of these treatises readily owns himself indebted for his first rational ideas of grammar and language', *Works* (1841), p. 393. Harris used the 1st edn.

processes of the human mind revealed a universally valid rational structure found not only in language but in nature itself.

When Harris writes on the 'coalescence' which exists between adjective and substantive, adverb and verb, subject and verb, and verb and object, he sees a grammatical relationship as a reflection of a fact in Nature: 'Those Parts of Speech unite of themselves in Grammar, whose original Archetypes unite of themselves in Nature . . . the great Objects of Natural Union are Substance and Attribute' (pp. 263–4). His discussion of the verb includes the notion that the verb *be* is 'a latent part in every other Verb' (p. 90). On the other hand, as Ian Michael has noticed, Harris's 'tense-scheme is not distorted by logic. It respects the facts of the language and is more perceptive than any previously suggested. It is meant to apply universally, but is in fact based on Greek and moderated by English. Harris had a sufficiently sound understanding of the nature of language to question the universality of his own scheme.'[19] Harris's metaphysical theory of universals clearly recognized the intransigence of human practice (pp. 122–3):

It is not to be expected that the above Hypothesis should be justified through all instances in every language. It fares with Tenses, as with other Affections of Speech; be the Language upon the whole ever so perfect, much must be left, in defiance of all Analogy, to the harsh laws of mere Authority and Chance.

Sanctius has been described as 'the first modern grammarian', and *Hermes* as the first book to give *Minerva* the respect it deserved.[20] There was a peculiar appeal for Harris in a grammatical theory which combined Aristotelian categories with the contemporary Augustan demand for rational ordering within a comprehensive system. The precise reasons for Harris selecting Sanctius as his mentor rather than, say, Scaliger or Ramus, or even Ben Jonson's *English Grammar* of 1640 and Wilkins's *Real Character* of 1668, among many other possible models, are best illustrated by comparing a summary of Sanctius's three main contributions to grammatical theory

[19] Michael, *English Grammatical Categories*, pp. 34, 23.
[20] G. A. Padley, *Grammatical Theory in Western Europe 1500–1700: The Latin Tradition* (Cambridge, 1976), 109.

with a brief outline of analogous points in Harris's work. The Spanish Mercury and the English Hermes have much in common.

G. A. Padley notes that Sanctius's fundamentally rationalist approach is revealed in his insistence on underlying 'causes', in the 'logical substructive of language', and particularly in the difference between this 'underlying logical structure' and actual speech.[21] Such a procedure anticipates the modern distinction between deep and surface structure.

It was Harris's purpose to show that logical categories were indeed the deep structure underlying not only grammar but also *literary language*. For Sanctius, 'every figurative usage is an invitation to the linguist to demonstrate the existence of an underlying logical level'. When Harris looked at the opening scene of *Macbeth*, and the witch's interrogative sentence 'When shall we three meet again, | In thunder, lightning, or in rain?', etc., he found the predicaments of *when* and *where*. Ultimately, all the arts and sciences could be so arranged. As a summary of this lifelong conviction, Harris states in *Philosophical Arrangements* his 'Wish to persuade his Readers, of what he has long been persuaded himself, that every thing really elegant, or sublime in composition, is ultimately referable to the Principles of a sound Logic; that those Principles, when Readers little think of them, have still a *latent* force, and may be traced, if sought after, even in the politest of Writers.' Harris expands this conviction to a world-view, a logical key to all arts and sciences:

history, natural and civil, [arises] out of Substance; Mathematics, out of Quantity, Optics, out of Quality and Quantity; Medicine out of the same; Astronomy, out of Quantity and Motion; Music and Mechanics, out of the same; Painting, out of Quality and Site; Ethics out of Relation; Chronology, out of When; Geography, out of Where; Electricity, Magnetism, and Attraction, out of Action and Passion.

The second characteristic feature of Sanctius's grammatical theory is his introduction of the tripartite word-class system:

---

[21] The following remarks on Sanctius are taken from G. A. Padley, *Grammatical Theory in Western Europe* 1500–1799: *Trends in Vernacular Grammar I* (Cambridge, 1985), 269–75.

noun, verb, and particle.[22] *Hermes* proceeds on the division established in Aristotle's *Poetics*, i.e. nouns, verbs, articles, and conjunctions, for which terms Harris substitutes substantives, attributives, definitives, and connectives (*Hermes*, p. 36).

Thirdly, Sanctius's most important grammatical contribution to linguistic theory is 'the particular status he gives to syntax . . . seeing syntax as the final object of grammatical description: *cuius finis est congruens oratio*. The corner-stone of his syntactic doctrines is his theory of ellipse, and it is here that a revolution is operated that is of cardinal importance for the development of theories of universal grammar'.[23]

Harris discusses and illustrates the operation of syntactic ellipsis in chapter 8 of the first book of *Hermes*, and his concept of 'latent' structure also indicates that he was fully aware of the importance of this aspect of syntactic theory. The notion of ellipse was not new in Sanctius, but if Harris was alerted to it by *Minerva*, his annotations in *Hermes* indicate that he reread Quintilian's distinction in *Institutiones Grammaticae* in its light. The latter is referred to more than two dozen times in *Hermes*. The distinction is important because it places Harris in the centre of what Noam Chomsky was to call 'Cartesian Linguistics'. Without using such terms, Harris nevertheless differentiates between 'surface' and 'deep' structure, two levels of syntax, the one spoken or written on the surface (semantic) and another implied on a deeper level (logical). Such is Harris's point in *Hermes*, where he draws a distinction between what is 'expressed' and what is 'understood', as in the example of the interrogative sentence (p. 153):

Custom hath consulted for Brevity, by returning for Answer only the single essential characteristic Word, and retrenching by an Ellipsis all the rest, which rest the Interrogator is left to supply from himself. Thus when we are asked—'How many right angles equal the angles of a triangle?'—we answer in the short monosyllable, 'Two'— whereas, without the Ellipsis, the answer would have been—'Two right angles equal the angles of a triangle.'

Horne Tooke was later to attack Harris for attributing to Nature what was only attributable to language, but in the

---

[22] Ibid. 271.
[23] Ibid.

same chapter in the first and all subsequent editions of *Hermes* this idea is elaborated:

> Even when we speak such sentences as the following, 'I choose to philosophize, rather than to be rich . . . the infinitives are in nature as much accusatives, as if we were to say, 'I choose philosophy rather than riches.' (pp. 164–5)

In his earlier discussion of the pronoun in chapter 5 (pp. 78–9) Harris provides nothing less than a transformationalist account of the pronoun's role in the sentence:

> Suppose I was to say, 'Light is a body, Light moves with great celerity.' These would apparently be two distinct sentences. Suppose, instead of the second Light, I were to place the prepositive pronoun *it*, and say, 'Light *is* a Body; *it* moves with great celerity'; the sentences would still be distinct and two. But if I add a Connective, (as for example an *and*), saying, 'Light is a body, *and* it moves with great celerity'; I then by Connection make the two into one, as by cementing many stones I make one wall.

As if to enforce the authority and primacy of his *classical* sources for the distinction between 'expressed' and 'understood', Harris first cites an illustrative example from Appollonius' *De Syntaxi*, and then adds his only published reference to 'an ingenious French treatise, called Grammaire Generale et Raisonnee'.

There is no doubt about Harris's dominant influence and guiding purpose in *Hermes*. Although he is indebted to Plato's *Timaeus* for his discussion of time, to Ammonius for the idea that the soul's leading powers are perception and volition, and to Sanctius for organizing his interest in universal grammar in the first place, all these influences are subsumed under a celebration of Aristotle. His indebtedness to Aristotle is overt, pervasive, and part of the whole purpose behind the writing of *Hermes*. His definitions of the word and the sentence, notions of gender, the idea that sensation is limited to a single instant of time, and the idea that the meaning of words is founded on compact, are all deduced from Aristotle.

Paradoxically, Harris was most confident and most vulnerable when tracking language theories back to Aristotle. If we take Johnson's opinion of logic as representative of contemporary literary opinion, Harris's conscious opposition to the Lockian orthodoxies is even more clearly revealed as a deliberate attempt to reverse what he saw in the preface to *Hermes* (p. xv) as 'the bigotted contempt of every thing not modern'. Johnson contributed a preface to Dodsley's *The Preceptor* (1748), an early version of a home-university library which included the first appearance of William Duncan's important *Elements of Logic*. Under the subject of logic Johnson recommends the work of Crousaz, Watts, Le Clerc, Wolfius, and Locke, and then adds: 'If there be imagined [*sic*] any necessity of adding the peripatetick logic, which has perhaps been condemned without a trial, it will be convenient to proceed to Sanderson, Wallis, Crackanthorp, and Aristotle.'[24] In his *Dictionary*, however, there is no attempt to rank Aristotle, even at the bottom of the list of preferences. For authoritative logical examples Johnson adopted the moderns: Locke's *Essay*, but more particularly Isaac Watts's 2 volumes (*Logick*, 1725; *Improvement of the Mind*, 1741), citing more than 150 times from them in the first volume alone, and recommending both for Revd Daniel Astle's course of private education.[25] The reason for this preference can readily be inferred from the Dodsley preface: traditional Aristotelian logic as presented by the scholastic tradition had become 'a mere art of wrangling of very little use in the pursuit of truth'.

This was the commonplace judgement, and it was fully acknowledged by Harris. For the training of the modern mind, only modern logicians would do, because they gave 'an account of the operations of the mind, marking the various stages of her progress, and giving general rules for the regulation of her conduct'. In other words, Johnson's attitude towards logic was moulded by the psychologistic and ultimately ethical concerns of the moderns since Locke, rather than by the disputative tradition which Aristotle's logic had generated. There is also a sense in which Johnson would terrify the young student out of a study of logic altogether:

---

[24] *The Preceptor: Containing a General Course of Education* (1748), in *The Works of Samuel Johnson*, ed. A. Murphy, 12 vols. (1824), ii. 250. Duncan's *Elements* and contemporary logic in general are discussed in W. S. Howell, *Logic and Rhetoric in England, 1500–1700* (Princeton, 1956). Duncan's logic, recommended by Johnson, was abstracted to form the 20-page entry on logic in the 1771 *Encyclopaedia Britannica*.
[25] W. K. Wimsatt, *Philosophic Words* (New Haven, 1948), 71.

But how much soever the reason may be strengthened by logick, or the conceptions of the mind enlarged by the study of nature, it is necessary the man not be suffered to dwell upon them so long as to neglect the study of himself, the knowledge of his own station in the ranks of being, and his various relations to the innumerable multitudes which surround him, and with which his Maker has ordained him to be united for the reception and communication of happiness. To consider these aright is of the greatest importance, since from these arise duties which he cannot neglect. Ethicks, or morality, therefore is one of the studies which ought to begin with the first glimpse of reason, and only end with life itself. Other acquisitions are merely temporary benefits, except as they contribute to illustrate the knowledge and confirm the practice of morality and piety, which extend their influence beyond the grave, and increase our happiness through endless duration.[26]

Johnson has clearly shifted the grounds of his argument away from logic to ethics: logic is a specialized intellectual pursuit and like all other specialisms must ultimately give way to a consideration of the final duties of self-knowledge and the practice of Christian piety. In this context Harris's apology for the priority of classical learning, and of Aristotelian logic in particular, is antipathetic. True, he informed Monboddo, as *Hermes* was being transcribed for the press, that in his next book 'ethics, imagery, and as many quotations from the Poets, as I could with any propriety collect together' were to be added not as fundamental parts of the argument but to make logic 'palatable'. But Harris was in fact in precise agreement with Johnson's ethical priorities. Like Shaftesbury before him, Harris stressed at the end of 'Dialogue on Happiness' that if this 'moral science' is lacking, then we can only 'value Logic but as Sophistry, and Physics but as Raree-show; for both, assure yourself, will be found nothing better'. The rationale of *Hermes* itself, after all, is not only the service of a narrowly philological purpose but the vastly more ambitious one of revealing in language the operation of timeless and universal truths.

Johnson would no doubt have agreed with Locke's epigram on the Peripatetic tradition: 'God has not been so sparing to men to make them barely two-legged creatures, and left it to

[26] *Works*, ed. Murphy, ii. 250–1. See further C. T. Probyn, 'Johnson, James Harris, and the Logic of Happiness', *MLR* 73: 2 (Apr. 1978), 256–66.

Aristotle to make them rational.' In a *Rambler* paper published in the same year as *Hermes* Johnson attacked those writers who 'think themselves entitled to reverence by a new arrangement of an ancient system' (*Rambler*, no. 129), a fair approximation of Harris's aims in *Hermes*, and an uncomfortably accurate prognosis of his intentions in *Philosophical Arrangements*. As for Locke, Harris sees him as one of Swift's Moderns in *A Tale of a Tub*. In a subsequent letter to Monboddo, 14 May 1773, he makes a rare reference to Locke by name:

I freely subscribe to your ideas of Mr Locke. Ignorant of all Ancient Literature, he had an inclination to spin out everything from his own brain, as if so stupendous a work as an Analysis of the Human Understanding could be raised by the effort of one unassisted man . . . Life is too short, and the labour too immense, for a single man to carry anything to perfection.[27]

The implications of that last remark provide the best insight into Harris's intellectual position. He saw culture as a synchronic and collective process, a continuing accumulation whereby the past gives strength to the present by means of memory in the individual and constant updating of cultural history for a whole society. Like Johnson, he was oppressed by an awareness of human limitations, of our tendency to forget where our intellectual resources originate. Like Imlac's vision of the business of a poet in chapter 10 of *Rasselas*, Harris's whole career as a writer was directed at re-marking and re-assembling the general properties and combating the narrow view.

For his contemporary world, stooped over the ocular proofs of empirical and experimental science, Harris offered the challenge of the 'sublimer part of Science, the Studies of Mind, Intellection, and the Intelligent Principles'. In place of Hobbes's pessimistic materialism Harris put the intellectually restorative and primary human faculty of imagination. Against Locke, he argues that the mind is stored with its own energies and perceptive powers, acting upon but in no way wholly determined for its operation by the sensible world, and he cites the example of mathematics to prove the mind's ability to generate abstract logical theorems without reference

---

[27] W. Knight, *Lord Monboddo and Some of his Contemporaries*, p. 79.

to the physical world. The progress of language acquisition thus moves *from* the merely sensible to the intellectual, as a child becomes an adult: 'the first Words of Men, like their first Ideas, had an immediate reference to sensible Objects, and that in the after-days, when they began to discern with their Intellect, they took those Words which they found already made, and transferred them by metaphor to intellectual Conceptions' (p. 269).

As in the individual, so in the nation: the chief intellectual error of his own time was its insularity and ahistoricity: 'we Britons by our situation live divided from the whole world . . . our studies are usually satisfied in the works of our own Countrymen; that in Philosophy, in Poetry, in every kind of subject, whether serious or ludicrous, whether sacred or profane, we think perfection with ourselves, and that 'tis superfluous to search further' (pp. xi–xii). England had acquired a political freedom the envy of Europe, but it was still in chains forged by a materialistic ethic and a cultural complacency. More narrowly, *Hermes* was to have a para-doxical effect: the aim to revive the 'decaying taste of antient Literature, to lessen the bigotted contempt of everything not modern' (p. xv) was to liberate the teaching of English grammar from the grip of Latin. Harris's book was to end the dominance of William Lyly's Latin grammar, the very text which probably initiated his linguistic interests at the Grammar School in Salisbury.

When William Hazlitt's uncle, Sir John Stoddart, wrote his book on Universal Grammar, originally planned for inclusion in Coleridge's projected *Encyclopaedia Metropolitana*, he did so under the influence of *Hermes*. Like Harris before him, Stoddart had been a pupil at the Close Grammar School in Salisbury, where Coleridge's older brother had been under-master, and where the mindless rote-learning of Lyly's Latin grammar had ruled the linguistic curriculum. Harris's protégé, William Benson Earle, had given *Hermes* to the 14-year-old Stoddart, thus opening his eyes for the first time to the 'many profound speculations of intellectual philosophy' and eventu-ally propelling him into a lifelong study of language. When Stoddart's book appeared it included not only a fierce defence of Harris—'whatever may have been the errors of Harris, they

were not a thousandth part so gross, or so injurious to the
science of grammar, as those into which Tooke himself had
fallen'—but also confirmed Harris's view of the priority of
mind, not an 'inert mass, receiving objects, and returning
them back . . . from the structure of the mental machine . . .
purely mechanical' but 'an active energy, a self-moving
power'.[28]

In his own mind Harris's role was to invite the reader to
collaborate with him in a sociable compact, effecting the
rescue of a cultural tradition from oblivion, extending
backwards the intellectual resources of his present time. His
method is to classify, compare, and amalgamate. His ultimate
aim is to discern patterns and systems for further exploration:
hence his gratification when he attended one of John Hunter's
evening classes on comparative anatomy in February 1773:

I could not but hear him with the highest satisfaction. Instead of
many strange tales about many strange things (the usual process of
modern philosophy) we had the sure antique method, that of looking
to the Whole; of tracing out *identities*, in comparing things
heterogeneous and dissimilar; of tracing out *diversities*, in comparing
the similar and homogeneous, of going from Means to Ends, and
from Ends back again to Means; of investigating, by this process,
general Theorems; and of employing those we discover, to help us to
the discovery of new ends . . .[29]

Harris might just as well be describing his own intellectual
methodology.

Just occasionally in the history of linguistics Harris has
been credited with insight, in spite of the fact that when his
work is measured against that of his contemporaries and
predecessors his achievements and his originality are
compellingly clear. Ian Michael's exhaustive study, *English
Grammatical Categories*, judges *Hermes* as 'by far the most
penetrating of all the works written in the name of Universal
Grammar'.[30] Harris's innovations are considerable. He is, for
example, the first grammarian to make the 'implication of

[28] Sir John Stoddart, *The Philosophy of Language; Comprehending Universal Grammar, or Pure Science of Language*, ed. W. Hazlitt (1830; 3rd rev. edn. 1854), pp. v–vi, 22–3.

[29] Knight, *Lord Monboddo and Some of his Contemporaries*, p. 75.

[30] Michael, *English Grammatical Categories*, p. 172. The following references are to pp. 359, 360, 173, 477, 422, and 517. André Joly (in his edn. of Thurot's trans., p. 55) claims that Harris was the first to use the word 'system' about language.

previous reference the essential function of the definite article';
he is 'one of the first English writers to point out that the
absence of a linguistic element . . . can have a positive
significance'; since his analysis of speech begins with the
sentence, Harris thereby shows 'a clearer understanding of
language than was to be found in any English grammar before
his day'; his classification of conjunctions is done with 'ardour
and freshness', his tense-schemes established 'on fresh criteria'.
He is, with William Ward, one of those 'rare individuals' who
can reunite logic with the idea of universal grammar, and
resist both educational utilitarianism and the crushing weight
of tradition, thus allowing 'the enlightening power of reason to
affect his grammatical categories'. R. H. Robins accorded
Harris's linguistic analysis in *Hermes* precisely the sort of value
Harris himself observed in Hunter's scientific method, and the
kind of instrumental role in the extension of intellectual
continuity he most sought, i.e. the construction of a general
theory from observed and comparative data which may then
be used as the basis for further refinements and definitions:
Harris's Aristotelian distinction between 'matter and form . . .
with reference to the phonic substance and the semantic
function of speech foreshadows the important doctrine of *innere
Sprachform*' later found in the work of von Humboldt, which 'in
turn may be likened in some degree to the *langue* of de
Saussure's later *langue-parole* dichotomy'.[31]

III

On 1 May 1778 Mrs Harris write to her son: 'Parson Horne
[Tooke] has writ a little pamphlet, called "A Letter to Mr

---

[31] R. H. Robins, *A Short History of Linguistics* (London, 1967), 155, 175. Joly argues
that Robins is mistaken in seeing Harris defending the concept of innate ideas against
the prevalent English empiricism. This may be technically true, but Harris does adopt
the phrase in *Hermes* (p. 393), and Harris's whole epistemological theory depends
upon innate powers of mind. He often scorns the idea of the mind's passivity. As for
Harris's 'hostility' towards empiricism, his praise of Baconian induction, his defence
of the experimental method (*Hermes*, p. 352), and his enthusiasm for mathematics, are
major if not fundamental characteristics of empiricism. Harris's position is that
intellect *begins* when it seeks the *why* in the *what* supplied by experience and
experiment. On p. 353 he remarks: 'while Experiment is thus necessary to all
Practical Wisdom, with respect to Pure and Speculative Science . . . it has not the
least to do. For who ever heard of Logic, or Geometry, or Arithmetic being proved
experimentally?'

Dunning". 'Tis the most strange stuff ever seen: he has presented it to your father, whom he has fallen upon most egregiously on account of his *Hermes*.'[32] In the whole history of linguistics, Tooke's pamphlet address to his friend and defending counsel John Dunning must rate as the one with the strangest provenance. Even though interpretation of the law often turns on the meaning of a word, the King's Bench prison is not the ideal place for philological theory. The prospect of being hanged on the morrow may concentrate the mind wonderfully, but in Tooke's case a twelve-month sentence for seditious libel was to concentrate his mind on the problems of the conjunction, and on Harris's linguistic theories, which he was to attack remorselessly.

In his own words, Tooke had been made 'the miserable victim of—*Two Prepositions and a Conjunction*' (i.e. *that*, *of*, and *concerning*). Tooke had accused the king's army of murdering loyal American colonists at Lexington and Concord on 19 April 1775, and through the agency of Benjamin Franklin, Tooke's Society for Constitutional Information had sent £100 to the widows, orphans, and aged parents of American soldiers killed in action. With wonderful historical irony, Tooke's trial opened on 4 July 1777, Lord Mansfield presiding, and Attorney-General Thurlow prosecuting. Found guilty, Tooke was fined, required to provide sureties for good behaviour, and jailed in the King's Bench prison. In the event, he was soon allowed to inhabit more comfortable accommodation outside the prison walls, where he was visited by his friends and permitted to dine once a week at The Dog and Duck in St George's Fields. Among his books was Harris's *Hermes*, whose arguments and reputation Tooke set out systematically to annihilate.

The comparison between the amiable Harris in his comfortable book-lined study in Salisbury, his mind roaming freely over classical culture, and the incarcerated seditious philologist whose vitriolic prose smouldered with a sense of political injustice, is startling. Both were linguistically expert, but the only thing they had in common was gout. Nothing delighted Tooke more than having to admit that *Hermes* had

---

[32] *Series of Letters*, i. 388. All references to *A Letter to John Dunning* (1778) are given in the text.

been 'received with universal approbation both at home and abroad', and that it had been quoted as undeniable authority on the subject 'by the learned of all countries' (*Diversions*, i. 120). If the whole world (except Bacon, Wilkins, much of Locke, and a few things in Johnson's *Dictionary*) was against him, then Tooke's triumph in his own certitude would be the sweeter.

The ostensible purpose of Tooke's pamphlet was finally to clear away all the confused nonsense in definitions of the conjunctives (if, unless, yet, still, though, but, without, since, and so on), as having no meaning. Harris had defined them as 'devoid of signification' or with 'a kind of obscure Signification, when taken alone . . . they appear in Grammar, like Zoophites in Nature, a kind of middle Beings, of amphibious character, which, by sharing the Attributes of the higher and the lower, conduce to link the Whole together' (*Hermes*, p. 259). Tooke achieves his purpose by some deft comparative etymology, but the crucial part of his argument is based on the English language, whose character is distinctly and properly different from that of Greek or Latin: he thus traces every conjunction to its Anglo-Saxon verbal root (thus *if* stems from *to give*, *though* from *allow*, *but* from *be-out*, and so on: p. 25). In the service of this narrow etymological purpose Tooke has a wider political argument about the nature of authority itself. Harris the gentleman scholar, linked with Lord Monboddo, is to the philosophy of language as Lord Mansfield is to the law. The first two, as it were, had gentrified linguistic philosophy but had only succeeded in substituting the fog of an elegant class dialect for Locke's demotic common sense. Mansfield, the Lord Chief Justice, had proved his legal incompetence by imprisoning Tooke through an exercise in mere word-play: 'if this reverend Earl's authority may be safely quoted for any thing, it must be for *Words*. It is so unsound in matters of Law, that it is frequently rejected even by himself' (p. 53). To Tooke, the whole trio seemed to act in a class tyranny against truth and justice. In a quite marvellous piece of rhetorical plea-bargaining, the ex-Wilkesite Tooke remarks (p. 20):

I can easily suppose that in this censure which I thus unreservedly cast upon Mr Harris, (and which I do not mean to confine to his account of the Conjunctions alone, but extend to all that he has written on the subject of language) I can easily suppose that I shall be thought, by those who know not the grounds of my censure, to have spoken too sharply. They will probably say that I still carry with me my old humour of Politics, though my subject is now different; and that, according to the hackneyed accusation, I am against authority, only because authority is against me . . . it is not my fault, if I am forced to carry instead of following the lanthorn; but at all events it is better than walking in total darkness.

If bad grammar and injustice are inseparable, then any philosophy of language will inevitably, if unconsciously, transmit a political ideology. Tooke therefore pointedly and mischievously transfers Harris's examples to show the ideological behind the grammatical. Whereas Harris defines *unless* as an 'adequate preventive' in the example drawn from classical history, 'Troy will be taken, unless the Palladium be preserved' (p. 256), Tooke illustrates the 'inadequacy' of Harris's definition by showing the conjunction in a political context, and, more specifically, in the context of contemporary constitutional theory:

'England will be enslaved, UNLESS the House of Commons continue a part of the Legislature.'
Now I ask,—Is this alone sufficient to preserve it? We who live *in these times* know but too well that this very House may be made the Instrument of a Tyranny as odious and (*perhaps*) more lasting than that of the Stuarts. I am afraid Mr Harris's *adequate Preventive*, UNLESS, will not save us. For though it is most cruel and unnatural, yet we know by woeful experience that the *Kid* may be *seethed in the Mother's milk*, which Providence appointed for its nourishment; and the Liberties of this Country be destroyed by that very part of the Legislature which was most especially appointed for their security.

For Tooke, those authorities, be they judges or scholars, who distort the meaning of words, not only degrade the English language; they also distort truth, restrict freedom, and perpetrate political tyranny. And although his chief target throughout the *Letter* is Harris's 'blind prejudice for his Greek Commentators' (p. 69), he also attacks the errors in Johnson's *Dictionary*, the occasional lapses of Locke himself (whom, in Ben Jonson's words on Shakespeare, 'I reverence on this side of idolatry', p. 45), and the blunders of Casaubon, Skinner, Junius, Wallis, Lowth, Sanctius, the 'sheer nonsense' of the '*common sense* of Oswald, Reid and Beattie' (p. 67), and even the lapses committed by 'a real Grammarian and Philosopher, J. C. Scaliger' (p. 69). Harris is at least placed in the best company, even though the combined efforts of all of them, according to Tooke, have produced a radical confusion.

In Tooke's view, the study of the human mind must be kept absolutely distinct from a study of words. Words are subject to human duplicity and are often misapplied to mask an evil purpose. It was therefore Locke's error to call his book 'an Essay on the Human *Understanding*' when it was really and only 'a *grammatical* Essay, or a Treatise on *Words* or on *Language*' (p. 70). Tooke was a nominalist, and therefore implacably opposed to the Aristotelian arguments of Harris and Monboddo. Words are 'the wheels of language', and if only nouns and verbs existed as absolute and natural linguistic objects, then everything else was simply a device for the abbreviated conveyance of thought, or signs for *signs*. The old grammarians were thus wrong to think 'all words to be immediately either the signs of things or the signs of ideas . . . in fact many words are merely *abbreviations* employed for despatch and are the signs of other words' (p. 21). Language does not reflect the operations of the mind.

The definitive answer to Tooke did not come from Harris himself, but from Coleridge, who took Harris's side. Coleridge pointed out that the contestants were actually talking about different things: Harris was talking about the 'philosophy, the moral and metaphysical causes' of language, whereas Tooke was talking about the formation of words only, even though he might have believed that etymology provided the key to the philosophy of language. Thus, Tooke's 'abuse of Harris is most shallow and unfair'.[33] In 1824 John Fearn claimed that both Harris and Tooke were equally in the dark in some respects, but nevertheless pointed out that 'the Theory put

---

[33] *Specimens of the Table Talk of Samuel Taylor Coleridge*, ed. H. N. Coleridge, new edn. (1858), p. 66 (7 May 1830). 'Tooke affects to explain the origin and whole philosophy of language by what is, in fact, only a mere accident of the history of one language, or one or two languages. His abuse of Harris is most shallow and unfair. Harris, in the Hermes, was dealing—not very profoundly, it is true—with the philosophy of language, the moral, physical, and metaphysical causes and conditions of it, &c. Horne Tooke, in writing about the formation of words only, thought he was explaining the philosophy of language, which is a very different thing.'

forth by the Author of HERMES, is far less unphilosophical, or gratuitously visionary, than that of the Philologer of Purley . . . [and] the reading community has labored under a profound mistake with regard to the nature of that triumph . . . which is generally supposed to have been obtained by Mr. Tooke, over Mr. Harris'.[34]

Tooke's attack on the technical misconceptions in *Hermes* is much more damaging than Johnson's rather vague objections to its style: the whole basis of universal grammar had been undermined. If the English language could only properly be explained in terms of its derivation from early *English* origins, what then was the point of a comparative study of Greek, Latin, and French? In his *Dissertations on the English Language* (1789), dedicated to Benjamin Franklin, Noah Webster remarked that 'Grammars should be formed on *practice* . . . The business of a grammarian is not to examine whether or not rational practice is founded on philosophical principles; but to ascertain the national *practice*.' Accordingly, Harris is listed here among other writers of eminence, such as Adam Smith, Beattie, Blair, and Condillac, 'but the discovery of the true theory of the construction of language, seems to have been reserved for Mr Horne Tooke, author of the "Diversions of Purley" '.[35] In the first volume of the *Diversions*, Tooke remarked that 'Hermes, you know, put out the eyes of Argus: and I suspect he has likewise blinded philosophy' (p. 15). For the Utilitarians Tooke was an undisguised blessing, and they leapt upon Tooke's scorn for Harris's work. Jeremy Bentham's *Fragments on Universal Grammar* adopted Tooke's point of view and his disdainful rhetoric: 'In the days of those ancients [i.e. the Greek and Latin grammarians] the star of Locke had not risen. In the days of their idolater Harris, that star had risen, but his idolatry had shut his eyes against its light.'[36]

Hans Aarslef has remarked that the year 1786 was 'a crucial year' for the study of language in England.[37] It was the year in which Horne Tooke brought out the first volume of his

---

[34] J. Fearn, *Anti-Tooke; Or an Analysis of The Principles of Language*, 2 vols. (1824–7), i. 300.

[35] N. Webster, *Dissertations on the English Language* (Boston, 1789), p. 204 and preface, p. xi.

[36] *The Works of Jeremy Bentham*, ed. J. Bowring, 8 vols. (1838–43), viii. 357.

[37] Aarslef, *Study of Language in England*, p. 3.

Diversions of Purley. More accurately, the crucial year was eight years earlier, when the pamphlet addressed to Dunning appeared. This contained the essence of Tooke's linguistic research. But whichever date is chosen, it is no exaggeration to claim that neither of Tooke's publications could have been written at all without Harris's *Hermes* in front of him. The advertisement prefixed to the second edition of the first volume of the *Diversions* (1798) remarks:

> The three following chapters (except some small alterations and additions) have already been given to the public in *A Letter to Mr Dunning* in the year 1778; which, though published, was not written on the spur of the occasion. The substance of that Letter, and of all that I have farther to communicate on the subject of Language, has been amongst the loose papers in my closet now upwards of thirty years; and would probably have remained there some years longer, and have been finally consigned with myself to oblivion, if I had not been made the miserable victim of—*Two Prepositions and a Conjunction*. (p. 74)

Both of Tooke's volumes are specifically directed at refuting *Hermes* and the powerful influence it exerted, on Monboddo and Lowth in particular, and it is impossible to imagine what Tooke would have written if *Hermes* had not provided him with a constant target and an organizing principle for his litigious criticism. If *Hermes* had been smouldering in his mind for more than thirty years, it was Harris's bad luck that Tooke's leisure to write down his vituperation came in the form of a prison sentence.

Tooke's materialist argument asserts his hero Locke's view of the essential passivity of the mind, which 'extends no farther than to receive Impressions, that is, to have Sensations or Feelings. What are called its operations are merely the operations of Language.' Moreover, Harris's idealist theory of language was boxed in by other followers of Locke. Condillac's *Essai sur l'origine des connoissances humaines* appeared some years before the first edition of *Hermes*, in 1746, but Harris nowhere mentions it. This is not surprising. Even if he had known the book, Harris would not have been impressed either by its advocacy of Locke's argument or its reduction of Locke's idea of the origin of knowledge in sensation and reflection to the *single* origin of sensation. The subtitle of the first English translation of the *Essai*, by Thomas Nugent (1756) was 'a supplement to Mr Locke's Essay on the Human Understanding'. *Hermes* was thus surrounded on one side by Locke himself, and on the other by pro-Lockians, and, when the French translation of *Hermes* by François Thurot appeared in 1796, Locke was again used against him, for Thurot was a disciple of Condillac. It is ironic, however, that for the most detailed, sympathetic, and 'modern' discussion of *Hermes* in the context of historical linguistics, one must go to the French reprint of Thurot's translation which appeared in 1972. Thurot's errors are here corrected, and Harris emerges as a pioneer of modern structural linguistics, 'un rénovateur et un précurseur', offering strikingly similar ideas (pp. 42–4) to those of de Saussure (*langue/parole*) and Chomsky (*competence/performance*). *Hermes* finally was seen as 'Le premier grand ouvrage de grammaire générale en englais [qui] joue donc un rôle comparable à celui de la *Grammaire* de Port-Royal au siècle précédent'.[38] Harris had to wait two centuries for such recognition.

To be sure, *Hermes* often anticipates modern linguistic theories; but to see value in *Hermes* only because it does so is to attribute to it a specious modernity as well as a false representativeness. Harris's purpose, after all, was to advance his subject by reconstructing classical and Renaissance sources in the light of his own understanding, and in attempting this he expected difficulties, knowingly courted unpopularity, and predicted charges of epistemological heterodoxy. In view of this last point it is a pity that Harris has given us few clues to his reading of contemporary texts on language. He seems to have avoided citing modern authorities on principle and apart from two passing references to the Port-Royal grammarians the most surprising omission is Bishop Wilkins's *Essay towards a Real Character and a Philosophical Language* (1668), which was certainly in Harris's own library. Moreover, on occasions Harris attributes too much to the wisdom of Aristotle and too little to his own skill as an interpreter of Aristotle. Just as he never mentions his targets by name (referring once to Hobbes as 'the philosopher of Malmesbury' and to Locke as the author of an essay), so he

[38] Joly, *Hérmès*, p. 9.

deletes any suggestion that specific contemporaries may be of some value. The Platonism of Petvin's book is an exception, as is the handful of acknowledgements to Bacon, one to Boerhaave, and one to the bilingual grammar of his friend's grandfather, Samuel Hoadly, for his *Natural Method of Teaching* (1683). At the philosophical core of *Hermes*, chapters 4–5, on general or universal ideas, Harris dismisses the possibility of specific reference to individual materialists among the moderns and sees 'our latter metaphysicians' struggling to replace the abstract mysteries of the occult with an equally spurious mechanism: 'For my own part, when I read the detail about Sensation and Reflection, and am taught the process at large how my Ideas are all generated, I seem to view the human Soul in the light of a Crucible, where Truths are produced by a kind of logical Chemistry' (p. 404–5). Against this material-ism, which he sees as a reduction of the power of the mind to that of a passive receptor of chance primary and external stimuli, Harris opposes a concept of intellectual priority which is proactive, dynamic, radically metaphysical, and the product of design (pp. 393–4):

the *intellectual* Scheme, which never forgets Deity, postpones every thing corporeal to the primary mental Cause. 'Tis here it looks for the origin of intelligible Ideas, even of those which exist in human Capacities. For though sensible Objects may be the destined medium to awaken the dormant Energies of Man's Understanding, yet are those Energies themselves no more contained in Sense, than the Explosion of a Cannon in the Spark which gave it fire.[39]

It is therefore the materialist argument itself, not the sources of its multiple authorship, with which Harris chooses to

[39] Harris was not, of course, breaking new ground in opposing the idea that all mental operations are dominated by the reception of sensible reality. Herbert of Cherbury's *De Veritate* (1624) describes the mind's innate powers as 'Common Notions' or 'hidden faculties which when stimulated by objects quickly respond to them'. The remarkable parallels between *Hermes* and *De Veritate* are noted by Chomsky (*Cartesian Linguistics*, pp. 60–3), but not discussed. Their common source is in Platonist writings (for further discussion, see ch. 7 below). Cf. C. G. Jung's note in *Alchemical Studies*: 'the character of the classical Hermes was faithfully reproduced later in the figure of Mercurius . . . [who] consists of all conceivable opposites . . . both material and spiritual . . . *the process by which the lower and material is transformed into the higher and spiritual* [my italics], and vice versa . . . on the one hand the self and on the other the individuation process . . . God's reflection in physical nature': see *The Collected Works of C. G. Jung*, trans. R. F. C. Hull, xiii (1968), 230, 237.

engage. Dr Burney's quip about *Hermes* being the 'Pourquoi de pourquoi' was and still is of precise relevance to Harris's aims and method, and it is to such metaphysical inquiries that Harris turns in his next book, *Philosophical Arrangements*. As a subsequent chapter will demonstrate, it is an error to think that Harris's admiration for the classical writers and his method of deleting contemporary references imply that his interests were out of step with his own times. More generally, his interest in the systematization of language coincides very precisely with contemporary attempts in Europe to take the universal view. The year 1751 saw the publication not only of Harris's work on universal grammar, but also Hume's *Enquiry Concerning Principles of Morals* (and the third edition of his *Philosophical Essays Concerning Human Understanding*), Henry Home, Lord Kames's *Principles of Morality and Natural Religion*, the first volume of Diderot's *Encyclopédie*, and D'Alembert's *Discours preliminaire de l'encyclopèdie*. Questions about the ordering of things universal were the vital contemporary questions, even though the answers to those questions were diverse. There was the mechanistic response of Condillac's *Traité des systèmes* (1749) and *Traité des sensations* (1754), the associationist response of David Hartley's *Observations on Man* (1749), or of Helvetius's *De l'esprit* (1758), and there was Kant's first major book, *Allgemeine Naturgeschichte und Theorie des Himmels* (1755), as well as Holbach's anti-clerical *Système de la nature* (1770).

IV

The reception of *Hermes* by Harris's inner circle of friends was generally gratifying to him. As we have seen, both Fieldings and John Upton were warm in their approval. There were similar responses from Benjamin Martyn, William Petvin (brother of the deceased John and custodian of his papers), Robert Stillingfleet, Wadham Knatchbull, Henry Newcombe of Hackney (who knew Scaliger, Sanctius, and the *Grammaire raisonnée*), Thomas Robinson, and several others. Joseph Highmore, who had painted Harris's three-quarter length portrait, said he set 'a peculiar Value' on Harris's 'token of remembrance'. Thomas Harris informed his brother that

Isaac Hawkins Browne, Floyer Sydenham, Edward Hooper, and Bishop Warburton had all approved the book, and that Thomas himself had personally delivered *Hermes* to its dedicatee, Lord Chancellor Hardwicke. He added that Lady Hardwicke 'was saying she was afraid there was too much Greek in it for her', whereupon the Archbishop of Canterbury borrowed it and took it away to Lambeth Palace.[40]

Apart from social grace, there is little expression of critical engagement in this group with the issues raised by *Hermes*. On 21 January 1752 John Hoadly wrote from Winchester Close that he had not had time to read it, but nevertheless wished to compliment Harris on its binding, and the author's benevolence in sending it:

I don't know that I have had two Hours to myself since ye Receipt of your Packet—and yet, I have read Amelia—(*not* Mr Fielding's Chef D'ouvre) but it is not *all such Reading* that will serve *Hermes* . . . He is too crabbed, to be wound round one's Finger, like a Withy.

Arthur Collier found another way of ducking the challenge of *Hermes*. On 9 December 1751 he wrote from Doctors Commons:

The last memorial of your friendship, my dear Friend, I have received . . .
    Had I considered my Self in the first class, once reading would have sufficed; If in the Second, twice would do for a nod of approbation; But as I rank myself in the last to which I am sensible I most properly belong, have resolved, as I said before, to go through it a third time, before I deliver it over to any Shelf-ranged quietude,

Samuel Richardson also adopted a face-saving approach, in a letter from London on 12 January 1752, measuring his scholarly insufficiency against his very considerable reputation as a novelist:

Good Sir,
    You have done me equal Honour and Pleasure in the Present you have made me of your truly excellent Book. I have had Leisure only lightly to run thro' it; and find, that I shall, on another Perusal, be greatly instructed, and entertained, tho' I pretend not to be Scholar

---

[40] All the preceding and following references and extracts from Harris's correspondents are taken from Harris Papers, vol. 40, part 4, 'Letters chiefly on Literary Subjects from divers friends', and are identified by date only.

enough (Hence my Grief!) to be benefited by the Learning, with which it abounds. I have often wished since I had the Pleasure of looking into it, that I had some Years ago had such an Instructor. If I had, I should either not have dared to appear in Print, or should have produced more accurate Pieces than those I have obtruded upon the World.

I ought, Sir, to have acknowledged this Favour before now: But I was willing to have found an Opportunity to reperuse it: But that not having been yet lent me, I would no longer defer my Acknowledgements.

I hope, Sir, that your Lady and You and Yours are in good Health. May you for many, very many Years, enjoy this Blessing, and every other that your Hearts can desire, because you have Hearts that may be trusted with your own Wishes, always . . .

Very few were able to take up the actual argument of *Hermes* but some could see that it had taken up an ideological position against Locke. John Upton's own fascination for classical learning prejudiced his mind in favour of his friend's book, and he may have overvalued its cultural effect. Nevertheless, he described accurately enough one of Harris's intentions, before attaching some additional interpretations which have nothing whatsoever to do with Harris's book:

I wonderfully praise you for yr Endeavours to bring your country-men back again into the roads of antient Philosophy and learning, for they have been set aside, and are like to be so. Ist: see what impressions are early made in us by New Philosophers, such as *Locke*, who comprehends all things relating to Logick and Meta-physicks. IIdly: our youth are sent to France, and Frenchified. IIIdly: Mothers have the whole and sole direction of the no-studies of their sons. IVly: University-studies are disregarded. Vly: the University disregards itself. VIly: Parties, Preferments, parliaments &c &c—Had it not been for Ld Shafts[bury] (chiefly) an end had been put ere this to Plato and Aristotle; they had been lost in *Locke*, as Aratus and Manilius are in our Modern Astronomers. Go on therefore and prosper . . .

Harris may have been amused by Upton's cultural dyspepsia, but when Charles Yorke (later Solicitor-General, and the first Earl of Hardwicke's son) wrote to him on 2 September 1752, it was to register not only a pointed criticism of Harris's ideological position but also to question his objectivity. This letter must have worried Harris, not because Yorke had carefully *read* the book, nor because this criticism came from the son of the book's dedicatee, but because Yorke was gently reproving him for carrying a family prejudice beyond the bounds of scholarly objectivity:

May I say one word for the Metaphysicians; especially for that great Master, and indeed Author of sound Metaphysical Reasoning, Locke? You will laugh, if I should hint my Suspicions, that one of your strong and exact manner of thinking could only be seduced from his school by your affection for a very noble and ingenious Relation, who had conceived a prejudice against him. Where has Locke given *Body* the precedence to Mind, either in the order of God's works, or in the scheme of universal Knowledge? He expressly proposes *Reflection* as one original Source of our Ideas . . . Is not the notion just, beyond the possibility of a question, that to know the value of our abstract Ideas, the mind ought frequently to review its own process? . . . It requires therefore all my friendship for you . . . to forgive what you say in p. 404 of *a crude account of the method how we perceive Truth, passing for an account of Truth itself,* and *logical Chemistry*; which the Reader can apply only to that incomparable work, the Essay on Human Understanding . . . as one of those who honour you, I am anxious that you and I should be of the same sentiment on all Subjects of moment . . . because, when I consider the antients, whom you study, and Lord Shaftesbury, whom you love, I consider your prejudices against Locke, as some of those *idola tribûs,* which Lord Bacon (whom I left you reading with due admiration) mentions, as one great impediment to the general advancement of Truth.

Through his bookseller John Nourse, Harris sent copies of the second and revised edition of *Hermes* to many of his friends and acquaintances, and also to writers whom he did not know personally. In Salisbury his friend Dr Jacob, a member of a social club which met at different homes each Sunday night, put the book in his best glass-fronted bookcase along with the leather-bound works of 'Duck, Dodwell, Rutherford and Hanway'. In the wider world of scholars, one of the more gratifying responses came from Dr Robert Lowth on 14 February 1760. Responding to a letter from Harris praising his forthcoming popular *Short Introduction to English Grammar* (1762), Lowth remarks that 'I should hardly have made any such attempt, had I not had the advantage of having Hermes before me, the most perfect Creation that I am acquainted with on any subject. It was of great use to me.' Henry Home, Lord Kames, initiated a correspondence with Harris on 8 October 1761 by saying that 'this work is universally admired in Scotland'. He also reprimanded Harris for his modesty:

Vanity leads most men to stand a tip-toe and to make the greatest figure they can. You on the contrary, assuming nothing to yourself, affect to borrow all from the ancients and even to hide your talents under the cover of Greek Grammarians, some of whom were never heard of, till they were brought to light in your work. The great and important *work* of Education is so much relaxed, that we have a sort of tremor even at the sight of a Greek book.

Some months later Kames again praised *Hermes* as 'the most complete Grammar ever the world saw so far as it goes', and then asked Harris to read and correct his own *Elements of Criticism* for a second edition (20 February 1762).

Nourse's list of recipients includes the Earls of Hertford and Northumberland, Lord Bute, Richard Owen Cambridge, the Bath painter William Hoare, Thomas Burnett, and Voltaire (whose copy was not to be sent until Nourse arranged a meeting with Harris in London). Nourse himself gave Lord Moreton a copy for the Royal Society, and Harris kept letters of thanks from Ambrose Serle, Lord Bute, Grantham, Charles Morton (Secretary of the Trustees of the British Museum), Shebbeare, Thomas Randolph (on behalf of the Royal Society), Richard Warner (who reciprocated with his translation of the comedies of Plautus), Edward Hooper, Thomas Birch, the Earl of Hardwicke, George and Charlotte Grenville, his cousin the Earl of Shaftesbury, his brother George, his kinsman George Wyndham (Warden of Wadham since 1744), Soame Jenyns (who had read the first edition), Joseph Warton ('I reckon [it] among the most capital books of any age for true Taste and true Philosophy': 3 December 1765), Edward Fawconer (Harris's Oxford research assistant), Jonathan Toup, John Lockman, M. Carlotti in Paris, and Father Jacquier in Rome. Such letters do little more than express their recipient's gratitude and admiration for Harris's scholarship, but there are more interesting reactions, and at least one which was exceptionally well informed. On 3 December 1765

Edward Sandys placed *Hermes* in the larger European context of contemporary linguistic theory:

My situation with regard to Grammar was the same as with most of my acquaintance for many years after I left School, disgusted with the dryness and confusion of Lilly's Grammar I trusted for my further information to my own observation in reading the Classicks, I met with the Learned Johnson's Commentaries, from which dull and heavy as they are I found some entertainment, and information; but upon reading your Hermes which I did when it first came out, I was so much entertain'd that I was desirous of seeing whether any other nation had produc'd a rational Grammarian; I bought Monsr Vaugelas [*Remarques sur la langue françoise*, Paris, 1670] who has a great name in France, and thought him as dull as our Johnson; I afterwards procur'd the Grammaire Raisonnee with which I was much pleas'd and then began to have some curiosity to know whether Messrs de Port Royale had treated as elegantly upon the Gr[eek] Language as they had perform'd for their own; and was not disappointed, tho' I could not at that time procure their Jardin des Racines Grecques. They have a long note upon the two Aorists where they mention with applause Sanctius's Greek Gram. who seems to lean to your opinion about these Tenses, but *they* seem rather to favour Vossius & Bishop Sharpe's Notions. I am sorry they did not quote Sanctius at large, as his Book is extremely scarce & I could never no more than you obtain even a Sight of it.

Before her marriage on 11 October 1763 Mrs Thrale had regarded Harris as a mentor and friend. Until the deterrance of Arthur Collier and Samuel Johnson forced her to repress such interests ('Oh what conversations! What correspondences were these!', she recalled in her autobiography), she had studied *Hermes* in a specially interleaved presentation copy, 'proving my attention to philosophical grammar, for which study I had shown him [Harris] signs of capacity'.[41]

---

[41] *Autobiography, Letters, and Remains of Mrs Piozzi*, ed. A. Hayward, 2 vols. (1861), i. 261. Mrs Thrale had one major reservation about Harris's writings. On 19 Oct. 1790 she wrote: 'I have been reading Harris's Logical Treatise upon Happiness [i.e. the third of *Three Treatises*] again—'tis a fine Performance: one should learn Logic if it was but to defend oneself from the Logicians, as one should acquaint one's self with the Principles of Legerdemain that one may not be blown as the coarse Phrase is, by a Jugler. Mandeville's second Volume in Dialogue which he calls a Defence of the Fable of the Bees is an admirable Antidote against Shaftsburism the Principle James Harris means to establish, for as the Rt. Honourable Author of the Characteristicks endeavours to insinuate his sweet but subtle Poyson into the Veins, and stop the Circulation of Christianity—his Grandson or Nephew tries to burst even the Rock on which it stands by Vinegar—Vain Wisdom all, and false Philosophy!' (*Thraliana*, ii.

Harris's works earned a place in most public and many private libraries, including those of Henry Fielding, Joshua Reynolds, Gregory Sharpe, Samuel Johnson, Thomas Jefferson, Macaulay, and many others. Catherine the Great and George III, of course, were to receive presentation copies. *Hermes* in particular came to be regarded as one of the period's essential books. In his home town, sales of the second edition of *Three Treatises* and *Hermes* were sluggish, and the bookseller Benjamin Collins cannily advised Nourse to buy advertising space in Collins's *Salisbury Journal*.[42] Yet the book was not always seen as a work of merely abstract or speculative value for philosophers alone. Fanny Kemble recalled that around 1826 her friend Harriet St Leger had set her the bizarre exercise of copying out portions of *Hermes*, 'a most difficult and abstruse grammatical work, much of which was in Latin, not a little in Greek. All these I faithfully transcribed, Chinese fashion, understanding the English little better than the two dead languages which I transcribed.'[43] On the other hand, in 1840 an under-educated tradesman called John Young, a Sunderland chemist and Methodist lay preacher who had made no sense of school Latin, not only taught himself 'the principles of language' and 'the grammar of my own language' from *Hermes*, but also found that it unlocked the doors to Latin and other languages.[44] On the other side of the

784). Thomas Tyers recorded that 'The posthumous volumes of Mr Harris of Salisbury (which treated of subjects that were congenial with his own professional studies) had attractions that engaged [Johnson] to the end' (*Johnsonian Miscellanies*, ii. 344). Was Johnson's posthumous affection for Harris's work prompted by guilt at his earlier scorn?

[42] Collins wrote: 'As Mr Harris's Books don't seem to move here I would advise you to advertise 'em in the Salisbury Journal, otherwise twill be known to few in these Parts that they are published, you may see the Paper every week at the Chapter Coffee House, Paul's Church Yard, or at Mr. Newbery's' (2 Jan. 1766): see Mrs Herbert Richardson, 'Wiltshire Newspapers—Past and Present, Part III', *WANHS* 40 (June 1920), 60. For Collins's connection with the Newberys (with whom he published *The Vicar of Wakefield*) see C. Y. Ferdinand, 'Benjamin Collins: Salisbury Printer', in M. Crump and M. Harris (eds.), *Searching the Eighteenth Century* (1983), 74–92.

[43] F. A. Kemble, *Record of a Childhood*, 2 vols. (1878), i. 170. This incident took place around 1826, when Fanny was about 16 or 17.

[44] G. E. Milburn (ed.), *The Diary of John Young 1841–43* (Surtees Society, vol. 195, Leamington Spa, 1983), 47. Cf. G. L. Kittredge: '[*Hermes*] must be credited with no little influence in emancipating English grammarians from a slavish adherence to Latin', *Some Landmarks in the History of English Grammar* (1906), 10.

world, and almost a century after its first publication, the acting-chief justice and judge of the Supreme Court of Victoria in the colony of Victoria, Redmond Barry, requisitioned *Hermes* from the London bookseller Frederick Guillaume. It was part of the first consignment of books for the new Melbourne Public Library.[45]

[45] PRO Melbourne, Victoria, Outward Letter Books of Registered Correspondence 1853–73, 4 vols., VPRS i. Barry was chairman of the Board of Trustees, Melbourne Public Library, subsequently the State Library of Victoria.

6

Politics and London

My gloom was gone, but my spirit of dissipation still remained, so that I was not solidly happy. I dined at Lord Eglington's. We had an exceeding good company, amongst whom were Mr Harris of Salisbury, Dr. Blair and Mr Macpherson and Sir James Macdonald. I felt myself happy in such a set . . .

Boswell's London Journal, 9 May 1763

IN 1760 James Harris stepped out into the political world, and for the next twenty years his scholarly and artistic avocations were forced to compete with a career in national politics. On 24 June 1759 Harris wrote to Sir Thomas Robinson, the current Member for Christchurch, to say that the burgesses resident within that borough had 'almost unanimously encouraged me to offer my self as Candidate for a Seat in Parliament'.[1] Harris described himself as 'having been ever uniformly a Friend to that Government' in which Robinson himself had played such a conspicuous part (as Secretary of State and leader of the House of Commons in 1754). Harris's political moment occurred in the middle of the Pitt–Newcastle coalition, the latter being Secretary of State and the former First Lord of the Treasury. As Robinson well knew, the real force in the Christchurch constituency was Edward Hooper. He it was who controlled the borough, and when Robinson had first offered himself as a candidate in 1749 he had been told by Lord Portsmouth (19 December 1749) that 'my good friend Mr Hooper has taken care that it is morally certain that you will have no trouble there, however [I] shall write to some of my friends by this post'.[2]

Edward Hooper (1702–95) was the son of Lady Dorothy Ashley, daughter of the second Earl of Shaftesbury, and

---

[1] 'Letters of Thomas Robinson, 2nd Lord Grantham, 1758–1771', West Yorkshire Archive, Leeds, letter 12425.

[2] Ibid. 12433.

therefore Harris's first cousin. He had acted as the second
James Harris's executor in 1731, and had trained at Lincoln's
Inn before becoming MP for Christchurch in 1735. In 1748 he
was obliged to vacate his seat on being appointed to the Board
of Customs, where he remained until 1793. Harris and
Hooper were also linked by legal process: had the fourth Earl
of Shaftesbury died without heirs, as initially seemed likely
when his neurotically over-protective mother married him off
before his fourteenth birthday to his first bride, the 13-year-
old daughter of the Earl of Gainsborough, his fortune would
have descended either to Hooper or to the children of Lady
Elizabeth Harris as heirs-at-law. Malmesbury points out that
Harris and Hooper's characteristics 'were in all essential
points so congenial that tho' each followed their respective
objects with great earnestness [literature and politics respect-
ively], their friendship remained uninterrupted and un-
impaired'.[3] In addition to the support of Hooper, there was a
more substantial influence at work on Harris's behalf. On 16
July 1760 Harris was informed by the Duke of York's equerry
that his master had so much enjoyed his visit the previous
year that 'He will not leave this County without calling on
you.' The duke dined with the Harris family after a visit to the
Bishop of Salisbury on 1 September, and spent 3 nights (9–11
October) at the Harris house. A correspondent in *The
Gentleman's Magazine* remarked in 1781 that 'Mr Harris was
metamorphosed from one of the most amiable and independent country gentlemen that ever sat in the House of
Commons, into a pliable and modest courtier, by the interest
of the late Duke of York . . . [who] assisted at a private
concert, and supped at the house of Mr. Harris at Salisbury.
Pleased with his conversation . . . he recommended him to the
minister.'[4] With Hooper's unqualified support in the constituency and the Duke of York's influence at court, Harris simply
could not lose. He was returned for Christchurch on 31 March
1761 with Hon. Thomas Robinson (the eldest son of Sir
Thomas Robinson, who had vacated the seat on being created
Baron Grantham on 7 April 1761). Both commanded the

   [3] Malmesbury, 'Memoir', p. 15.
   [4] Harris Papers, Letter Book, vol. 31, part 1, pp. 22–4; *The Gentleman's Magazine*, 51
(1781), 24; see also W. A. Wheeler, *Sarum Chronology* (1889), 59.

unanimous support of the seventy-odd constituents qualified to vote, as the voters themselves were in turn commanded by Hooper.

Hooper's confident managership of the borough's political direction is best illustrated by a letter he wrote to Thomas Robinson junior on 27 September 1766, when the latter was facing re-election after his appointment as secretary to the Congress of Augsburg (which never in fact met). Hooper warns Robinson that his developing metropolitan career had weakened his electoral base and he therefore needed to ingratiate himself with the mayor and burgesses, some of whom believed that their representative had been tempted to 'slight them' in the past. Accordingly, he was to send each of them a personal circular letter, wait on both the burgesses and out-burgesses, and attend the election in person. Before communicating this confidential strategy to Harris, Hooper cannily pointed out his own difficulty in the matter. As a Commissioner of Customs, he was 'absolutely muzzled by severe penalties from attempting to persuade or dissuade in the choice of a Member of Parliament. However my friends perfectly well know my great respect for Ld. Grantham [Robinson's father] and you; and I have a friend at Hurn Court whose sentiments they know upon these occasions perfectly conformable to mine.'[5] Currying favour with the mere electorate was certainly more than a public ritual posture at election times for intending candidates; but clearly the determining factor of success or failure was rather more subtle. Political interest, family relationships, and patronage were the filaments of a powerful web which enclosed a select few and excluded many others. In the case of Harris's own son, the interconnections are even more startling. It was James Harris junior, again on the interest of Edward Hooper, who replaced his father's colleague Thomas Robinson as MP for Christchurch in 1770 when the latter was appointed ambassador to Madrid. But Harris junior was already in Madrid as secretary to the embassy and chargé d'affaires, his Commons representation of Christchurch being almost entirely nominal. Harris junior acted as the guide and mentor of his

[5] 'Letters of Thomas Robinson', letter 12427.

own ambassador.[6] A further stage in the intermarriage of
political interest and family connections was consolidated on
11 June 1785, when James Harris senior's daughter, Gertrude,
married Thomas's brother, the second son of Lord Grantham,
the Hon. Frederick Robinson.

Harris was to serve Christchurch until his death in 1780.
Promotions came quickly: two were short lived, and the third
was prestigious but of no political significance. In December
1762 he was appointed one of the Lords of the Admiralty (for
four months only, until the end of the Bute administration),
and then Lord of the Treasury from April 1763 to July 1765,
serving under Grenville. For the last six years of his life he was
secretary and comptroller to the Queen: 'They were two
offices, required no duty, were worth together £500 a year
(£25 deducted), were regularly paid, and the duty done by the
deputy . . . all I had to do was sign Receipts.'[7] As each
promotion came, he stood for re-election, and succeeded.
There was little news value in so predictable an event, and the
Salisbury Journal of 28 March 1768 lumped together the
announcement of Harris and Robinson's re-election (by this
time 'inseparable' political colleagues as Hooper had put it),
with news of Sterne's death, the election of Henry Thrale for
the borough of Southwark, and a note on the real political
anxiety of the day: 'Very considerable betts are now depending
on the success of Mr Wilkes canvass for Middlesex.'

Harris's achievements as an MP for Christchurch included
efforts to promote his own home town of Salisbury, and to that
extent were no different from the activities of any other
representative in the Commons with a seat to maintain:
improving civic amenities such as roads and turnpikes,
receiving petitions, pressing patrons and politicians for offices
on behalf of constituents, and occasionally entertaining his
corporation to dinners, and their wives to balls. Harris and
Robinson shared the responsibility of providing a buck (of
venison) for such periodic gatherings.

Since the 1740s it had been largely through the Shaftesbury

---

[6] Ibid. See letters from Harris to Robinson (11866 ff.) giving the latter advice on
protocol, housing, selection of servants, travelling arrangements, etc. Robinson was
officially appointed 24 May 1771.

[7] Malmesbury Papers, 'Account of my coming into Office', p. 1.

family and his brothers Thomas and William that Harris had been kept in touch with first-hand reports of life in London— Handel's health and private rehearsals of his latest oratorio, the progress of the '45 rebellion, the possibility of a French invasion, and Lord Lovat's trial (William sent Harris a copy of Hogarth's caricature of Lovat on 28 August 1746, and Thomas attended the trial in person on 9 March). All these first-hand accounts were supplemented by Harris's own reading of the gazettes and newspapers at a period when he was unravelling the meaning in more scholarly commentaries, either in the Close or at St Giles. He also found time to read some contemporary literature. On 13 June 1749 he wrote to Samuel Richardson thanking him for the present of *Clarissa*, in terms which suggest that Richardson's family was already known to him.[8]

Harris's own active involvement in politics is not notable for doing things as much as for witnessing and recording the actions of others. In the latter area he was remarkable. In the judgement of John Brooke,

Harris was a first-class reporter: brief, accurate, and amusing . . . the most important source yet discovered for debates in the Parliament of 1761 . . . Harris was intelligent and accurate; he tried to summarize the arguments of the speakers fairly, and often succeeds in capturing typical phrases and turns of speech . . . he had a lively mind, graced both with wit and humour, and his learning never obtrudes . . . Together with those of Horace Walpole, [Harris's memoranda] are the best that have been found for the Parliament of 1761–1768.[9]

Unfortunately, only fragments remain, i.e. for the period 1761–6, and a partial record for 1779–80. Yet there is

[8] Facsimiles of 3 of Harris's letters to Richardson were reproduced in A. L. Barbauld's *Correspondence of Samuel Richardson*, 6 vols. (1804); for the 1st see i. 161–2. A 2nd, 19 Jan. 1752, indicates that Richardson had received a copy of *Hermes* from Harris, the latter remarking, 'the sordid views of trade have not (as usual) been so far able to engross you, as to withdraw you from the contemplation of more rational, more ingenious subjects' (i. 162–3). The 3rd letter, 10 Nov. 1753, thanks Richardson for another book (presumably *Sir Charles Grandison*), and congratulates the novelist for successfully combining 'public utility and improvement' (vi. 288). It was characteristic of Harris to initiate or, in this case, consolidate literary relationships by sending presentation copies of his books.

[9] Sir Lewis Namier and J. Brooke, *The House of Commons 1754–1790*, 3 vols. (1964), ii. 588, i. 522–3 (app. 2).

sufficient evidence in what survives to show some of the characteristics of Harris as a parliamentarian, and his own diaries record a greater involvement than has so far been recognized.

Harris took his seat in the Commons on 2 November 1761, already an experienced local magistrate, publicly celebrated for his scholarship, and known as the author of two books, *Three Treatises* and *Hermes*. Looking back to this event sixteen years later, Mrs Thrale recorded in her diary:

I have at last been in Company of old James Harris of Salisbury with whose Name and Character I was very familiar in my younger days. There is a famous joke of Charles Townshend's concerning him which must be mentioned—when he was introduced into the house of Commons—who is this Harris says the witty Charles— why James Harris replies somebody—the great Logician who has written one book about Grammar and one about Virtue: and what brings him here then enquires Townshend—he will find neither Grammar nor Virtue in this house.[10]

Townshend's quip was, in the long run, to be proved correct. On 19 August 1779 Harris was to write to his son, by then his country's ambassador to Russia, that 'I think patriots and Opposition have ruined us . . . Prating is a mere habit, which is possessed now as it was by the old demagogues without a grain of virtue, or even political science.'[11] During Lord North's Government an income tax had been proposed, to fund 'our unhappy broils with the puritanical Americans': it would have exempted the chancellor, the judges, Speaker, foreign ministers, officers of the army and navy, and all place-holders with salaries of not more than £200 a year. Harris observed of the universal Commons approval: 'A noble patriotism this, a heroic generosity, to dispose of money *not your own*; yet this produced on the faces of all the patriots present a most broad unmeaning grin; they all expected to be in a few days Lords of Trade, Admiralty, Treasury, &c.'[12] And on 18 February 1779 he was to make a final estimate of the Commons' claim to grammar and virtue: 'Tired with

[10] K. C. Balderston (ed.), *Thraliana: The Diary of Mrs Hester Lynch Piozzi 1776– 1805*, 2 vols. (Oxford, 1942), i. 107.

[11] Malmesbury (ed.), *A Series of Letters of the First Earl of Malmesbury His Family and Friends from 1745 to 1820*, i. 424–5.   [12] Ibid. i. 380–1 (10 Mar. 1778).

Tautology and futile Patriotism, I have not deemed our trifling debates worth recording.'[13]

On the second day of his first session in the parliament which ran from 2 November 1761 to 11 March 1768, Harris was immediately impressed by George Grenville's speech on the choice of a Speaker. His loyalty to Grenville, in and out of office, was to remain constant until Grenville's death in November 1770. On 19 May 1763 he was to record in his diary, 'I meet no one, whose sentiments and Speculations of every sort appear more candid, just, and constitutional, than those of Mr Grenville';[14] and 2 months before Grenville handed in his seals of office (10 July 1765) Harris recorded his continuing esteem for Grenville, as a member of about eighteen independent MPs who stayed with him until the end. In May 1765 he noted 'great encomiums of Mr Grenville by all persons. The persons of the first rank and credit at the Cocoa Tree declared for him. On Tuesday last he had a very full levee, when there were Sir James Dashwood, Sir Charles Tynte, Sir Robert Burdett, Sir Walter Bagot, and many others.'[15] Although Harris appeared to have no strong party affiliation at all when he entered parliament—and was eventually to cease recording the debates altogether because of his dislike of their specious factionalism—his initial leaning was towards Hardwicke and his administration. His interventions in parliamentary debates and the detailed political diary he kept are both marked by an independence of thought and, in the latter case, a wry scepticism of all political rhetoric. Harris's parliamentary deeds and words reflect personal characteristics and moral convictions rather than political expediencies, and for this reason his loyalty to Grenville, once given, remained strong. In many ways the

[13] Malmesbury Papers, 'Parliamentary Memorials from October 31 1780', 476B–482B.   [14] Malmesbury Papers, 707B.

[15] Cited in Namier and Brooke, from the Malmesbury Papers, *The House of Commons*, ii. 543. For Harris's account of the formation of the Grenville administration, see Sir Lewis Namier and J. Brooke, *Charles Townshend* (1964), 98. The most recent study of Grenville's career is P. Lawson, *George Grenville: A Political Life* (Oxford, 1984). Harris regularly dined at Grenville's house in 1763–4. A diary entry for 11 June 1764 states: 'I was with Mr Grenville at his house, in which he opened himself to me with his usual Candour and Friendship', and on 19 Oct. 1764 Harris writes, 'He talked, when we were alone, as usual; nothing but Principles of Freedom and Whiggism.'

career of Harris represents the characteristics of the majority of independent Whig members and place-holders.

On 29 March 1762 Harris opened the debate on the Game Bill with a lawyer's objection to a clause which would have enacted an entirely unprecedented concept of voluntary trespass. It seemed to him a charter for manorial feudalism, an offence against 'all the Laws of common and natural Justice'. Before the Bill was lost Harris went up to the House of Lords to observe its passage, and his memorandum indicates that the new MP for Christchurch entertained no illusions about the dignity or disinterest of parliamentary debate:

Their Lordships made no better work of it, than we had done—Had an Indian viewd so many grave men and been told they were the chiefs of the nation assembled in Senate he would of course have imagined them debating on Peace and War, or other Matters of an importance equal to their rare dignity—what had been his Surprize had he been told, 'twas whether Grouse should be killed the 1st of September or the 20th of August? Mon Dieu, qu'est que ce Grouse?[16]

On 20 April he was the third speaker, after the Secretary of War and Lord Barrington, to object to a clause in the Militia Bill which proposed naturalization of foreign officers serving Britain in America. Again, he was objecting to an extension of legal jurisdiction into dubious and unprecedented areas: this time he invoked the Act of Settlement, which debarred the Crown from conferring military honours on foreigners. Harris was then appointed to a committee chaired by the Secretary of War to consult with the Lords on the matter.

His representation of Christchurch notwithstanding, for his 'Friends in Salisbury' Harris also 'solicited and carried through' the additional Toll Bill for the Fisherton turnpike; successfully presented a petition for the repair of Ham House Hill; was partially successful in reducing the labour and tolls imposed on the Shaftesbury Road; successfully petitioned the king, through Halifax's under-secretary, Robert Wood, to allow a respite on the sentence of transportation passed on a

---

[16] Malmesbury Papers, 995B. 37–8, of Harris's political diary. The clause was lost by a majority of 14.

Mr Scammel; and was instrumental in getting a military camp sited near Salisbury. Though such achievements were modest, they were real. Among the political leaders, on the other hand, fantasy appeared a necessary ingredient of their performance: 'Mr Pitt I call an Inigo Jones in Politics, a Man of great Ideas, a Projector of noble and magnificent Plans— But Architects, tho they find the Plan, never consider themselves as concerned to find the Means.'[17] Grenville, by contrast, was no great orator; many found his exhaustive attention to detail stultifying and his manner tedious, but he was thoroughly expert in the procedures of parliament and an economic realist. His diffidence when eventually faced with the possibility of leading a government was a function of his initial domination by Lord Temple, his wealthy and tempestuous brother (to whose interest Grenville owed his seat in the Commons), and after 1754, by William Pitt, his erratic brother-in-law. Both kept him down. Temple moreover stopped making any financial provision for his younger brother, and thus the highest office found for him before November 1761 had been at the Admiralty, as Treasurer of the Navy. The prospect of leadership which Bute held out to Grenville and Fox in 1761 elicited only a deep and so far justified political insecurity: 'Mr Fox has a great number of friends in the House of Commons, attached strongly to him; Mr Fox has *great connections*,' Grenville complained, 'I have none; I have no friends; I am now unhappily separated from my own family.'[18] In spite of this pessimism, Harris was tipping Grenville for either Secretary of State or, more likely, Head of Treasury, in May 1762. In the following month Grenville was appointed Secretary of State, Northern Department, and Henry Fox became Leader of the Commons.

On 26 November 1762 Harris was offered, through his Wiltshire neighbour Lord Shelburne, a seat on the Board of Admiralty. Harris consulted his friends. His cousin Hooper advised him to accept, Philip Lord Hardwicke, the Lord High Chancellor and the dedicatee of Harris's *Hermes*, advised declining, and his brother Thomas could not make up his

---

[17] Ibid. 54.
[18] Namier and Brooke, *House of Commons*, ii. 539. See also Lawson, *George Grenville*, pp. 124 ff.

mind. Harris at first declined the offer, but after a sleepless night returned to Shelburne to ask if he could reverse his decision. Shelburne said he could, and Harris gave as his reasons 'the contrariety of opinions among my friends and my own diffidence'. His resolve finally to accept was principally shaped by considerations of protocol and prestige. Political appropriateness came second: 'I reflected how great an honour had been made me and in how generous a manner, unsought, unsolicited . . . I reflected too that even though I refused a place, I so far approved the present measures of Government that I should certainly vote for them.'[19] Consequently, Harris's first major parliamentary role in the Commons was to move the Address to the King on the Preliminaries of Peace, on 9 December 1762. His speech was a formality, a routine performance expressing loyal respect, gratitude for the Peace, and particular satisfaction at the consequent expansion of British interests by the restitution of Minorca, new acquisitions in Africa, possession of the sugar islands of the West Indies, and enormous trading possibilities between Florida, Canada, and the north American colonies. It was a recommendation of the Peace based on practical economic considerations and the *realpolitik* of a negotiated settlement:

. . . it must be remember'd that France and Spain, tho' humbled, are not annihilated. How then would they have treated a resolution to cede them nothing? They would have consider'd it, Sir, as a condition not of Peace but of war, the war then must have gone on for one year, for two years, for more . . . in one thing we are sure they would have been successful, in augmenting our great our formidable debt. How singular in such case had been our situation?—Triumphant abroad and approaching to bankruptcy at home.

. . . there is nothing more easy than for lively imaginations to form to themselves plans of peace, the most plausible and specious. But, Sir, those lively minds would do well to distinguish between what is imaginary and what is practicable . . . Peace is a thing not dictated [by] One but mutually agreed to by Many.[20]

Grenville himself had led the attack on the war policy of Pitt and the immense debts caused both by the war and by the subsidy to Prussia. His view of Britain's future was based on

---

[19] Namier and Brooke, *House of Commons*, ii. 588.
[20] Malmesbury Papers, 209A. 4–6.

the expansion of a sea-borne empire, but he vacillated over certain of the peace terms. For this he was demoted by Bute, and it was the unpopular Henry Fox who carried the Peace to the Commons without a dissenting voice raised by the Government. Grenville needed a ministerial salary to support his family and accepted demotion to the post of First Commissioner of the Admiralty. On 19 December it was Grenville who informed Harris that he, Harris, was to kiss the king's hand the following day before taking up his appointment as a Lord of the Admiralty on 1 January 1763.[21] Harris's promotion cemented his friendship with Grenville at a time when Grenville was in the political wilderness and in great need of loyal supporters. As Philip Lawson has remarked, Grenville's six-month period at the Admiralty was a kind of political 'purgatory', dramatically ended when Bute offered him the highest office in government, on 25 March 1763, as First Lord of the Treasury.[22] When he moved to the Treasury in April (thereby assuming leadership of the Commons), he took Harris with him.

Harris was already working closely with Grenville, pausing only briefly to be re-elected for Christchurch on 27 December 1763. Their political relationship had naturally extended to regular social contact in which members of each family shared. While James Harris was professionally occupied (dining at Lord Halifax's with 'all the *grandees*', as Mrs Harris put it), on 10 November Mrs Grenville and four of her children had assembled at the Harris house in Whitehall to watch the Lord Mayor's show. In a letter to her son, studying at Oxford, Mrs Harris continues: 'Your sisters are invited to see the King go to the House, at Mr Grenville's, which pleases them greatly.' Five days later, not even the grandest civic pageant of the year could match the furore caused by the incendiary writings of John Wilkes:

Your father returned from the House at two this morning: great debating. Lord North moved to enquire into No. 45 of the 'North

---

[21] The full list of Commissioners for executing the office of Lord High Admiral (27 Dec. 1762) was George Grenville, George Hay LL D, Thomas Orby Hunter, John Forbes, Hans Stanley, John Lord Carysfort, and James Harris. See J. Redington (ed.), *Calendar of Home Office Papers of George III 1760–1765* (1878), 236.

[22] See Lawson, *George Granville*, pp. 114–53.

Briton.' Mr Pitt was the chief and almost only manager for Wilkes. They were pretty smart at times, and the House often called them to order. They divided twice; the first was 300 to 111, the second 273 to 111. The House came to a resolution that No. 45 'North Briton' was a false, scandalous, and seditious libel, and ordered it to be burnt by the hands of the common hangman; they are upon the same business today, that is, to make the seizing of his person no breach of privilege. Tomorrow he is to speak for himself, and Friday the affair may end, and he may be expelled the House. The House of Lords is engaged with Wilkes also, for among his papers was found the most blasphemous, profane, and obscene thing ever written, in Bishop Warburton's name . . .

This instant a note has come from your father to inform me that Wilkes fought a duel, this morning, in Hyde Park, with Mr Martin; he has received two balls in his body, which are extracted, and the wound thought not dangerous . . . I think 'tis better he is not quite killed, for I should wish he might be made an example of; for a more wicked wretch never lived in any age. His blasphemy with regard to our Saviour is enough to shock those who never think of religion.[23]

Wilkes was to be disposed of neither by lead nor paper bullets, however, and Harris was to speak in the protracted Commons debate on the question of parliamentary privilege on 23 November, 'remarkably well; at least it was said so in the House of Lords, this morning', Mrs Harris reported. If

---

[23] *Series of Letters*, i. 99–101. Elizabeth Harris refers, of course, to no. 45 of the *North Briton* (23 Apr. 1763) by John Wilkes (1725–97). Wilkes was a member of the notorious Hell Fire Club of Medmenham Abbey (whose blasphemous antics were presided over by Sir Francis Dashwood, the chancellor of the Exchequer), and had been a member of the House of Commons since 1757, following his purchase of the seat of Aylesbury. 'No. 45' of Wilkes's newspaper had insulted the king's mother and attacked the praise of the Peace (which concluded the Seven Years' War in Mar. 1763) in the king's speech as 'the most abandoned instance of ministerial effrontery'. Wilkes was thrown into the Tower of London, and on 15 Nov. the House of Commons declared no. 45 a seditious libel. One of the 13 copies of Wilkes's obscene poem *An Essay on Woman* (1762; privately printed on his own press) also reached the House of Lords in Nov., and at the king's request it was prosecuted as 'a most scandalous, obscene, and impious libel; a gross Profanation of many Parts of the Holy Scriptures; and a most wicked and blasphemous Attempt to ridicule and vilify the Person of our Blessed Saviour' (*Journals of the House of Lords*, xxx. 415). Wilkes had attached Bishop Warburton's name to the mock notes in the *Essay*, hence breaching the privilege of the House of Lords. Wilkes was banished (in his absence in France), and on his surrender in 1768 suffered equal fines and concurrent prison sentences for the seditious *North Briton* 45, and for the obscene *Essay*. He became Lord Mayor of London, 1774–5, and City Chamberlain in 1779. See further G. Rudé, *Wilkes and Liberty* (Oxford, 1962), and I. R. Christie, *Wilkes, Wyvill and Reform* (1962).

Mrs Harris's view of Wilkes was only partly coloured by her husband's opinion, it is no surprise that Harris voted for the expulsion of Wilkes on 3 February 1769, even though this went against the judgement of Grenville.

Harris's first significant duty, particularly appropriate now he was in the Admiralty, was joining Grenville and others on a committee to investigate the petition received from John Harrison, the inventor of 'the Ingenious Watch' (or chronometer) which could calculate the longitude at sea, and which would work a revolution in navigation. Harrison was asking for official recognition of his invention and for the balance of the £10,000 reward first offered in 1713. The Board of Longitude refused certification and simply paid him some more money on account. In spite of the fact that Captain Cook was to circumnavigate the world in the *Endeavour* in 1768–71 with the aid of an exact copy of Harrison's fourth 'watch', the balance of £8,570 was not paid to Harrison until 1773. Before listening to the expert opinion of the Astronomer Royal, Nevil Maskelyn, on 6 March, Harris's eye was caught by a bizarre example of parliamentary deference:

Then came on the Longitude, and Harrison's Time-Keeper. Lord Barrington opened it very properly and told us a noble Lord was ready to give more accurate information, if called upon. Lord Moreton was called in, with great Ceremony—a large Elbow-Chair placed for him within the Bar—he approached, attended by the Serjeant at Arms, with three Reverences to the Chair-man—then took his Seat in the Chair, and put on his Hat—then being questioned, told us twas too dark for him to refer to his Papers, and that he could not read by Candle; and so his Lordship departed, and we adjourned.[24]

Harris himself was not always quick off the mark. On one occasion during debate on the Militia Bill he failed to get to his feet in time to speak about the Army. A memo survives among his papers which notes the increase in personnel from 7,000 in King William's time to 8,000 in that of George I. Harris apparently intended to argue for an increase in expenditure. The note is described as 'Heads of a Speech

[24] Malmesbury Papers, 'Parliamentary Memorials from 16 January 1764 to 4 April 1765', p. 58.

intended to be spoken on the Army, could I have got up in time'. On Tuesday, 22 March Harris spoke in favour of the Cider Tax, 'and had all the Reward I wished, that is, to be heard'; and on 13 April 1763 he again kissed the king's hand as a Lord of the Treasury, an event which occasioned another successful re-election for Christchurch. He took up his appointment on 15 April 1763 and attended his first meeting three days later. Harris attended the Treasury Board meetings assiduously from April 1763 to 10 July 1765, when a new Board was appointed. Together with that of Grenville and North, Harris's signature appears on several Royal Warrants in 1763. One gives permission to the inhabitants of the Portland Garrison to quarry stone (6 July), another appoints Henry Fane a Commissioner of the Salt Duties (29 August), and a third appoints Thomas Baron Hyde and Robert Hampden to the position of Postmaster-General (12 September).[25]

Harris's third and fourth sessions of Parliament (15 November 1763 to 4 April 1765) were dominated by the Wilkes affair, although his own duties included membership of a parliamentary committee on postage, a continuing involvement with the Harrison petition, a defence of the Salisbury turnpike, an inquiry into smuggling in the Isle of Man, and a report on the navigation of the River Tyne. Each such activity formed part of the extraordinary transformation of England's commercial character and its imperial expansion. Harris was aware of this slow revolution, but he was also quick to note, with healthy scepticism, the political bluster which accompanied it. Particularly during the Wilkes debates, he notes in his diary the distortions of argument and the inflation of statistical evidence on the debates over the legality of general warrants. A scrap of verse scribbled down during these debates speaks of 'A motley, party-colour'd Rout | Of Wastlings in, and Worthy's out'. A longer poem (undated, untitled, and unfinished, but evidently written about the Harris family's regular visits to the family of Richard Owen Cambridge at their Twickenham home, and probably dating

---

[25] BM Add. MSS 36133, Hardwicke Papers, vol. 785, pp. 35 ff. The Treasury Commissioners were Frederick North, Sir John Turner, Thomas Orby Hunter, and James Harris (13 Apr. 1763): see *Calender of Home Office Papers . . . 1760–1765*, p. 360.

from 1765) enacts the Horatian ideal of retreat as a literary fiction contrasted with the whirligig of politics:

> While at your Twickenham, pleasant Seat,
> You prove the Bliss of sweet Retreat,
> And calmly view the rising Day
> In peace and order pass away,
> Me Fortune gives to busy Life,
> Where Faction seizes, and party-ship,
> Where lucid Order ne'er was known,
> And frightened, Peace far off is flown;
> Where now we're in, and then we're out
> A Triumph now, and then a Rout.[26]

Wilkes's *North Briton* 45 touched Harris in a particular way. It had attacked the praise of the Peace in the king's speech as 'the most abandoned instance of ministerial effrontery'. Halifax, as Secretary of State, had issued a general warrant for the arrest of the authors, printers, and publishers of Wilkes's paper. Wilkes's consequent imprisonment was interpreted by the Chief Justice as an infringement of parliamentary privilege, since he had not committed acts of treason, felony, or a breach of the peace, and he was released. Wilkes republished his libellous attacks, sued Halifax and his under-secretary, Robert Wood, and was awarded £1,000 damages. On 15 November the Commons retaliated, declaring number 45 to be a 'false, scandalous and seditious libel' which would cause 'traitorous insurrections against his majesty's Government'. When his *Essay on Woman*, with notes ironically attributed to Bishop Warburton (for whose scholarship Harris had little respect) was declared obscene and pornographic, Wilkes was also denounced in the House of Lords. Wilkes withdrew to France, where he learnt that he had been expelled from the Commons in January 1764. Even among those hostile to Wilkes, there was anxiety about the political use of general warrants, which had effectively lapsed with the demise of the Licensing Act of 1694. The Commons narrowly voted that

---

[26] Malmesbury Papers, 818B, slightly repunctuated. The Harris family often spent Easter week at the Cambridge home in Twickenham. To Mrs Harris it was 'the seat of the Muses, who are to me more edifying than the politicians' (*Series of Letters*, i. 225: 5 Apr. 1771). For letters between the 2 families, see Newnham Box 35, 'Miscellaneous Letters', pp. 377–504.

general warrants were in fact legal on 15 February 1764, but
Colonel Barre and General Conway were both dismissed from
office for opposing the administration's policy towards Wilkes.
Charles and John Yorke resigned in protest, the London mob
attempted to stop the public hangman burning number 45,
and mocked the whole issue by symbolically burning a pair of
jackboots (for Bute) and a petticoat (for the Dowager
Princess). Harris kept among his papers a memorandum of
November 1764 defining the kinds of offenders covered in the
general warrant (journeymen and devils, the real printers and
publishers of number 45, Beardmore and others), and notes
on the libel law going back to 1555. A letter to Harris from
Grenville in Downing Street for 30 October 1763, which may
have been Harris's draft of the Commons reply to the king's
speech, suggests that Harris had been preparing papers for
Grenville's approval for the previous 12 months.[27]

Harris's opinion of Wilkes the man, as opposed to Wilkes
the public incendiary, is only apparently contradictory.
Clearly, Harris saw him as the puppet of some rather more
sinister and anti-monarchist forces in the City. Long after the
crisis had blown over, in 1775, Harris described Wilkes as 'a
pleasant social companion, a lover of letters, and (I believe) in
his heart, a most complete contemner of his illiberal puritanical associates in the City, whom, notwithstanding, he must
endeavour to please, as if he were as earnest as themselves'.[28]
Four years later Harris suggested Wilkes as the part-author of
'Anticipation', on the grounds that 'no one but a perfect
master of our eloquence could so well delineate it'. Neither
was he moved to passionate denunciation of Wilkes on 22
March 1769, when he witnessed a mob of Wilkes supporters
besieging St James's Palace. Harris's diary entry is cool and
minimal—'About noon or soon after being informed the Mob
were gathering round the Palace (which I could easily see as I
lived in St James's Street) I went in consequence into St
James's Coffee House, where I had a complete view of
everything that past.' Mrs Harris's account, in marked
contrast, is more personal, circumstantial, and declamatory,
even though much of its detail could only have been based on
her husband's report:

[27] Malmesbury Papers, 220A.   [28] *Series of Letters*, i. 384–5.

The Duke of Northumberland was severely pelted, as he went to the back door of the Palace; the ammunition of these rioters consisted chiefly of dirt, but many stones were seen to be thrown, and one glass bottle. The Riot Act was read without any effect, Lord Talbot harangued the mob, and whilst he was haranguing at the gate, one Mr Whitworth (not Sir Charles) was haranguing from St James's coffee-house, and a drunken woman in a third place: they each had their audiences, but the Wilkism, and obscenity of the woman proved the greatest attraction. The tumult still continued at its height, when from the Palace yard issued the Horse Guards and Horse Grenadiers, with their swords drawn, and commanded by three officers. The rabble, whose spirit of mischief is only equalled by their timidity, immediately retired . . .

. . . The Guards patroled the streets that afternoon and evening. It was said that, amongst the mob, there were men of better appearance, supposed to be their leaders, but this is not certain. Your father was in St James's coffee-house all the morning, so saw the whole. Your sisters and I were at Clapham in the morning, and came down Pall Mall in the midst of the mob: we let down our glasses, they cried *Wilkes and liberty* enough to us, but did not insist on our joining them, so we got safe home, though I was a great deal flurried at the time. Many of the mob cried *Wilkes and no king*, which is shocking to think on.[29]

The London mobs, sometimes spontaneous but more often 'hired' for a specific occasion and purpose, inspired more indignation than fear amongst unaffected witnesses after the event. Harris seems always to have kept a cool head during such disturbances, and his subsequent accounts show his great skill at reporting first-hand events of national importance without opinionated prejudice, and with some compassion for the inarticulate. This is best illustrated by an extract from a carefully written-out memorandum dated 13 May 1765. The occasion for this impartial account was the first reading in the Lords of the Silk Bill, a measure designed to protect the weaving industry but which was thrown out of the Lords without a division after the Duke of Bedford had spoken against it. Harris observes:

This was an unlucky Step. The Spital Fields weavers came down some thousands the next morning to St James's, and not finding his Majesty there, went on to Richmond, and made their Complaints.

[29] Ibid. i. 177–8 (24 Mar. 1769).

The King gave them a gratious answer. The next day (Thursday)
they beset the Avenues of the House of Lords, and with black Flags
flying filled Old Palace Yard. They addresst every Lord, as he got
out of his Coach. The Lords had a Witness at their Bar, who
deposed that he heard them say, *if they caught him, they would use him
worse, than ever he had been used in Ireland*—on being asked whether he
heard them explain whom they meant he answered, No—Whom did
he believe?—He supposed they meant the Duke of Bedford—Justice
Fielding and other Justices were also brought to the Bar of the
House, and received Instructions to be active in their duty—Lord
Marchmont with the Chancellor even the principal Directors—The
same Evening, they gathered about Bedford House, and Guards
were sent to protect it—The next day (Thursday: [*sic*]) they again
beset the Lords House, and after that Bedford House—The Guards
Horse and Foot were differently disposed, and a Squadron of . . .
Dragoons brought in—on Friday (the fourth day) they made a
formal attack on Bedford House, and began pulling down the Wall
of the Court—on this the Guards opposed their Bayonets, but did
not fire—the Mob stoned them, and bid them defiance—the Horse
at length rode in upon them, and dispersed them, but fortunately
did no harm—The Proclamation was redde, and some were seized,
when upon examination they appeared to be a motley Crew, a
Rabble of various Professions, and from every Quarter. Twas
commonly believed, had they carried their attack, the House had
been plundered, and the Duke murdered. The Duke had Mr
Grenville, Lords Halifax, Sandwich, Gower, General Waldegrave
and many others dine with him—He would have gone out to talk to
them, but his Friends dissuaded. His Company went out, and
discoursing with those who were taken, could only learn that they
had large Familys—the same day in the Evening they demolished all
the Windows of the House of Car the Mercer, Ludgate Street,
notwithstanding my Lord Mayor with his officers endeavoured to
oppose them. Not only, many of them were no Weavers, but others
confesst they were hired.

Mori mallem dente Leonino, quam morbo pediculari.[30]

Harris understated the official verdict on the role of Sir
John Fielding, Henry's half-brother. A week later, 2 May
1765, the Lords committee published on account of its
investigation into the tumult, and placed the blame squarely

---

[30] Malmesbury Papers, 'Debates, 19 April 1765–4 March 1766', pp. 35–7. Harris
paraphrases an old proverb, to the effect that 'I would rather be killed by a lion than a
fox'.

on the inefficiency of the magistrates. Harris's friend was censured as the chief culprit:

That Sir John Fielding is particularly blameable; having (as he himself acknowledged) thought, that this was not such a Mob as, by their Insolencies, authorized him to read the proclamation, though he well knew that the Duke of Bedford had been assaulted and wounded in his Way from the House, and could not attend his Duty in Parliament, without Danger of his Life.[31]

The king's temporary illness in 1765 persuaded him that a Council of Regency was needed, and on 23 April Harris attended Downing Street, along with North, Pitt, Townshend, and thirty others, to hear Grenville read over the king's speech on the Regency Bill. Harris reported that 'Nothing was said by any one'. It was what Harris called the 'late occurrencies in America' that brought members to their feet and to the divisions. These debates elicited from Harris himself some characteristic interventions against sentimental, hypocritical, and subversive rhetoric. Grenville's position on the insurrections in the North American colonies was unequivocal, maintaining the British parliament's right to impose taxes on the American colonists, denying the principle of no taxation without representation, and treating American resistance as simple rebellion. There was much more to this than 'a meer Mob, nothing more than the late Spital-Fields Riot', and Harris's speech, declaring that the Americans were 'in actual Rebellion', was in complete support of Grenville. When Grenville was dismissed, in July, Harris followed him into opposition to the Rockingham administration. Among those who kept him informed (notably John Shebbeare, QC on Grenville's instructions), there was Edward Hooper from the Board of Customs. In October Hooper was explaining to Harris the more sinister domestic ramifications of the uprisings at Boston and Rhode Island: 'The American example will certainly operate strongly with our common people, and as there does not seem to have been any care taken to suppress the riots, which have been in great measure foreseen, even from the first, the flame may now, perhaps, have got so much ahead, that the mischief may be irreparable.'[32] The Commons

[31] Resolutions printed by order of parliament, 22 May 1765.
[32] *Series of Letters*, i. 132 (26 Oct. 1765).

debate on 15 January 1766 was dominated for some hours by Pitt and Grenville, 'those two great Masters, and the Ministry stood by, like the Rabble at a Boxing Match,' Harris observed: 'I spoke as a lover of Freedom, but that twas the Freedom of Great Britain, and of the British Parliament.' Accordingly, when Lord Cavendish moved the Commons address to the king, using the innocuous phrase 'the late important occurrence in America' to describe the rebellion, Harris was the only member to vote against it: 'I gave them a Negative and prevented the nem. con.'[33] His most revealing remarks about the relationship between America and the mother country were made in his report of the examination of Benjamin Franklin in the Commons:

Thursday the 13th [Feb.], more Examinations, among other that of Dr Franklyn, the celebrated Electric Philosopher, who answered boldly and explicitly, but appeared a most complete American, a perfect Anti-Briton, denying not only the authority of our Legislature, but that our two last Wars were undertaken on American motives— the Capture by the Spaniards, which occasioned the first, were, he said, Capture of Goods belonging to British Merchants—the Disputes about Limits, and the Forts erected on the back of the Settlements by the French had nothing to do with the Americans; the Ground where those Forts stood were no American's Property; it was the ungranted Land of the Crown—This last Answer was rather weak, in others he was sufficiently acute.[34]

When Conway's speech, seconded by the Secretary of the Treasury, Cooper, on a motion to repeal the Stamp Act was delivered, Harris remarked, 21 February 1766: 'I never heard ever since I have sat in Parliament two Speeches, where America was so completely set above Britain, and Commerce above the Constitution and the British Legislature. Burke, the other Secretary, much later in the day, talked much the same, and if possible, carried the last Preference still further.'[35]

After five years as a member of the Opposition, Harris's reports of the debates get shorter as his indignation increases. By 6 January 1770 his attendance and interest in the Commons had declined steeply. Mrs Harris remarked to her

[33] Malmesbury Papers, 'Parliamentary Memorials from 17 December 1765 to May 1766', pp. 9, 12.
[34] Ibid. 20.   [35] Ibid. 20–1, 25.

son, in Madrid, 'I fancy he will not make much stay this year in town.' Ten days later, and twelve days before Grafton's resignation (to be succeeded by Lord North), Grenville urged Harris to attend the new session as a patriotic duty, beginning with a debate on the state of the nation and an examination of the Middlesex election:

many questions of the greatest moment are in agitation, and will certainly come on in a few days. It is impossible for me and many others of your friends, who both love and esteem you highly, not to lament your absence at such a conjuncture, both for your own sake and that of the public. Neither of these considerations, at such a crisis as now presents itself, will admit of that appearance of a total indifference, which your absence seems to hold forth, and which, I am convinced, you are far from feeling in any instance wherein the happiness and constitution of your Country are so nearly concerned.[36]

Harris's state of mind may be judged from the fact that he arrived in London only after the divisions had been taken. His private views were expressed to his son, and they suggest that he had found the debates over the Middlesex election profoundly demoralizing. On the Cabinet resolution to expel Wilkes, he commented: 'We talked much upon this, and mounted the high patriotic horse upon the unconstitutional mode of this practice.'[37] Doubtless also depressed by the recent deaths of Mrs Grenville and Hardwicke's brother, the Lord Chancellor Yorke, Harris continued to find hypocrisy and disorder in debates, 'full of that patriotic commonplace which no one believes that talks it, nor anyone else but a few dupes in the provinces'. When both Government and Opposition collaborated on a motion that the Wilkes decision had been both proper and constitutional, Harris had to resort to metaphor in order to convey his dismay: it seemed 'as strange and heterogeneous a thing as ever was engendered in the mud of the Nile'.[38] As for the general opposition to the king's Government, and in particular the City of London's remonstrance demanding a dissolution and removal of his ministers (presented by the mayor of London, Alderman Beckford), he

[36] *Series of Letters*, i. 186–7.
[37] Ibid. 191 (6 Feb. 1770).
[38] Ibid. 192–3 (13 Feb. 1770).

could predict nothing but anarchy and confusion; and his letters search out absurdities for some temporary relief:

The violence, I may say madness, of the opposition, have done Administration as much good as all their own sagacity. A man may wish to see a house altered or cleansed, who does not wish to see it blown up.

General Armiger, aged sixty-five, was married at eight o'clock last Saturday evening to a lady between thirty and forty, went to bed, and was dead by one in the morning.[39]

In common with many others, Harris was also deeply troubled by the Government's clumsy but nevertheless vigorous attempts to exert its control over the printed word, an attempt just as wrong-headed as the abuse of the limited 'freedom' of the press which provoked it. When some City printers were arrested for publishing somewhat less than accurate or respectful versions of the parliamentary debates early in 1771, one of them was freed by John Wilkes himself, now an alderman. The ministry bungled the whole affair, even adjourning parliament for 10 days so that the order requiring Wilkes himself to appear on a charge of breach of privilege should lapse. Harris predicted that 'We shall lose the most valuable privilege under heaven, by the rascality of the lowest of scoundrels.'[40] But he also ridiculed the Commons' arbitrary and illiberal posturings: 'we clamour as if our oppressions were ten times greater. Perhaps our clamours prove against us, for, when tyranny is complete, no one dares think of complaining, being well apprised of its danger.'[41] In November Harris's political mentor, Grenville, died; by the end of March 1771 the mayor of London was in the Tower, Lord North was attacked by the mob in his chariot, and the same happened to Charles Fox. As Harris's own political commitment waned, his son's conduct over the dispute with Spain and the Falklands brought some compensation, reflected glory, and congratulations from all sides of the House. He dined with the influential Shelburn, and his second daughter, Louisa, was presented to the king and queen on 13 February 1771. On a few occasions he was able to get away from politics to the

---

[39] *Series of Letters*, i. 199 (20 Mar. 1770).
[40] Ibid. 202 (3 Apr. 1770).
[41] Ibid. 189 (2 Feb. 1770).

family estate of Great Durnford, where the conversation 'was chiefly of grass and dogs . . . a place of no events, we enjoy ourselves and the fine weather, we eat, drink, sleep, read, work, walk, ride, laugh, play cards, and grumble every day, and we are so satisfied with the uniformity of our living that we have fixed no time for quitting it' (29 June, 7 September, 1771). Apart from home-grown musical concerts, the Salisbury news concerned innoculation against smallpox.

Lord North's style of government was conciliatory, broadly based, financially cautious, and moderate. These characteristics attracted independents such as Harris who were looking for order and stability, and he transferred his allegiance to him with no evident strain. North significantly reduced the national debt and managed not only to keep the land tax at the same level from 1771–4 but also to pay for increased defence forces. On 14 December 1770 Harris reported his satisfaction to his son: 'The land tax has passed at last, so that our whole expense of six millions sterling is actually raised, our whole augmentation of our fleet and army as good as paid for, and our navy as forward as ever was known, and a formidable one moreover. Britannia need not tremble, whatever the event may be, whether peace or war.'[42] Within two years Harris was soliciting North for office.

Another issue in 1772 brought a renewed involvement in parliamentary debates: the Dissenters Bill. This required subscription to the Thirty-nine Articles by dissenting ministers, and in this debate Harris joined company with Burke, Soame Jenyns, Thomas Pitt, Solicitor-General Wedderburn, Lord North, Fox, and Sir George Savile. The Egerton transcript records three Harris contributions to the Opposition's case:

I shall only say a few words, to give my sentiments before I vote. I am myself a friend to toleration. I am free to declare that toleration is no such thing, is not worth having, that depends upon the will of a Prince, or sovereign. If it was, let us go into Africa for toleration under their grand Emperors; to Rome for toleration under the Pope; let us go to Spain. These are my reasons why I am a friend to toleration, why I wish this bill may not be postponed. I hope after the declaration of the Honourable Gentleman over the way [i.e. Captain Phipps] has given it will not be thought anti-Christian.

[42] Ibid. 208.

Mr Harris concluded with a short comparison between ignorant monks, the crude, barbarous Inquisitors, the German divines, and Bacon, Newton, Locke, and Clarke.[43]

On May 14 1778 Sir George Savile proposed a Bill to repeal one of the seven punitive laws which prevented Catholics from legally either owning or inheriting land and which provided a sentence of life imprisonment on priests exercising their functions. As a 'friend to toleration' Harris naturally applauded the proposal, adding in a letter to his son in St Petersburg, 'I wish religion would make us love one another; 'tis certainly high time it should no longer make us hate one another.' In the debates preceding Savile's proposal on 10 March Harris spoke against a petition signed by 2,000 Dissenters ('our Puritans', as Harris put it) who *objected* to the Bill. Harris regarded the spirit of the petition as expressing 'a diabolical principle inconsistent with the humanity of this House, inconsistent with common sense, therefore I hope we shall pay no regard to it'. Burke listened to Harris's remarks and responded in the same vein, but more coolly: 'I should be glad that some church of England counsel were to give reasons for men wanting toleration [for] themselves, [and] why others should not be tolerated.'[44] In his last intervention Harris invoked historical precedent to enforce the idea of humane progress achieved since the Smithfield fires burnt heretics in the reign of Queen Mary, now only remembered for its foolish superstitions about witches and broomsticks ('these foolish women were cruelly executed by men far more silly than themselves'). In these more enlightened days, the extension of toleration was the proper duty of reasonable men. Moreover, even the despotic French king tolerated the Huguenots in the large towns of the southern provinces, although this was a spurious toleration because it depended on royal whim: the king of France could still hang a malefactor or send him to the galleys, whereas in England malefactors could only be fined or imprisoned. It was another four years before the Sacramental Test Act was finally to be abolished.

Harris's judgement of parliamentary speaking was based,

[43] BM Add. MSS Egerton 233, p. 267. For Harris's 2 additional interventions, see MS 243, pp. 261, 263.   [44] Ibid. 264.

not surprisingly, on the model provided by the classical rhetoricians. Accordingly, he found Burke's speeches 'long, vehement, and florid, without order, and therefore not to be remembered'. The oratory of Charles Fox, on the other hand, when speaking in the American debate on 2 February 1778 was 'orderly, able, and masterly, and prefaced with a good exordium, like an oration of old', and again, on 22 November 1778, he 'had a plan; his speech had an exordium, a narration, and a conclusion, regular, (though as it ought to be,) concealed with care'. The distinction Harris draws is not only formal, but also on the extent to which the passions were engaged, legitimately or otherwise. In the same debate, 'Burke was diffuse and eloquent. I, who am used to order, and study no books but such as have a beginning, a middle, and an end, cannot relish a rhapsody, however exquisite, and when the sufferings of the Americans, their wives and children, are attended with exclamations and tears, I smile at the mock tragedy, when perhaps at Drury Lane I should weep.'[45] His increasing disillusion with the Patriots' rhetoric fed his desire to escape the Commons and coincided with solicitations to Lord North for office. By October 1775 he could write freely of his increasing reluctance to attend the House, now characterized by 'faction and rebellion, weeds which are too common in free states', because since the end of January he had escaped to a more dignified and entirely congenial office as secretary and comptroller to the Queen. Harris wrote the following account of his 'coming into office':

January 12 1774. I was at the Cockpit, to hear the King's Speech and the Address redde over, the day before the Session. When they were done, and the Company were going, Lord North took me aside and told me, he had long been concerned at not being able to answer my Letters but that now he could offer me something, if I approved it, it was the office of Secretary and Comptroller to the Queen—that General Greene, on Lord Guildford's accepting the office of Treasurer to the Queen, on the death of Stone, had resigned, and made the vacancy. I thanked and accepted, and we parted.

January 14 1774. I saw Lord North in the House. He told me he had mentioned my name to his Majesty, who perfectly approved the Nomination—that I was to kiss the Queen's hand first, and then the

---

[45] *Series of Letters*, i. 396 (23 Nov. 1778).

King's. I should have noted that on January 12 he said they were two offices, required no duty, were worth together £500 a year (£25 deducted), were regularly paid, and the duty done by the deputy Vincent Matthews, whom he recommended me to continue—that all I had to do, was to sign Receipts.

On my first interview I had asked Lord North, if I was at liberty to mention this affair—he replied, better not. Things continued in this state for some days. I communicated the affair to none, but one or two confidential Friends, and expected every day to be sent for to kiss his Majesty's hands.

On the 18th the Queens Birth-day, my self and Family were at Court. Then in a crowded Drawing Room, before all the Company, Lord Chamberlayne (Lord Hertford) wished me Joy, then Lord Holderness and others. In telling Lord Holderness I did not know whether I was to acknowledge the thing, he smiled and said he had heard it the night before in the company of 150 people.

In the Evening I attempted to attend the Ball, but the Crowd was too great; however in the narrow passage leading to the Bath Room, as her Majesty past by, she honoured me with particular notice.

Next morning, Wednesday the 19th Lord Hyde and Lord Bruce calld on me. The latter said he had heard it from Mr Jas. Brudenell, who heard it from Lord North.

Since that time to this present Monday January 24 I have had Congratulations from people of all Ranks, both at Court and in Parliament, but no kissing of hands. General Fitzroy told me yesterday he waited a month in the same circumstances. This not pleasing—the Queen's pregnancy makes her confine her self to her House, and tis not expected she will be at St. James, till after her delivery.

Harris's suspense came to an end a week later, when the seals of his office were handed over. The queen then detained him for a private conversaton which lasted at least twenty minutes. Harris's record of what he calls 'this extraordinary Conference, in which I had the strongest proofs of the Queen's excellent Understanding and Goodness of Heart' (entirely conducted on her part, he remarks, in excellent English) also revealed his own abiding passion for the sister arts:

we talkd of Bach, Abel, Sacchini, and many others. I told her Majesty that for the honour of Germany it had the best Scholars, the best soldiers, the best Musicians, and the best Painters now in Europe—her Majesty hesitated about saying Painters—on this I quoted Mengs and she readily assented—on this we past to

Painting, to the Guidos, the Cartoons, and other such Pictures that were in the House—thence to Writers, when she honoured me with mention of my works—I lamented that the subjects being philosophical were some of them difficult—she politely said there were parts intelligible—that she saw the King reading a Book at Richmond—what Book is that Sir (said the Queen)—Mr Harris's (said his Majesty)—what Harris, said the Queen. Harris of Salisbury said the King, you know him—He then redde what she much approved—in singing she talkd of young ladys who sing by Rote, but could not sing the easiest thing laid before them—that there was no knowledge when they could not do that—we then past to the Fire and Winterslow House, and the sad ordeal of waters—this lead us to Water Expeditions—the Queen said she did not love Water much, and Sea especially, since her Voyage to England . . .[46]

The actual duties of Harris's new office were, as he admits, slight. The work of preparing warrants for appointment and pensions, and compiling household accounts, was done by his deputy, and the bills were immediately settled by Lord Guildford, the Treasurer. Harris was otherwise free to engage his masters with discussions about 'Gardening and the English Taste', the absolute contrast between German and English landscape design (28 June 1774), Handel, 'Books, Stile, Grammar, Literature, Poetry, Painting, . . . the Royal Academy', and his own work: 'His Majesty asked me which of the three polite arts (meaning Music, Painting and Poetry) I thought had most force—I hesitated for a moment, and replied I thought Poetry, for that Poetry excelled in Sentiments—His Majesty smiled and said an Author, who had written on the works had not clearly so decided it, meaning by that author my self.'[47] This was at least the second such occasion. On 27 November 1772 at the king's levee the king asked Harris about the progress being made on his new book (*Philosophical Arrangements*), and 2 days later the king remarked:

whatever may be the Effect of Music and of Poetry, when taken *singly*, they had never so much Force, as when they were united—I replied it was an undoubted Fact—I have learnt this, says his Majesty with a smile, from a certain Book I have redde—your Majesty, sayd I, does that book too much Honour—your own

---

[46] Malmesbury Papers, 632A, the start of Harris's 101-page political diaries, pp. 1–4.   [47] Ibid. 32.

Genius is abundantly sufficient for *that Information*—His Majesty in a gratious manner past on to the next person. (The Book alluded was the second Treatise of my first Volume).[48]

On 24 June he was asked to play the harpsichord for the king, and did so, choosing a sonata and the overture from *Alcina* by Handel. Harris was later informed that the king's solicitude for Harris's gout may have outweighed his enthusiasm for Handel; 'I did not understand his goodness in making me play, till I found from his Page, twas to give me an opportunity to sit, which he knew my weak feet wanted.' In subsequent discussions Harris gave his opinion of Gray's letters (some of which were 'too juvenile and hardly worth notice', an opinion with which the king agreed), and Lord Chesterfield's letters ('the morals of which he by no means approved').[49] At the concerts performed in the queen's house, Harris met Abel and J. C. Bach, signs of the queen's musical preference, since she had been discouraged on her first arrival in England to find that musical taste was dominated by Handel, Corelli, and Geminiani. On 21 May 1779 Harris watched Reynolds painting the king's portrait.

Harris's royal appointment might seem to merge, for the first time, his apparently separate and parallel lives as public politician and private dilettante. In fact, Harris's ability to combine aesthetic interest with political activity, though unusual, was certainly known from the very beginning of his career. When Manzolini's clothes and finery were seized and taken to the Customs House, their owner was obliged to petition the Lords of the Treasury to get them back. Mrs Harris reported (25 October 1764) that 'This event diverts Lord North, as he says not one of the Treasury know a note of music, or care one farthing what becomes of Manzolini, *except Mr Harris.*'[50] This intense and absorbing musical involvement was by far the most significant *personal* aspect of his metropolitan career (see Chapter 7), but Harris's political career sprang nevertheless from a serious commitment both to his constituency and to the national interest. Towards the end of his career, while in attendance at court, he may have

---

[48] Malmesbury Papers, 632A, 3 of sheets following diary.
[49] Ibid. 44, 52, 55.
[50] *Series of Letters*, i. 117; Mrs Harris to her son.

enjoyed the social opportunities of his various offices far more than the humdrum and eventually disagreeable round of parliamentary duties: who would not? But at least under Grenville he had demonstrated the valued loyalty of an independently minded political lieutenant. Whether Harris owed his political preferment to a dinner at Salisbury provided for the Duke of York on 9 October 1761, his subsequent achievements in parliament and at court were due to rather more than social elegance and literary prominence.

7

Music and Theatre in Salisbury and London

He, whose Judgement isn't cloid
With admiring Vicar Floid,
May an equall Pleasure earn
From the Voice of Vicar Hern;
From the pleasing Bray of Asses
From Naish or Collier's Tunefull Basses.

By Notes of Greece was rais'd a spacious Town
By Jewish Notes a City fair fell down:
If by mens Tastes their Country we discern
From Greece came Hull, from Jury Floid & Hern.

HARRIS's interest in music was lifelong, and his musical proficiency included both theoretical and practical skills. He performed, composed, arranged, conducted, and managed public musical performances for most of his adult life. It is not surprising that the opera, the musical genre which contains all others, was to absorb his interests, nor that Handel was to remain his great musical hero, overshadowing his earlier enthusiasm for Purcell. Harris was born into a musical family in a cathedral city with a 500-year-old musical tradition, largely the responsibility of the choristers and the Vicars Choral (some of whom are satirized in the above Harris poem written in July 1730). The 'Hall of the Vicars Choral' was on the site of number 12 in the Close, next to the Harris home, and the mid-fourteenth-century 'Chorister's House' nearby still partly survives.

His musical skills were developments of those already practised by the Harris family, friends, and home city. On the one hand there were the catches and glees composed by his father and performed by the family for domestic amusement; on the other, the exemplary patronage of Handel by the fourth Earl of Shaftesbury. We have already seen that Handel represented no less than the fate of Harmony itself in *Three Treatises*, but Handel's death came in 1759, approximately half-way through Harris's own musical career. In later life, Harris was to range far and wide, becoming one of the research assistants to the greatest musicologist of the day, Dr Charles Burney. Connecting every stage of his career, and at the centre of it, was the brilliant annual musical festival in Salisbury.

From the mid-1740s onwards, Harris's musical taste and enthusiasm engineered a conjunction of the metropolitan repertoire of the London operatic and concert stages and the musical diet of his home city. London and Salisbury did not compete, except in the opinion of some understandably prejudiced Wiltshire residents, but they did share the same aims and tastes in the business of musical promotion. James Harris was the impresario who not only brought London music to Salisbury but also stimulated a traffic in the other direction.

Harris's library contained almost a thousand folio pages of manuscript scores drawn from Italian composers of the late seventeenth-century as well as the best Elizabethan, Restoration, and early eighteenth-century books on the theory and practice of music. In addition to this scholarly resource, which he put at the service of the Salisbury repertoire, his emotional devotion to music was such that, according to his daughter Gertrude, he refused to hear any music for the last three or four weeks of his life lest it should overpower him. Readings from *Joseph Andrews* were substituted as a less stressful consolation.[1]

Music was indeed Harris's most passionate experience. Along with his wife and two daughters, James attended a concert in the Burney household in May 1775, in the house

---

[1] Gertrude Harris, 'Portrait . . . of my Mother' (1806), PRO 30/43/1/4, p. 18. The sections chosen were bk. I, chs. 15–16 (Parsons Adams and Barnabas on the law and vanity), and bk. II, chs. 12–13 (romantic comedy, 'high and low people', and the debate on Charity between Adams and Trulliber)—each resonant with memories of Fielding's days on the Western Circuit (Harris and Fielding were both, in Fielding's words, 'of the same Trade'), and of William Young's scholarly absent-mindedness. Harris's musical library contained T. Morley, *A Plaine and Easie Introduction to Practicall Musicke* (1597/1608); J. Playford, *A Breefe Introduction to the Skill of Musick* (1654/1660); W. Holder, *A Treatise on the Natural Grounds and Principles of Harmony* (1694); C. Simpson, *A Compendium: Or, Introduction to Practical Musick* (1667/1706); C. Avison, *Essay on Musical Expression* (1752); J.-J. Rousseau, *Dictionnaire de musique* (1768), and a 17th-cent. MS, *Traicte des modes, ou tons, de la musique*.

once occupied by Sir Isaac Newton in St Martin's Street. Fanny Burney described the presence of James as 'the pride of the evening': 'His looks, indeed, are so full of benignity, as well as of meaning and understanding; and his manners have a suavity so gentle, so encouraging, that notwithstanding his high name as an author, all fear from his renown was wholly whisked away by delight in his discourse and his countenance.' The hesitant Louisa was encouraged to sing an unpublished air by her teacher Sacchini to her father's accompaniment on the piano, and Fanny observed that during a harpsichord duet played by Dr Burney's son-in-law Charles Rousseau Burney and his eldest daughter Hettina, 'Mr Harris was in an ecstacy that played over all his fine features'.[2]

I

Harris's enthusiasm had been nurtured in peculiarly interesting but still somewhat obscure Salisbury origins at the very beginning of the eighteenth century. On St Cecilia's Day, 22 November 1700, the musically inclined subdean of Sarum, Thomas Naish, gave his first sermon to the Society of Lovers of Musick in the cathedral on the themes of Harmony, Love, and Charity. This sermon inaugurated Salisbury's progress towards becoming the oldest as well as the best non-metropolitan centre for musical festivals, pre-dating those of Wells, Oxford, Winchester, and Devizes, and that known as the Three Choirs Festival which still exists. Naish's sermon is also the first essay on the theoretical principles of music to be published in Harris's own time in Salisbury. The Society had doubtless grown from a strong local tradition of private and civic musical activity (in 1660 the silver chains belonging to the 'Town Musicians' were stolen from the Council House). It was to be Harris's role to transform ideals into performance. Both the success and the decline of the festival coincided exactly with the term of his involvement, from about 1744 until 1780, when the death of Harris and that of Dr John Stephens, the cathedral organist, occurred within a few weeks

---

[2] M. d'Arblay, *Memoirs of Dr Burney*, 3 vols. (1832), ii. 13, 17. Fanny Burney remarked that Harris's 'soul seems all music, though he has made his pen amass so many other subjects into the bargain' (ii. 16).

of each other. Thereafter, the Salisbury annual musical festival went into a steady decline.

Naish's first sermon laid down high-minded guidelines for the Society's secular and religious purposes. As Harris was to point out to Monboddo in 1773, Harris himself was no 'theologue', and thus his own essay on the theory of music published in *Three Treatises* was to have no dealings with analogical arguments about worldly and other-worldly musical harmonies. Even so, and with this important difference, Naish's 1700 sermon, as well as those delivered on St Cecilia's day in 1726 and 1727, provide clear anticipations of Harris's own thinking about the role and purpose of music (and, for that matter, art in general) as an unmediated expression of man's idealistic nature and needs.

To be sure, Naish stoutly defends the inspirational and sacramental role of music; but he begins by stressing its non-rational and affective properties: music is thus 'the happiest Means of expressing and representing any kind of Affection to our Mind, in a very secret Manner, but with very subtle Effects . . . [and is] a happy Means to open and enlarge [men's] Hearts, to bring fresh Fire to the Sacrifice, and new Flames to their Devotion'. Accordingly, he charges the new Society to 'strengthen the Bonds of brotherly Love, and . . . maintain perfect Harmony in your Affections to each other'.[3] By the mid-1720s, the Society's annual meeting had adopted a regular and tripartite format of a sermon, followed by a supper, and then a concert. This format was probably borrowed from the St Cecilian tradition in London, which dated from perhaps 1683 until a few years beyond 1700. By 1726 the Society had at least one patron (in that year there were two, Sir John Crisp and John Wyndham), six additional stewards, and each was appointed on an annual basis.

A month and two days after Swift's *Gulliver's Travels* launched its message of subversive misanthropy, Naish was

---

[3] T. Naish, *A Sermon Preach'd at the Cathedral Church of Sarum, November 22 1700 Before a Society of Lovers of Musick* (Salisbury 1701), 14, 19, 26. The order of proceedings seems to have been a cathedral sermon first, followed by a dinner elsewhere, and then the concert, ticket-money from the latter being used for charitable purposes. Naish remarks (p. 27) 'I hope you are not come together purely to eat and drink, and rise up to play . . . now I shall dismiss you, recommending the Saints in Bliss as a pure Pattern for your Musick and Love.'

5. Title-page illustration for Thomas Naish's *Sermon* to the Society of Lovers of Musick, Salisbury (1726).

once more preaching the beauties of harmony to the Society. The subdean congratulated the Society on a long and well-established tradition of musical excellence allied to charitable as well as aesthetic purposes. He remarked that 'through the Course of many Years I have found your Meetings so regular and harmonious, your Designs so honourable and good', and once more developed his analogical argument that music provides not only an innocent intellectual pleasure but also a near-mystical experience of religious rhapsody:

Here is the Beauty of Harmony in its Perfection, when 'tis inspir'd with Devotion, then, like *Moses's Bush*, it will flame and burn continually, without consuming. This is the Glory of Musick, that it enables us to praise God with more fervent Passions, and helps us to excite in us a Love and Joy, and Reverence towards him; and when we feel the greatest Transports of these Divine Passions, we most resemble the heavenly Choir, and come nearest to their exalted Way of praising God.

'Tis but a poor and mean Office of this noble Science to play upon the Ear only, or to raise the Fancy to a chearful Sprightliness; But when its Subject is holy and sublime, it darts itself through the Organs of Sense, warms all the Powers of the Soul, and fills the Mind with the brightest and most ravishing Contentments. It warms the Affections as with a Coal from the Altar, and fills the Soul with blissful Breathings of divine Love. In Musick so directed we may expend all our Powers, and yet feel fresh Repletions; here is no Danger of Excess, and we may be perfect Voluptuaries without Sin.[4]

Such a comment provides the perfect anticipation, if one were needed, for the reception of Handel's biblical oratorios in the cathedral city. Appropriately, then, the annual St Cecilia's day concert, the high point in year-round musical activity, was moved from the cathedral to the Assembly House (in the evening, and followed by the customary ball), in 1740. This freed the cathedral for a performance of Handel's *Te Deum* and two of Handel's anthems in the morning of the

---

[4] T. Naish, *Sermon Preach'd at the Cathedral Church of Sarum, November the 30th, 1726. Being the Anniversary Day Appointed for the Meeting of the Society of Lovers of Musick* (London, Sherborne, and Sarum, 1726), 13. The Salisbury printer was Edward Easton. Naish remarks on pp. 9–10: 'all Nature is harmonious, and musical Proportions are to be found in all well disposed Things: There's something of the Power and Virtue of Harmony that lies hid in every Part of the Creation, such as truly affects every tuneful Mind.'

same day. Thereafter, Handel's music dominated the repertoire until well into the nineteenth century, particularly, the *Dettingen Te Deum, Messiah* (at least twenty performances between 1750 and 1782), one of the 'Coronation Anthems', parts of *Judas Maccabaeus, Joshua*, and the overture from *Saul*.[5] The scores of *Judas, Messiah, Joshua*, and the *Te Deum* were acquired from the extensive library of Handel scores owned by the Earl of Shaftesbury, Harris's cousin.[6] By 1752 the musical talents of the 'town musick', augmented by players from other centres, comprised sixteen violins, two hautboys, two violas, a bassoon, a harpsichord, four violincellos, two double basses, French horns, trumpets, and drums.[7] The splendid four-manual organ designed and installed by the London organ-builder Renatus Harris in 1710 was probably the first of its kind in England. With such musical resources at his disposal, Dr William Hayes, Professor of Music at Oxford since 1742, successfully enlisted Harris in his efforts to acquire Handel's scores for performance in Oxford and Salisbury in 1756.

At least eight performances were given in Salisbury of Harris's own musical pastoral *Daphnis and Amaryllis* from 1761 onwards (and at least one performance at the Winchester Festival of 1779), and two performances of his pastoral *Menalcas*. On 6 July David Garrick, already a visitor to Harris's Salisbury home, solicited the pastoral (then called *Damon and Amaryllis*) for performance at Drury Lane 'the next winter'.[8] A third letter from Garrick eventually acknowledged its arrival (7 September) in a form suitably revised for performance. On 22 September 1762 the pastoral, technically

---

[5] Information about the concert programmes at Salisbury is drawn chiefly from D. J. Reid (with B. Pritchard), 'Some Festival Programmes of the Eighteenth and Nineteenth Centuries: 1. Salisbury and Winchester' *R.M.A. Research Chronicle*, 5 (1965), 51–67, and the addenda and corrections by A. D. Walker in 6 (1966), 23, and B. Matthews, 8 (1970), 23–33.

[6] See B. Matthews, 'Handel—More Unpublished Letters', *Music and Letters*, 42 (1961), 127–31.

[7] It should be added that all these instruments were illustrated in the engraved frontispiece to Thomas Naish's 1700 sermon to the Society and formed a framework to the words 'A Sarum Consort'. I suspect that this lavish plate may have been used as a cover for the Society's concert programme, but I have no proof that this was so.

[8] The following letters from Garrick to Harris may be found in *Series of Letters*, i. 85–7, and Harris Papers, Letter Book, vol. 31, part 1, pp. 26–31, 32–3 (Stephens to Harris). The standard edn. is *The Letters of David Garrick*, ed. D. M. Little and G. M. Kahrl, 3 vols. (Cambridge, Mass., 1963): i. nos. 289, 291, 295.

speaking a *pasticcio*, with words by Harris and music by Handel, was performed at Drury Lane in order to introduce the Salisbury chorister Thomas Norris, a pupil of Arne, in the title-role. Garrick had required substantial cuts for Drury Lane; an afterpiece lasting two hours was far too long. The process of rewriting was transacted through Dr Stephens, the Salisbury cathedral organist since 1746 and one of Harris's closest collaborators. Stephens delivered the score to Garrick on 7 September and wrote back to Harris with an account of its London rehearsal, urging Harris to compose and score an additional 'fine song' in order to alleviate 'too many serious Songs together'. There was also some encouraging news: 'I find the King will be at the 2d. performance of it. If this gay Song cou'd be ready to clap in at the 2d. performance if necessary, it wou'd make Mr Garrick extravagantly happy . . . Grimaldi is to conduct the Dances, and to perform himself' (12 October). Finally, Stephens reported four days later, 'Tenducci has the Parts that are missing, all but Daphnis, which you was so kind to lend Norris to study . . . I have rec'd the Song, and think it will do charmingly.'

It changed its title again for the London stage, becoming *The Spring*, and was included as an afterpiece at Drury Lane on 20, 22, 23, 25, 26, and 29 October, and again on 19 and 22 November 1762. (It reappeared at Covent Garden on 10 February, and at Oxford the following year.) On 29 November it was the afterpiece to *Cymbeline* (Garrick played Posthumus), and on 19 November it followed Garrick's *Lear*. A puff for the piece specifically commended Norris's singing but kept the author's anonymity. It also remarked that 'this Pastoral was not originally designed for the theatre'; that although the choruses and airs had been selected from Handel and 'other Eminent Masters', the connecting recitative had been composed by its author; and that 'the words were of necessity composed in perfect subservience to the music. The piece has been several times performed at Salisbury and greatly admired by many of the first Rank.'[9] Harris's own remarks on the pastoral indicate clearly that it was written for 'private' rather than public performance, and since this is the only

[9] Quoted in G. W. Stone, *The London Stage, 1660–1800. Part 4: 1747–1776* (Carbondale, Ill., 1962), 948.

surviving printed evidence of a once thriving and well-known tradition of performances at his own private theatre above St Ann's Gate, it is worth quoting Harris's own estimate (to Garrick) of its purpose and function:

For the Music, it is wholly ye choicest Italians taken from pieces *little* known in England, of other pieces well known. For length, it consists of 2 Acts which may be finished rather in less than two hours. As all Shepherds & [illegible] are Simple, you will find their Words & Sentiments to be simple likewise. These & ye Music of ye Recitativos are all, for ye merits or demerits of wch I am my self to stand responsible. You herewith receive the Words, the better to judge of them.

I have only one thing to add, which is that I cannot spare the piece, till after ye end of September, because we propose at that time to perform it here at our annual musical festival.[10]

The Salisbury performance went ahead under the new designation of the 'Salisbury Annual Musical Festival' on the evening of 27 August. Moreover, the pastoral was only one part of a wholly 'Harris-arranged' evening, including his setting of Milton's 'Hymn' from *Paradise Lost* and with excerpts from Handel's *Jephtha*, conducted by Harris. Finally, this Miltonic operatic masque of 16 pages was published, without Harris's name, at Exeter in 1766.

II

In Harris's time, from 1742 to 1780, the annual musical festival in Salisbury lasted for 2 or 3 days and was held at slightly varying times in the months of August, September, or October, no earlier than 22–3 August and no later than 28–30 October. Morning performances usually began at eleven or twelve o'clock, almost without exception in the cathedral, and continued into the afternoon. Up to 1750 all evening performances took place in the Assembly Room in New Street, and thereafter in the New Assembly Room in High Street. Given his parliamentary duties after 1761, as well as the necessary research and writing of his next book, *Philosophical Arrangements*, Harris could not have been the only director of

[10] Harris Papers, Letter Book, vol. 31, part 1, p. 27.

The Annual MUSICAL FESTIVAL,
AT SALISBURY,
WILL be celebrated on the thirtieth Day of
September next, and the firſt Day of October following,
At the ASSEMBLY-ROOM, on the firſt Day,
Will be performed an ORATORIO, called,
The TRIUMPH of TIME and TRUTH,
Compoſed by Mr. HANDEL,
And never before performed in the Country.
And on the ſecond Day,
The celebrated Hymn in Milton's Paradiſe Loſt,
AND
A Paſtoral called DAPHNIS and AMARYLLIS,
Both adapted to Muſic of the greateſt ITALIAN Compoſers,
JOMELLI, PERGOLESE, GALUPPI, and others; with the ad-
ditional Songs and Choruſſes, as at the laſt Performance.
In the CATHEDRAL CHURCH will be
A TE DEUM and JUBILATE.
The Muſic from the beſt Church Compoſers of Italy; and two
Anthems by the ſame, and one by Mr. Handel.
The principal Vocal Parts by Signior TENDUCCI, Meſſrs.
NORRIS, HIGGINS, and CORFE. The Inſtrumental Parts by
the beſt Performers from Bath and Oxford.
A BALL each Night, as uſual.

6. Advertisement from the *Salisbury Journal* (16 August 1762) for the annual festival: music by Handel and Harris.

the festival. But he was certainly involved in choosing the programme, acquiring and arranging scores, as well as conducting and rehearsing the musicians.[11] He could fairly be described as its consultant and impresario. Another important role, and one which he shared to some extent with his wife, was that of the festival's social contact with the star performers of the London operatic stage, particularly at the King's Theatre. The precise degree of his overall management remains difficult to quantify, since his name does not appear

[11] Gertrude Harris recorded in her 'Memoir', p. 2: 'He had frequently Music at his House and he constantly attended (and in a great measure directed) the subscription Concert and from his procuring for it, some of the choicest pieces of Music then performing in London, it acquired a greater degree of celebrity than usually belongs to provincial centres.'

in any of the surviving lists of Society stewards. Nor are there detailed records of the festival's organization. Even so, there is no doubt that he was the driving force behind its continuing high quality. Apart from evidence in the family's correspondence and notes in the *Salisbury Journal*, there is the remark by Handel's librettist, Dr Thomas Morell, who wrote to an unidentified correspondent in 1764 about the first complete performance outside London of *Jephtha*, on 2 October 1760:

> My own favourite is *Jephtha*, which I wrote in 1751, and in composing of which Mr Handell fell blind, I had the pleasure to hear it finely perform'd at Salisbury under Mr Harris; and in much greater perfection, as to the vocal part, at the Concert in Tottenham Court Road.[12]

There is little doubt that the introduction of Italian music into the Salisbury festival for the first time in 1749 (Geminiani and Corelli in particular) was Harris's doing. Thereafter, the frequent recurrence of the phrase 'selected from capital Italian Masters' in the concert publicity, together with performances of Milton's 'Hymn of Adam and Eve' and *Daphnis and Amaryllis* (1761), identify Harris as the arranger and supplier of such scores.[13] We could not have been sure of this had not Harris's musical protégé, the cathedral organist Joseph Corfe, published two elegant volumes of *Sacred Music* around 1800 dedicated to the Earl of Malmesbury. Its title-page describes the contents as 'a Selection of the most admired Pieces of Vocal Music from the Te Deum, Jubilate, Anthems & Milton's Hymn, Adapted to some of the Choicest Music of the greatest Italian and other Foreign Composers, Jomelli, Pergolesi, Perez, Martini, Scolari, &c . . . By the Late James

---

[12] Historical Manuscripts Commission, 15th Report, app., part 2, *The Manuscripts of J. Eliot Hodgkin* (1897), 93. Admission tickets to morning performances in the cathedral generally cost 3*s*. 6*d*., and to the evening performances in the Assembly Rooms, 5*s*. In his *Handel: A Documentary Biography*, p. 622, Deutsch claims that 'Handel may have visited the Harris family at Salisbury repeatedly'. Nothing is more likely, although I have found no supporting documentary evidence. There is, however, some evidence to support the equal probability that Handel visited the Shaftesbury's home at St Giles. Betty Matthews ('Handel', facing p. 131) reproduces a musical fragment from the Harris family papers, a Larghetto by Handel, annotated in what may be Harris's hand: 'The above Air was composed at St. Giles by Mr. Handel extempore, & afterwards by Desire or the Company writt down in his own hand writing as above.'

[13] See Gertrude Harris's comment, n. 11 above.

Harris, Esq.' These were the only Harris 'compositions' to be published, and clearly represented the favourite and most performed pieces in the festival repertoire. They include thirty-six settings; recitatives (two by Harris himself, one scored for instrumental accompaniment), airs, choruses, duets, and a quartet, all drawn from Giovanni Pergolesi, Giuseppe Scolari, Gioacchini Cocchi (musical director of the Haymarket Theatre until replaced by J. C. Bach, and also in charge of Mrs Cornelys's concerts at Carlisle House), Johann Hasse (German born, but a composer of operas in the Italian style, and married to the Italian prima donna Faustina Bordoni), Nicolo Jommelli (Italian born but a composer of operas in the German style), Torquato Tasso, Giovanni Chinzer (trumpeter, composer, and impresario), Leonardo Vinci (stylistically a follower of Scarlatti and 'the first opera composer', remarked Burney, 'who . . . without degrading his art, rendered it the friend, though not the slave to poetry'), David Perez (all of whose operas existed only in manuscript), Giovanni Battista Martini (the friend of Burney, and the tutor of Jommelli, Mozart, and J. C. Bach), Beretti, and Francesco Geminiani (violinist, composer, and Corelli's pupil).[14]

Corfe's two volumes in one contain twenty arrangements and eleven recitatives. The first group contains eleven settings by Harris of music by Pergolesi, two arrangements of Cocchi, two by Jommelli, two by Padre Martini of Milan, two of Pergolese, and one each of Scolari, Hasse, Chinzer, Vinci, Perez, Beretti, and Geminiani. The remaining pieces, eleven recitatives, are by Harris himself. The first volume is arranged as a sequence of musical settings of words taken from *Paradise Lost*, Book V, lines 137–45 ('Milton's Hymn'), set for four part strings, two flutes, two oboes, bassoons, two horns, two trumpets, and organ continuo. Adam and Eve, represented by soprano and alto, awake, the latter having dreamed a premonition of betrayal by Satan; Adam soothes her anxiety

---

[14] For information on composers, instrumentalists, and singers, see Burney, *A General History of Music*, S. Sadie (ed.), *The New Grove Dictionary of Music and Musicians*, 20 vols. (1980), and P. H. Highfill *et al.* (eds.), *A Biographical Dictionary of Actors, Actresses, Musicians, Dancers, Managers and other Stage Personnel in London 1660–1800*, 12 vols. (Carbondale, Ill., 1973–87). The reference to 'Milton's Hymn' is to the hymn of Adam and Eve, *Paradise Lost*, v. 137–206, words which Harris freely adapted, including a duet between Adam and Eve (v. 205–6).

and both begin to praise God as the sun rises over Eden's landscape. After an opening Gloria in C major, the words of the first recitative are: 'But see from under shady arb'rous roof, the new created pair come forth, in sight of Day Spring and the Sun.' This is scored for treble solo, and is followed by a duet between Adam and Eve. The linking recitatives are scored for both treble and bass (a figured bass used by the organist to improvise an accompaniment) and would have been sung.

Quite clearly, Harris's first volume is an entity. What we have is a self-contained musical drama, an oratorio on Handel's model using the best Italian composers, designed specifically for the instrumental and vocal resources available in Salisbury at the time. We may assume that with the excellent Dr Stephens playing the organ, Harris himself would have conducted the oratorio. The leading roles of Adam and Eve would have been well within the musical competence of local talents such as Norris, Corfe, Louisa Harris, and others, but equally, there is the possibility of two star roles for visitors from the London scene. The second volume, by contrast, contains occasional settings of the *Te Deum*, Jubilate, and anthems, a part of the regular ceremonial music in the cathedral, any one of which would have been performed individually at any appropriately festive occasion (orchestral instruments would not have been used at normal services). The recitative composed by Harris himself (p. 20) introducing his setting of Pergolesi's music to 'O Lord save thy People', is scored for a treble voice, a chorus, and organ accompaniment. Other pieces are scored for four voices. The final chorus (pp. 63–80), a Gloria again, is set out for trumpets, kettle drums, violins, viola, oboes, bass strings, and continuo in D major, with alto, tenor, and bass. The instruction to play and sing everything together (*tutti*), with the bass line comprising cellos, bassoons, and organ (left hand) all sounding together makes a particularly rich, even magnificent climax.

With such musical skills, and with a centuries-old tradition of church music on his doorstep, Harris enjoyed the perfect outlet for that combination of aesthetic pleasure and piety which he praised in his mentor Handel, but which was cast into doubt by Johnson's response to his philosophical writings alone. We know that Harris (unlike Johnson) could be 'transported' by music as by no other art form, and his argument for the pre-eminence of poetry and music when combined in opera or religious oratorio (*Three Treatises*) was thus achieved in practical terms. Moreover, Harris's own music (including Milton's 'Hymn' and the pastoral *Daphnis and Amaryllis*) was regularly performed alongside Handel's at the annual musical festival. The metropolitan and the provincial combined in the festival of 1761, for example, with a performance of Handel's last work, the pastoral oratorio *The Triumph of Time and Truth* ('never before performed in the Country', according to the *Salisbury Journal*), Harris's 'Hymn', *Daphnis and Amaryllis*, *Te Deum*, and Jubilate. The principal singers for this Handel–Harris festival were Giusto Tenducci (a Mozart family friend from the King's Theatre), Norris, a Mr Higgins, and Corfe himself, whose obligations to Harris are commemorated in the preface to his edition of Harris's *Sacred Music*. The instrumentalists on this occasion were supplemented with 'the best Performers from Bath and Oxford'.

The *Salisbury Journal* announced that the annual musical festival for 22–4 September 1784 would contain 'the most striking and finest pieces selected from the late commemoration of Handel' and 'a Jubilate and Anthem compiled from music of the most eminent composers by the late James Harris'. Thus Harris's 'local' talent lived on in the music of distinguished company, and for many years after his death his arrangements continued to enjoy both performance and prominence. In 1792 the two-day festival opened with Handel's Overture to the *Occasional Oratorio* followed by Harris's arrangement of the *Te Deum*, and then on the second day came Harris's adaptation of the Jubilate. One may be reasonably sure that it was Harris who enabled J. C. Bach's harpsichord recital at a Miscellaneous Concert (possibly in Harris's own home, or in his Chapel Room above St Ann's Gate) on the evening of 8 October 1773, but only probability supports the claim that Harris was the source for the 'M.S. Serenatas of Mr Bach' performed at a similar event after his death in 1782.

Harris's detailed acquaintance with all the leading Italian

singers of the day in London was an enormous asset to the Salisbury musical festival. Between 1750 and 1780 Salisbury regularly echoed to the sound of Italian voices. Many of the brightest stars of the operatic stage, singers, instrumentalists, composers, and managers, appeared in the cathedral or in the New Assembly Room. Among the singers there were the soprano trained by Handel and Burney, Giulia Frasi (performing each year in Salisbury from 1758–61 and again in 1765); the contralto castrato Gaetano Guadagni (1750 season) from the Haymarket (a pupil of Handel and Burney, tutored in acting by Garrick, and the singer for whom Handel created the role of Didimus in *Theodora*); the soprano Christina Passerini from the King's Theatre (who appeared at Salisbury with her violinist husband Giuseppe in 1754 and 1755); the castrato soprano Giusto Tenducci for the seasons of 1762, 1763, 1764, and 1770; the soprano Cecilia Grassi (the wife of J. C. Bach, also from the King's Theatre and regular performer in the Bach–Abel concerts from 1773–6) in 1773; the male soprano, composer, and harpsichordist Venanzio Rauzzini (King's Theatre) in 1777; the Czech-born violinist and composer Anton Kammell whose first London appearance was at the Bach–Abel concert in Almack's Rooms, 6 May 1768, and who then appeared in Salisbury every season from 1771 to 1776; and less well-known figures such as Andrea Pacini, Clara Polone, Tasca, Giuseppe Giustinelli (1768), and the Venetian tenor Gabriel Piozzi, Burney's friend and eventually the husband of Mrs Thrale (in 1780).

Among the musicians, to make an unreal distinction between singers and instrumentalists—Rauzzini, for example, played the pianoforte as well as he sang—there was the violinist-composer Gaetano Pugnani in 1769; the Piedmontese virtuoso violinist and opera-composer Felice de Giardini (who managed the Italian Opera at the King's Theatre for about forty years, and whom Burney described as 'the greatest performer in Europe'), in 1769; Giaccobbe and James Cervetto, father and son, both cellists, the latter, in Burney's word, 'matchless', in 1768; and in 1780 Johann Salomon, the German-born violinist, one of Europe's greatest musicians and for whom Haydn named his twelve 'London' symphonies. Among the home-grown talents were Dr John Stephens, Dr

William Hayes, and Harris himself as conductors, outstanding choristers from the cathedral such as Thomas Norris (who composed an anthem for Linley and Tenducci to sing on 5 October 1770, and who later became the organist of Christ Church, Oxford), and James and Joseph Corfe (the latter a lay vicar and organist from 1796–1804 until succeeded by his son Arthur Thomas, a pupil of Clementi). Other English performers included Dr Thomas Hayes, Jane Mary Guest (a child prodigy as a pianist, one of J. C. Bach's last pupils, a composer, and later Princess Charlotte's music teacher), and the soprano Elizabeth Linley (who eloped with and later married Richard Brinsley Sheridan in 1772), in the 1769 and 1770 festivals, and Cecilia Davies ('L'Inglesina', who had studied in Vienna and became the first English woman to be accepted as a prima donna in Italian opera houses), in the 1774 and 1775 festivals, and the organist-composer Thomas Dupuis of London's Chapel Royal, one of the directors of the Handel Commemoration of 1784.

One measure of Salisbury's Handelian fame is Thomas Gray's letter to Norton Nicholls (24 June 1769) promising that he will soon be able to hear Gray's 'Installation Ode' sung by 'Mr Norris, & Mr Clarke, the Clergyman and Mr [Thomas] Reinholt, and Miss Thomas, great names at Salisbury & Gloster musick-meeting, and well-versed in *Judas Maccabaeus*'.[15] For a first-hand experience of the frenetic musical, domestic, and social activity generated by and around the Harris's home in the Close, however, Elizabeth Harris's report written the day after the end of the 1770 festival (3–5 October), to her operatically indifferent son in Madrid, merits extensive quotation. Rehearsals had taken place in their own home, as well as in the room in the annexe over St Ann's Gate:

Our festival is just ended, and we have brought things to a happy conclusion. I never remember so much good company, or a more numerous appearance. We had a rehearsal here Tuesday evening, and a most crowded room. . . . Wednesday morning we rehearsed [Jomelli's] Passione, Pergolesi [Stabat Mater], &c. here, to another crowded audience. That day all the Professors dined with us, we ladies were obliged to assist at the dinner, as Madame Tenducci and

---

[15] Gray's letter to Nicholls, 24 June 1769.

Mrs. and Miss Lindley were of the party, so, from the heat of the breakfast-room we got into a far greater heat in the dining-room; it was literally out of the frying-pan into the fire. The music began here that morning at ten, and never ceased till three. We eat, drank, dressed, and went to the Oratorio of Hercules, which went off charmingly. Tenducci is amazingly improved; in his part the old Handelian songs were left out, and some fine Italian ones smuggled in, in their places. Thursday Lady Pembroke called on us to go to the Messiah, which went off divinely . . . Lord Pembroke and the Mr. Herberts were obliged to assist at the Mayor's feast at Wilton, they very humanely let your father off. In the evening we had some quartetts till after eight, and then adjourned to the card assembly. Yesterday we attended the music in the church, and had only gentlemen at dinner . . . The music in the evening was truly fine, the first act part of the Passione, the second the Stabat Mater, in which Miss Lindley and Tenducci sang like two divine beings, the third act miscellaneous. Your father set out this morning for Heron Court . . .[16]

Once established as a key provincial centre, of course, Salisbury attracted artists almost automatically. As early as October 1764, for example, Tenducci was asking Mrs Harris to use her influence to postpone the Salisbury annual musical festival to September of that year so that he could fit it in on his return from the Continent and after his London season. Presumably, she exerted such influence through her husband, although there is no reason to suppose that her almost constant attendance at operas and concerts in London meant that she was any less well informed or less influential on such matters than James.

III

After his election to parliament in 1761 Harris regularly transported his whole family to their London residence, first in Pall Mall, then in White Hall, St James's Street, and Charles Street, usually in the second or third week of January, where they stayed until the end of the parliamentary session, returning to Salisbury generally between the end of May and mid-June. Music and politics were always unequal competitors for Harris's interest, his passion for the former easily

[16] *Series of Letters,* i. 204–5.

7. St Ann's Gate, Salisbury: the 'Chapel Room', east wall of the Close.

8. Pen and watercolour sketch of the Harris home in the Close, Salisbury:
from Katherine Gertrude Harris's Day-book.

outweighing scepticism of the latter. In a letter from Salisbury, 27 November 1766, to his fellow Grenvillite and MP for Oxford University, Sir Roger Newdigate, Harris speaks of laughing at men 'calling themselves Whigs, avowing even in Whig days Principles, that every rational Tory would in *any* days have been ashamed of'. Harris had just acquired a house in St James's Street, its great advantage being proximity to Spring Gardens, and in this same letter to Newdigate he gives a revealing self-portrait of his musical single-mindedness:

Just as I was getting into my Chariot with my laced Coat on &c &c a hefty Walker in the Streets ran foul of me, and (to ye mortification of human Vanity) down we tumbled both together into the dirt. However this might affect my Pride, it did not abate my Love of Music. Up I mounted undrest into ye Gallery, and heard Gaulducci with admiration. Such chaste, elegant & simple singing I have hardly heard before. The Tenor too is meritorious, & Savoya improved, since last year. For the Women—alas! neither Voice for the Connoisseurs, nor Beauty for the Gay & young.[17]

The London household account books kept by Mrs Harris from 1764 to 1780 provide much information about the family's social and musical activities, and record their earliest contacts with musical celebrities later to appear at the Salisbury festivals. The children's education was placed in the hands of tutors: Mr Planta was paid at the end of December 1764 to instruct all three Harris children in the Italian language and, before matriculating at Oxford, the young James in French (Mrs Harris herself being already fluent in the latter); Mr Thomas was hired as a writing master; and Tedeschini was hired to teach the pianoforte to Louisa. There are, as we might expect, payments to their opera box keeper, records of payments for exhibitions (sometimes in the company of the family of Richard Owen Cambridge, whose home at Twickenham the Harrises knew well), and concerts,

---

[17] Sir Roger Newdigate Papers, Warwickshire County Record Office, CR136/B1703–6 (1705). I owe my knowledge of these 4 Harris letters to the kindness of Dr Fred Ribble, of Charlottesville, Virginia. For the political allegiances of Harris and Newdigate, see Lawson, *George Grenville* pp. 222, 257, 276–7. Sir Roger and Lady Newdigate had been 'early Protectors of my Son', Harris wrote, and when the future first Earl of Malmesbury was appointed secretary to the embassy at Madrid (Aug. 1768), he wrote to acknowledge the 'early Conversation with People of Fashion & Worth, such as his good Friends at Arbury' (CR136/B1706).

such as Manzolini's benefit on 5 March 1765, and Tenducci's benefit of 2 April.

As a devotee of the performing arts, Harris could hardly have chosen a better time to move to London. Musically speaking, it was a moment of brilliance and transition. Harris's idol Handel had died at the peak of his fame, on 14 April 1759, commemorating his friendship with the family by leaving Thomas Harris £300 in his will. James, of course, had closely followed every turn of Handel's career (and health), either in person as a concert-goer or as an onlooker corresponding with those inveterate Handelians the Earl and Countess of Shaftesbury throughout the 1730s and 1740s, and thereafter by sponsoring performances of Handel's music in the annual festival in Salisbury. His eulogy of Handel in *Three Treatises* has already been mentioned. In addition, as when his close friend Henry Fielding died, Handel's death called forth an act of commemoration. Harris compiled something for which he was peculiarly suited, a 9-page catalogue of Handel's works for the first biography of a musician in any language, *Memoirs of the Life and Writings of George Frederic Handel* (1760), by John Mainwaring. Although Harris's contribution is anonymous, the catalogue bears the imprint of Harris's taxonomic style, and the essay which followed (attributed to Richard Price) pointedly cites Harris's *Three Treatises* for its expert and judicious commentary on the imitative power of music. Harris classifies Handel's works into three genres (Church, theatrical, and chamber music), and then into ten 'lesser classes' (anthems, oratorios, operas, concertos, sonatas, lessons for the harpsichord, chamber duettos, terzettos, cantatas and pastoral pieces, and occasional or festal pieces). He then lists the operas with place and date of first performance, sometimes with the librettist's name, and very probably recalled his own attendance at many of them.

Johann Christian Bach had come to London in 1762, and was soon to establish a concert partnership with C. F. Abel. In particular, Bach was to delight all those who attended Vauxhall pleasure gardens, where the illuminations turned night into day. Its imitation Greek temples and statuary, alcoves set with supper tables, its Grove, Grand Walk, Cross Walk, and Lovers' Walk, were all set off with Hogarth's paintings in the loggias and (to James Harris's undoubted delight) with Roubiliac's magnificent statue of Handel. It was a place to parade, and a place to see London's musical stars. Dr Arne was Vauxhall's official composer, and his wife Cecilia Young was one of its leading performers, together with two equally celebrated castrati, Manzuoli (Mozart's singing teacher) and Bach's intimate friend Tenducci. All three singers were soon to appear in Salisbury. Equally astonishing was Ranelagh, west of the Royal Hospital, Chelsea. Its vast Rotunda could hold 3,000 people in reasonable comfort. This was the heyday for such palaces of the performing arts, and the Harris family were, of course, regular patrons at these and other concert venues.

On 9 April 1764 Leopold Mozart moved his family from Paris and arrived fourteen days later at their first London lodgings, the house of Mr John Couzin, haircutter, in Cecil Court, off St Martin's Lane. After Nannerl and the 9-year-old Wolfgang had played twice at court, their first public concert was given on 5 June. During their fourteen months stay in England, the price of hearing the Mozart children perform was to drop from half a guinea to 5s., and then, at the Swan and Harp Tavern in Cornhill, to a half-crown. Thus, on 30 April 1765, Mrs Harris paid £1 for tickets to hear her favourite Tenducci, and a half-guinea for two tickets to the 'Mozart Concert'. Unfortunately, there is no record of the family's response to this latter concert, nor of its programme; but one might speculate that it could have been a repeat performance of Nannerl and Wolfgang's benefit concert of 21 February at the Little Theatre, Haymarket (where the K.16 and K.19 symphonies were performed). Equally, it could have been a performance of the later programme of 13 May at Hickford's Guest Room in Brewer Street, which included a 'Concerto on the Harpsichord by the little Composer and his sister, each single and both together' (possibly the virtuoso Sonata for keyboard duet, K.19d). Unfortunately, this is the only mention of Mozart in the Harris papers.

The subsequent surviving section of the Harris London account book relates to the period January 1769 to 14 January 1780, and is sketchy. The firm of Jacob Kirkman and Rock continued to rent and tune the Harris's harpsichord (the former famous for the rich and powerful tone of his instruments); there were more trips to Ranelagh, theatre and opera visits, including Bach's concert on 26 February 1771, and on 9 June 1777 there was a very large payment of £105 for a new coach.[18] While Mrs Harris and her daughters were enjoying the opera season, particularly the subscription concerts of Almack, Mrs Cornelys at Carlisle House, and those of Bach, Cocchi, and Abel, often with their Salisbury friends, James himself, now a Lord of the Treasury, was closeted in the Commons, or dining with Grenville, First Lord of the Treasury. On 17 March 1764, for example, Mrs Harris wrote to her son in Oxford: 'Your father dined in Downing Street after the debate [on the plans for building London Bridge]; they are all in very high spirits. I was at the oratorio of "Nabal" last night with Lord and Lady Shaftesbury. The words are Dr. Morell's; the music taken from Handel's old operas and oratorios—charming fine things, but played stupidly.'

Mrs Harris persuaded both Manzolini and Tenducci to sing for their breakfast at the Harris's house in Whitehall on 20 October 1764, and when Manzolini's clothes and finery were seized by the Customs officers it was James Harris who interceded for him with Grenville. Almost certainly, his cousin Hooper, as Commissioner of Customs, would have arranged the restitution.

Seven months of each year were spent largely in Salisbury, which could offer neither the range nor the frequency of professional entertainment as the metropolis. Certainly, the Harrises attended the Salisbury professional theatres.[19] There was the small converted Assembly Room at the Vine Inn on the corner of Market Place, and the regular winter base, from 1766 to 1786, of Samuel Johnson's 'Salisbury Company of Comedians' at the Sun Inn by Fisherton Bridge (the latter opened in January 1765), as well as Collins's theatre opened in New Street in 1777. Mrs Harris, in common with other members of the local gentry, occasionally sponsored plays: the

[18] 'Household Account Book [kept by Mrs Harris] of James Harris 1764–80', HRO 27M 56/1.

[19] See A. Hare, *The Georgian Theatre in Wessex* (1958), 59–119, 120–40 (on the amateur companies).

1767 performance of *The Suspicious Husband*, the 1768 performance of *The Conscious Lovers*, and the 1770 performance of *Much
Ado about Nothing*, by the Johnson–Collins–Davies companies.

But the Harrises were also able to supplement the local
theatrical diet with material and talent from their own
amateur, if sometimes unexpectedly eminent, resources. Once
more, the records are scanty, but from the late 1760s the Hon.
Stephen and Lady Mary Fox ran a series of amateur
productions, possibly in a converted barn, at nearby
Winterslow House: Dryden's *All for Love* in August 1768,
Aaron Hill's adaptation from Voltaire, *Zara*, in October of
that year, the theatre being filled each night with 'the Nobility
and Gentry in the neighbourhood, the City and Close of
Salisbury', according to the *Salisbury Journal*, and Hughes's
The Siege of Damascus and Mrs Centlivre's *The Wonder* in
January 1769. In November 1770 David Garrick was in the
audience to watch *The Distressed Mother*, and Mrs Harris's
letter on the occasion implies that he had acted previously at
Winterslow and would do so again, in *The Clandestine Marriage*.
The Harrises were spectators rather than performers in
several seasons of these Winterslow productions. Winterslow
House burned down on 9 January 1774, however, and the
days of aristocratic barnstorming at the Fox's estate were
over.[20]

The Harris women had formed their own theatrical troupe
by 1770 and the Winterslow disaster removed their chief
alternative resource. They responded by increasing both the
quality and frequency of their performances. They produced
and acted *Elvira* and *Florizel and Perdita* in June 1774, and one
of their would-be recruits was a member of the now defunct
Winterslow troupe, the Earl of Pembroke. He wrote to James
Harris in November offering to 'snuff candles . . . scrape on
the violincello . . . I will perform gratis. P.S. My wife says she
can thrum the harpsichord or viol-de-gamba.'[21] *The Earl of
Warwick* and *Cymon and Sylvia* were performed in the second

---

[20] Confirmation (not found by Hare) that there was a theatre at Winterslow is
provided by the *Salisbury Journal* of 10 July 1775, which announced the sale of
household furniture and effects of Winterslow House on 19 July: 'N.B. In the above
sale will be sold, the complete Habits, Dresses, and Decorations of a play-house.' For
a reconstruction of the barn theatre, see Hare, *The Georgian Theatre in Wessex*, p. 129.

[21] Cited by Mrs Harris, *Series of Letters*, i. 283–4.

9. James Harris in 1769: engraving by William Evans.

and third weeks of October 1776, using home-made manuscript playbills and their own prologues and epilogues, some of which were corrected by James Harris himself. For a detailed account of such productions we must turn to the earlier performance in January 1770 of the poet laureate Whitehead's *Creusa, Queen of Athens*, a marmoreal adaptation of Euripides' *Ion*, followed by *Daphnis and Amaryllis*. This was perhaps the model for subsequent plays, and it is clear that great efforts went into such productions. Gertrude's 'antique' costume was designed by James Harris and Joseph Warton, the headmaster of Winchester, editor of Virgil, and Pope's biographer. Louisa played Thyrsus, Miss Wyndham took the title-role, and the painted scenery included the temple of Delphi and a laurel grove. As usual, the theatre was over St Ann's Gate, and Mrs Harris provided a detailed account of the rehearsal to her son:

The Chapel room makes a good theatre, the stage is near three feet high, there is room for between forty and fifty spectators, giving a good space for the orchestra, which consists in a proper band for the Pastorale of 'Daphnis and Amaryllis;' I have heard them rehearse, and I must say it was infinitely better performed than at Drury Lane. The dancing chorus is delightful, and the dance is all their own composition. Dr Stevens leads the orchestra; he is highly pleased. To fill up the choruses properly we are obliged to take two small choristers, and they make pretty shepherds: Parry sits in the orchestra, and sings the bass part in the chorus. Gertrude ventures to sing a song, which she does very sweetly and in tune, but rather too softly; Louisa says, she sings like a piping bull-finch.[22]

The celebrity of the Harris productions, particularly the questionable propriety of young women playing male roles, reached the *Bath Journal*, which printed a stuffy protest-poem 'On the ladies of the Close of Salisbury, now acting "Elvira" ' (17 November 1774):

> In good Queen Elizabeth's reign,
>   In a decent and virtuous age,
> That they ne'er might give modesty pain,
>   No female appeared on the stage.

[22] Ibid. 184–5.

But lo, what a change time affords!
The ladies, 'mong many strange things,
Call for helmets, for breeches, and swords,
And act Senators, Heroes, and Kings.[23]

Such theatricals alleviated the rural ennui of Mrs Harris in particular. She rapidly tired of the family country residence at Great Durnford when there was no company and the conversation 'chiefly of grass and dogs', although sometimes the rest afforded was anodyne: 'This is a place of no events, we enjoy ourselves and the fine weather, we eat, drink, sleep, read, work, walk, ride, laugh, play cards, and grumble every day, and we are so satisfied with the uniformity of our living that we have fixed no time for quitting it' (7 September 1771). When the family was thrown back on itself to provide musical entertainment, the 22-year-old Louisa could impress with her skill in Italian recitative and on the harp, a recent version of the single-action pedal harp invented around 1720. On 31 March 1775 Mrs Harris claimed that 'the harp is so new in England . . . Louisa is the only lady who performs on it'.[24]

The return to the London season was, therefore, eagerly awaited each year. When not attending Mrs Cornelys's soirées, or masquerading as a domino, she and James, it seems, rarely managed to attend concerts together. He was, after all, fully tied up with parliamentary business, and had to find time to research and write his third book, *Philosophical Arrangements*, in 1771, and another ambitious study of medieval literature, *Philological Inquiries*, which remained unpublished at his death. But he could occasionally sneak away to particularly significant concerts, as when he and Gertrude attended Bach's opening concert in his spectacular new Hanover Street concert room, lit by transparent paintings by Benjamin West, Gainsborough, and Cipriani during the first week of February 1775. The whole family seemed to enjoy an open house at the Cambridge's villa at Twickenham, which Mrs Harris described as 'the seat of the Muses, who are to me

---

[25] *Series of Letters*, i. 285. Mrs Harris enclosed a copy to her son in Berlin adding: 'You may easily imagine that the following verses were sent from some vinegar merchant in Salisbury, who could not get admitted to the performance.' For other poems and 2 reviews of the Harris theatricals, see *Salisbury Journal*, 27 Mar. 1775, and Hare, *The Georgian Theatre in Wessex*, pp. 124–8.   [24] *Series of Letters*, i. 299.

more edifying than the politicians', on 5 April 1771. On at least one occasion, Salisbury musicians were in town. At a performance of one of Handel's oratorios at the Haymarket (9 March 1775) Mrs Harris singled out 'my countrymen Corfe and Parry' as the better performers, and went on to describe a rehearsal of Sacchini's *Miserere* in their own Piccadilly house, the singers being Rauzzini, Savoye, Pasini, and Louisa.

Music continued to sustain Harris until a few months before his death. Thomas Twining is the only assistant Charles Burney acknowledged in the published *General History of Music* (1776–89: i, p. xix), but Harris also helped him, not only in his mammoth research in the university libraries in Cambridge, but also by turning over his own private music library in Salisbury. In February 1779 Burney thanked Harris for letting him see 'the Musical MSS wch. I consulted in your Curious Liby. wch. I cannot help wishing were more particularly described in your Liby. Catal., & assigned to the true Authors'. Burney then sent Harris the results of his research into the particular tenth-century manuscript (a copy of which Harris had made from Benet College library, Cambridge). Burney must have copied most of his letter to Harris from his manuscript of the *General History*, since the latter is virtually a verbatim printing of the former.[25] A fragmentary letter of 15 March 1779 suggests that Harris also read parts of Burney's *History* in manuscript. As we might expect, the international, national, and provincial musical establishment was well represented in Burney's subscription list. The names include C. P. E. Bach, Joseph Corfe, Diderot, James Harris, Adolpho Hasse, Professor William Hayes, the Kirkmans, Metastasio, Rousseau, and figures less known for their musical interest, such as Thomas Birch, Garrick, Samuel Johnson, Reynolds, Henry Thrale, Horace Walpole, and Joseph Warton. Harris's *Three Treatises* is cited twice (i. 498–9, and in Burney's chronological appendix of contemporary musical works). Harris wrote to Burney on 14 October 1779 to

[25] Burney to Harris (Feb. 1779), Beinecke Rare Book and Manuscript Library. Burney here thanks Harris for 'the many civilities and kind offices wth. wch. you honoured me during my late visit to Cambridge', and discusses the attributions in J. Nasmith's *Catalogue* of the music manuscripts given by Archbishop Parker to Corpus Christi library (1777). For the published version of this letter, see Burney's *General History of Music*, ed. F. Mercer, 2 vols. (1935), i. 492–3, 546.

thank him for a presentation copy of his essay on the infant musical prodigy William Crotch. This had been written for the *Philosophical Transactions of the Royal Society* (69: 1, 1779), and it also appeared in the *Annual Register*, where, in Harris's opinion, it had been 'unfortunately buried in rubbish'. Of much greater importance to Burney's *magnum opus* were his friend's distinctions between the sister arts in *Three Treatises*. These were recalled and developed in the *General History of Music*, where Burney described his own endeavour to 'point out the boundaries of music, and its influence on our passions; its early subservience to poetry, its setting up a separate interest, and afterwards aiming at independence' (i, p. xvii). Harris's treatise on music published back in 1744 had provided a milestone in this process, a major theoretical breakthrough in liberating music from a slavishly mimetic function.

Thirty years after Harris's death in 1780, Charles Burney and his daughter Fanny were each to recall the Harris family's practical musical contributions to social gatherings. In her *Memoirs of Doctor Burney*, begun around 1807 but not published until 1832, Fanny reconstructs the occasion when she was fortunate enough to sit next to James Harris ('whose soul seems all music', she observed), enjoy his conversation, and then listen to Louisa, 'a distinguished lady musician', singing an unpublished air by Sacchini, accompanied by her father on the piano, followed by a rondeau from Rauzzini's opera *Piramis and Thisbe*: 'She has very little voice, either for sound or compass; yet, which is wonderful, she gave us all extreme pleasure; for she sings in so high a style, with such pure taste, such native feeling, and such acquired knowledge of music, that there is not one fine voice in a hundred I could listen to with equal satisfaction . . . She is extemely unaffected and agreeable.'[26]

During the last five years of his life Charles Burney kept up an occasional correspondence with Louisa Harris, who was then in her late fifties. Signing himself, in 1807, 'an old broken down Chelsea Pensioner' with a failing memory, he nevertheless assured her that 'there is nothing it so well retains as that time wch I had the honour to spend in your revered

<hr>

[26] Mme D'Arblay, *Memoirs of Doctor Burney*, ii. 13, 16–17.

family and the happiness of meeting you at my dear Ld. Clarges, where Sacchini & Rauzzini were of the party, & where I cd. hear Duranti's Duets from Cantatas of Aless. Scarlatti, of wch. I am in possession of the original MS. in his own autograph.'[27] In particular, Burney remembered one anecdote, although he seems to have conflated the musical talents of Louisa and her older sister Gertrude. The latter had sent Burney some specimens of Russian musical compositions during her visit to Russia (August 1777 to September 1779), where she had been staying with her brother, the British envoy at the court of St Petersburg from 1777 to 1783.[28] It was Louisa who had been the better singer and more accomplished instrumentalist on the harp and the harpsichord. Burney had confused the two sisters, and also forgotten which instrument it was that Louisa had actually sent down to the Twickenham home of Richard Owen Cambridge:

my old story, wch does not seem to quadrate with what you remember much more accurately than myself—yet I am unwilling to allow the whole of my narrative to be unsound & apocryphal—I think that Time & an infirm memory have in my anecdotic tale jumbled 2 Harris together—If you, dear Madam, never sent a P[iano] F[orte] to Twickenham—you sent a *Harp* thither—& that wd do as well for my story as a Harp[sichor]d—which is only a horizontal Harp.—As to Mr C[ambridge] come down to desire a L[ad]y to continue who was at the Harp[sichor]d when he came into the room—& who got up in a fright, knowing what an *amousos* he was [person lacking a sense of harmony], an acquaintance of mine then in the same room, heard him say to the terrified female musician—'go on, go on—I don't mind it'—and it used to be a jocose anecdote with the musical part of my family—yet Mr. C was always highly respected at my House, as a personage of wit, learning, & genius extremely remote from the common road.[29]

At that time, during the first decade of the nineteenth century, and a generation after the golden years of the

[27] Draft ALS, Burney to Miss Louisa M. Harris [1807?] from Chelsea College; Beinecke Rare Book and Manuscript Library.

[28] For the definitive account of the 4th James Harris's diplomatic career during this period, see I. De Madariaga, *Britain, Russia, and the Armed Neutrality of 1780* (New Haven, 1962).

[29] Draft ALS, Burney to Louisa M. Harris [15 Feb. 1809 or later] from Chelsea College; Beinecke Rare Book and Manuscript Library.

Salisbury festival under Harris's direction, Louisa was living alone at Great Durnford, 6 miles north of Salisbury. She had never married, but the continuing reputation of her father deflected some casual interest towards her on the part of some visitors to the cathedral city. Some time after 1801 a visitor copied down William Batt's Latin inscription on James Harris's monument in the cathedral, recorded the local opinion that Bacon's medallion of Harris was a good likeness, and then passed by the Manor House at Great Durnford. He or she annotated a copy of James Easton's *Salisbury Guide* as follows: 'I passed it at a little distance, its appearance is respectable and situation delightful, surrounded by the sweet stream of Avon, and thick plantations. This lady, having experienced a slight from her love (Mr Bowles) devotes herself to celibacy and good works—Her benevolence and charity have endeared her not only to those who partake of her bounty, but to all who can appreciate such virtue.'[30]

[30] Manuscript annotation in a copy of the 22nd edn. of J. Easton, *The Salisbury Guide*, (Salisbury 1801), 25; Library of Congress.

8

The Archaeology of Knowledge and Modern Memory

There is a very remarkable inclination in human nature, to bestow on external objects the same emotions, which it observes in itself; and to find every where those ideas, which are most present to it. This inclination, 'tis true, is suppress'd by a little reflection, and only takes place in children, poets, and the antient philosophers.

D. Hume, *A Treatise of Human Nature* (1739)

Every man is born an Aristotelian, or a Platonist . . . They are the two classes of men, besides which it is impossible to conceive a third.

S. T. Coleridge, *Specimens of the Table Talk* (1830)

HARRIS's penultimate book, *Philosophical Arrangements* (1775), reveals more clearly than any other that the intellectual foundations of his whole career as a writer lie in a series of dialectical relationships between the past and the present, and between two broadly opposed ways of thinking. In each of his books he sustained an unbudging rearguard action in defence of a priori Idealism in an age which had sold out to an inductive scientific method. The impression he has left in the minds of many readers is that Idealism can only be reconstructed through literary archaeology, that it does not exist in modern philosophy, and that its permanent well-spring may only be found, in its pure state, in the culture of ancient Greece. As Harris wrote: 'I have drawn all my Sentiments from ye pure Sources of Classical Antiquity. To approach these Sources Litterature is as necessary, as the Golden Bough to the approaching of Elysium' (to Monboddo, 15 April 1775).[1] This reverential tone, together with a discursive method heavily dependent on quotations from authorities, might leave the impression that his whole output as a writer is an eccentric and wilful indulgence in anti-

---

[1] Harris Papers, vol. 40, part 1, 'Miscellaneous Correspondence of Mr Harris', a series of 11 letters from Monboddo dating 8 Dec. 1768–23 May 1779: hereafter referred to by date only.

modernism, a nostalgic retreat into scholarly antiquity. But Harris was perfectly well adjusted to his own times, and his books are his own achievements, not synopses of Aristotle. Moreover, there was already a distinct English tradition of Idealism of which he was well aware and to which he had access. His mission, after all, was neither to bring forward the past, nor to roll back the present, but to synthesize both. If the modern memory was shrinking into itself, the priority was to see the achievements of the present in the enriching context of history. The last thing he desired was that the present should be regarded as a derivative assembly of classical parts. That is why Harris wrote in English and not in the scholar's dialect of Greek or Latin. His learning was put to the service of those whose horizons had been limited by the smoke of their own chimneys, and in his last book, *Philological Inquiries* (1781), he was to commend those of his contemporaries, not a few of whom had been his close personal friends, who had shared his mission to inscribe English mid-eighteenth-century culture in the context of a continuous classical tradition. As we have seen, *Hermes* goes much further than the application of logic to linguistic categories: it treats *general grammar*, in words used by Foucault in another context, as 'the spontaneous form of science—a kind of logic not controlled by the mind—metaphysics'.[2] Harris's next book turned once more to the metaphysical causes of thought, but this time above and beyond the level of the sentence.

I

Philosophical Arrangements is a reactive rather than a reactionary book: its reconstitution of classical philosophy is written very precisely for the consideration of the contemporary world, and in this aim he recognized one staunch ally and pupil, his Scottish friend James Burnet, Lord Monboddo. Monboddo's *Of the Origin and Progress of Language* (1773–92) and *Ancient Metaphysics: Or the Science of Universals* (1779–99) constantly interrelate to Harris's *Hermes* and *Philosophical Arrangements* (1777). Both men knew that what confronted them was more

---

[2] M. Foucault, *The Order of Things: An Archaeology of the Human Sciences* (Paris, 1966; New York, 1973), 83.

than a choice between the ancient and the modern world. What was at stake, particularly for Harris, was the proper philosophical method. The moderns may have conquered the world of matter, but they had erred grievously in assuming that everything valuable and knowable in the world was determined and explicable by a mechanistic philosophy.[3] In assuming that the modern world was the creation of modern men, they found themselves on a cooling, utterly material, and therefore hostile planet.

Thus George Campbell, in *The Philosophy of Rhetoric* (1776), argued that the discredited but nevertheless still widely used method of traditional scholastic dialectic had failed to produce anything but 'an artificial and ostentatious parade of learning, calculated for giving the appearance of great profundity to what in fact is very shallow'. Moreover, in the area of moral philosophy the method of arguing by syllogism was simply obfuscatory and irrelevant:

this method of arguing has not the least affinity to moral reasoning, the procedure in the one being the very reverse of that employed in the other. In moral reasoning we proceed by analysis, and ascend from particulars to universals; in syllogizing, we proceed by synthesis, and descend from universals to particulars . . . [the latter] more resembles mathematical demonstration, wherein, from universal principles, called axioms, we deduce many truths, which, though general in their nature, may, when compared with those first principles, be justly styled particular, whereas in all kinds of knowledge, wherein experience is our only guide, we can proceed to general truths, solely by an induction of particulars.[4]

Having fallen 'greedily' upon his presentation copy of the second edition of Harris's *Hermes*, especially its discussion of

---

[3] Harris to Monboddo, 14 May 1774: 'Mind and soul are terms acknowledged by our language, and most other modern languages have terms analogous. But, alas, tho' they pass very well in common conversation, our philosophers are well satisfied with Body and its attributes. These last are abundantly sufficient for their purposes, and solve every part of *their* systems to their intire content', *Sixth Report of the Historical Manuscripts Commission*, part 1 (1877), 676. Monboddo agreed: 'the Ancients not only held *Mind* to be the first cause of all things, but the *immediate* cause of the chief operations in Nature. I used the word Mind in a large sense, so as to comprehend not only *Intelligence*, but *Vitality*, and whatever other principle there is in Nature that produces *Motion*' (Knight, *Lord Monboddo and Some of his Contemporaries*, p. 93: 21 June 1776).

[4] G. Campbell, *The Philosophy of Rhetoric* (Edinburgh, 1776), i. 165–6 (ch. 6).

General Ideas in Book III, Monboddo wrote to its author on
26 March 1766. He urged Harris to move on from grammar to
'something upon Logic, to show an ignorant age that the
greatest discovery in science ever made by any one man is the
discovery of the syllogism by Aristotle'.[5] Although Monboddo
may have first suggested the idea of Harris's next book, the
fact is that Harris's final volume is a natural and inevitable
continuation of *Hermes*, and was conceived by Harris himself
as exactly that. *Philosophical Arrangements* is usually referred to
in their correspondence as a book on the 'Categories', but only
5 of its 485 pages are specifically devoted to the syllogism. Its
larger purpose was to examine the operations of the mind,
whose active, infinitely copious, and prior energy could be
demonstrated by means of the Aristotelian method of classification, and thereby illustrate the universal validity of Reason
itself. His previous book had looked at universal grammar;
this one was to look at the way the operations of the mind
shaped the syntax of the world.

Further discussions ('sublime speculations', as Monboddo
called them) took place during Monboddo's visit to London
early in 1769, and for the next ten years their consultations
generated a series of letters describing their mutual progress
as authors. Monboddo set to work on what would become a 6-
volume study of language, repeatedly urging Harris to
complete his 'Categories', tutor him in philosophy, and join
together in the battle against Materialism (a term which for
Monboddo at least included Descartes, Hobbes, Locke, and
the 'Newtonians'). Monboddo shocked Harris only once, by
deducing the rise of man from the orang-outang: but even this
idea, the wounded Monboddo replied, had its authority in
Plato's *Theaetetus* and in Horace. In every other respect they
were of one mind, each bolstering the other's dismay at
Locke's epistemological blunders, and each feeding the other's
addiction to Aristotle. Harris's private letters, unlike his
published books, parade the Lockian subtext beneath *Philosophical Arrangements*. On 14 May 1773 Harris sent this
revealing and unrepentant message to Monboddo:

I freely subscribe to your ideas of Mr Locke. Ignorant of all Ancient
Literature, he had an inclination to spin out everything from his own

[5] Knight, *Lord Monboddo*, p. 50.

brain, as if so stupendous a work as an Analysis of the Human Understanding could be raised by the effort of one unassisted man. Euclid and Archimedes among the ancients, Copernicus, Kepler and Galileo among the moderns, preceded our illustrious Newton. 'Twas thus that Homer and Tasso pointed out the road to Milton. Life is too short, and the labour too immense, for a single man to carry anything to perfection.[6]

In other words, Harris's seething dislike of Locke was that same Baconian *idola tribûs* for which Charles Yorke had scolded him twenty-five years earlier when *Hermes* first appeared.

II

Harris compared the Mind to a library of books with one central classificatory system. This single system enables the mind to be 'furnished, like a good Library, with proper Cells or Apartments' into which we can file 'our Ideas both of Being and its Attributes', and where we can 'look for them again, when we have occasion to call them forth'.[7] The contemporary problem, as Harris saw it, was that ever since the time of Descartes, modern philosophers had relegated the work of their distinguished predecessors to a dusty store-room out of bounds to the general reader. The antidote to the moderns' intellectual imperialism and its consequential cultural amnesia was obvious: a regimen of classical reconstruction designed to expand the modern memory. Just as his own library in Salisbury contained busts of none but classical figures, so his third book contains not the slightest mention of the moderns. We know they were represented on his shelves, among thousands of other books, but we look in vain for the names of Hobbes (the philosopher of Malmesbury is noted only once, in *Three Treatises*), or John Locke (whose works Harris had pored over), or of Hume (whom he had met several times, and whose *Dialogues concerning Natural Religion* he regarded as 'a wicked Book calculated to do much Mischief').[8] And yet, even

---

[6] *Sixth Report*, p. 676.

[7] *Philosophical Arrangements* (1775), pp. 543–4. All subsequent quotations are from this, the 1st edn., and are given as page references in the text.

[8] Gertrude Harris, 'Memoir'. On 19 Mar. 1769 Harris recorded: 'With David Hume, as we went in his chariot to [Richard Owen] Cambridge's, spent the Night, and returned the day following': Malmesbury Papers, 933B, p. 37.

though their names are subsumed under such generic phrases as 'Modern' writers or 'Vulgar Philosophers', their anti-pathetic presences are everywhere in Harris's book. The 'modern' shelf in *Philosophical Arrangements* is only apparently empty. Ironically, it is the dead philosophers who are given body in Harris's book, whereas the modern materialists are reduced to textual phantoms—rather like a Battle of the Books with only one combatant actually inside the ring. The dead philosophers are revivified; the moderns are left to squabble irrelevantly outside the ring.

Harris evidently assumed that the modern writers whom he personally disliked were all sufficiently well known not to need particular nomination. Indeed they were, but the result of Harris's restraint is curious. Without the taste or perhaps the confidence for an eyeball to eyeball combat with Locke's writings and with those of his influential followers, Harris nevertheless provided a critique of modern philosophy of mind which was oblique and almost completely disguised. He fights Locke's shadow using the substantive words of Aristotle and his commentators, but it is the former which steers his reading of the latter. In this sense, *Philosophical Arrangements* is very much a 'modern' book.

As for the book's declared purpose of explaining Aristotelian logic, Harris knew as well as Thomas Reid that there were literally hundreds of such manuals currently available, and that it would be supererogatory to explain again what was already sufficiently familiar to the educated reader. Such an audience already had their Sanderson, Wallis, Crackanthorp, Aldrich, and Fell (to mention only the better known). Harris's book had a different purpose and a different audience in mind.[9]

[9] The most important Aristotelian logic books published in England were: H. Aldrich, *Artis Logicae Compendium* (1691), which enjoyed 'an astonishing popularity in England for the next hundred and seventy-five years' (W. S. Howell, *Eighteenth-Century British Logic and Rhetoric* (Princeton, 1971), 13); J. Wallis, *Institutio Logicae* (1687); R. Crackanthorp, *Logicae Libri Quinque* (1622); R. Sanderson, *Logicae Artis Compendium* (1615: 9 edns. before 1700, repr. in 1705, 1707, 1741, and 1841). For Pope's high opinion of Aldrich, see J. Spence's *Observations, Anecdotes, and Characters of Books and Men*, ed. J. M. Osborn, 2 vols. (Oxford, 1966), i. 181. Johnson cited and recommended the others in his preface to Dodsley's *The Preceptor* (1748), relegating Aristotle to the last; and see also *Tristram Shandy*, i. 19. W. Risse's *Bibliographica Logica*, 2 vols. (Hildesheim, 1973), lists almost 2,400 books wholly or partly on logic

More generally, this is one of the most remarkable examples of the eighteenth-century urge to systematize knowledge, and in this particular case to systematize the operations of the mind itself. We look for Harris's book quite properly on the same shelf as Locke's *Essay* (1690), Hume's *Treatise of Human Nature* (1739–40) and *Enquiry concerning Human Understanding* (1748), Thomas Reid's *Essays on the Intellectual Powers of Man* (1785), and Dugald Stewart's *Elements of the Philosophy of the Human Mind* (1792–1827). Each is concerned with defining the operation of the human understanding, and each of them in turn relates to a larger context. The context is taxonomy, the purpose of all Harris's writings about causes and principles.

In the fourth chapter of *Hermes* Harris had drawn the reader's attention to that most remarkable use of the binomial categories of genus and species in the botanical works of Linnaeus. Harris possessed *Classes Plantarum*, *Systema Natura*, and *Fundamenta Botanica* (1747), as well as the early volumes of Buffon's no less astonishing work on organic evolution, the *Histoire naturelle* (Paris, 1749–67). In the classificatory works dealing with the fine arts, he received and studied presentation copies of Joshua Reynolds's *Discourses on Art*; he owned Rousseau's *Dictionnaire de musique* (Paris, 1768), Diderot's *Encyclopédie* (1751–77), the Ur-text of Burlington's revival of classical architectural theory, Vitruvius's *De Architectura* (Venice, 1511); and of course he possessed the benchmark texts in science and philosophy—a first edition of Copernicus's *De Revolutionibus Orbium Coelestium*, Descartes's *Principia Philosophiae* (Amsterdam, 1650), Newton's *Principia Mathematica* (Geneva, 1739), a first edition of Hume's *Treatise of Human Nature* (1739), as well as dozens of mathematical texts and dictionaries such as Pierre Bayle's *Dictionnaire historique et critique* (Rotterdam, 1720), Johnson's *Dictionary* (1755), and those of Ainsworth, Skinner, and many more. In the early 1770s Harris's friendship with Monboddo involved lengthy

published in Europe between 1660 and 1800, including anonymous works, works in trans., and reprints. Numerically, a graph would show a steady decline in logic books during this period, but with resurgences in 1675, 1694, 1737, 1748, and with a minimum point (5 books) in 1784. The British pattern closely follows the European pattern. For further information, see R. Blakey, *Historical Sketch of Logic, from the Earliest Times to the Present Day* (London and Edinburgh, 1851), and W. and M. Kneale, *the Development of Logic* (Oxford, 1962).

epistolary discussions about errors in Locke's *Essay*, Leibniz's *Theodicée*, and of course Monboddo's own shocking evolutionary classification of 'men with Tails, Orangoutangs, and the natural State of Man'.

Often misleadingly understood as *only* a 'summary of peripatetic doctrine',[10] and consequently dismissed as an anachronistic defence of Aristotelian notions published at a time when Locke had allegedly won the day, *Philosophical Arrangements* is in fact an eclectic synthesis of Peripatetic logic, idealist philosophy, optimistic moral theory, and literary structuralism, organically and frequently linked back to Harris's previous books on aesthetic theory and universal language. Its prefatory advertisement is in part a disingenuous attempt to forestall criticism by camouflaging its contemporary relevance: it will examine the 'Speculations of ancient and respectable Philosophers'. If *Hermes* had uncovered the universal structures of thought in a theory of verbal discourse at the level of the sentence, the next book, using the same materials and synthetic approach, would now move on to examine the syntax of thought itself as a perception of latent structures in a providential Nature. Harris's notions about the philosophy of mind and the originality of his purpose are once again modestly passed off as the legacy of classical culture. The contentiousness of Harris's argument lay not only in its timing, but also, as we shall see, in its startling erasures.

III

Philosophical Arrangements is not itself a disingenuous book, but it is deceptive. It reads like a serenely untroubled account of universally agreed axioms. In some respects this was its desired effect, since all that Harris initially claims to do is to systematize the principles which underlie 'Natural Logic': thus 'The Vulgar can give Reasons to a certain degree, and can examine after a manner, the reasons given them by others.—And what is this but NATURAL LOGIC? If therefore these Efforts of theirs have an Effect, and nothing happen

---

[10] L. Lipking, *The Ordering of the Arts in Eighteenth-Century England* (Princeton, 1970), 87. To be fair, Lipking devotes ch. 4 largely to *Philological Inquiries*, but this dismissive remark is nevertheless a representative response.

without a Cause, this Effect must of necessity be derived from certain Principles' (p. 9).

But Harris knew perfectly well that he was walking into an epistemological minefield and sensed from the start that his book would cause trouble. One man's common sense is another's heterodoxy, and in returning to the 'first Philosophy', Harris was attempting not a refutation of empiricism, the most recent chapter in the history of ideas, but its relegation to an inferior status and a lower priority in the proper operations of the mind. His objection to it is the classic counter-argument drawn from mathematics: no empirical theory can explain our concepts of things which do not exist in Nature, such as a perfect triangle, but only in the mind. Harris thus addresses his philosophical 'Contemporaries' directly, and with some patrician mockery, in his fifth and sixth chapters on Essential Forms: 'I speak perhaps of Spectres, as shocking to some Philosophers, as those were to Eneas, which he met in his way to Hell . . . Yet we hope to make our peace, by declaring it our opinion, that we by no means think these Forms SELF-EXISTENT; things, *which Matter may slip off*, and fairly leave to themselves' (p. 90). Monboddo saw this attempt to placate Harris's opposition among the materialists as the most valuable aspect of his book, and told him so in a letter of 13 June 1775:

I am particularly pleased with the Explication you have given of two points of the antient Philosophy, which appear to almost all modern Philosophers to be altogether incomprehensible, and no better than mere Chimaeras; I mean *the first matter*, or *matter undressed* . . . and Essential Forms. These you have shewn to be no more incomprehensible than the Lines points or Surfaces of the Geometers. For a thing without parts—Length without Breadth, or both without Depth—are things that have no Actual Existence, but are Creatures of the Intellect.[11]

The names of Descartes, Hobbes, Locke, and Hume are thus erased from Harris's text, but no contemporary reader could miss the point of his careful and deliberate inversion of central materialist doctrines; and in retrospect we can see something that Harris, Monboddo, and their contemporary

[11] Harris Papers, 'Private Correspondence', vol. 31, part 3.

readers could not—Harris's apparently reactionary and minority argument for Idealism could hardly have been more timely. *Philosophical Arrangements* appeared only 6 years before Kant's *Critique of Pure Reason*. Though driven by an admiration for Newton and Hume and resisting extension into Harris's field of speculative metaphysics, Kant was nevertheless to describe twelve 'categories' of concepts known by a priori reason independent of experience. The route from Harris to Germany, and Kant in particular, has been traced by George ten Hoor, who sees each of the foremost pre-Kantian critics, Mendelssohn, Lessing, and Herder, as having 'based . . . their central doctrines . . . on a system of the arts derived from Harris'. For his attack on the *Metakritik* in the *Erstes Waldchen*, Herder made the energy–work distinction first announced in *Three Treatises* 'entirely his own. He not only applies it in his theories of art, but introduces it into the struggle with Kant. It permeates, in short, not only his aesthetic but his epistemology as well.'[12]

We can only speculate as to what Harris's opinion of Kant may have been, and in any case our concerns here are to establish the characteristics of Harris's philosophical argument in his own time. But we might note in passing that his aesthetic theories proved much more amenable to serious attention and further extension in continental Europe than in the anti-theoretical climate of his own country.

What were the doctrines, therefore, that Harris was responding against? In the first and second chapters of *Leviathan* (1651) Hobbes had argued for SENSE as the prior faculty of human perception and Imagination as 'nothing but *decaying sense* . . . an obscuring of it'. In the fourth chapter of the third book of *Hermes*, exactly a century later, Harris reversed this priority: 'SENSE [is] a kind of transient Imagination; and IMAGINATION on the contrary a kind of permanent

---

[12] G. ten Hoor, 'James Harris and the Influence of his Aesthetic Theories in Germany', Ph.D. diss. (Ann Arbor, Mich., 1929), 263–8. He adds that Herder 'stands indebted to Harris for . . . the definition of art as a human activity, the distinction between natural and arbitrary media, the theory of aesthetic senses, the view of the original union of poetry and music, and certain views on language, especially those dealing with time and the tenses of verbs' (p. 266). See also E. M. Szarota, 'James Harris: Die Bedeutung seiner *Three Treatises*', *Wissenschaftliche Zeitschrift der Ernst-Moritz-Arndt-Universität Griefswald*, 12 (1963), 351–61.

Sense . . . REASON and INTELLECT [cannot function] 'till IMAGINATION first fix the *fluency* of SENSE, and thus provide a proper Basis for the support of its higher Energies' (pp. 357–8); 'nothing can be more dissipated, fleeting, and detached' than the 'Perception of SENSE' (p. 365). The idea of sensory decay is replaced by a concept of the Imagination as the *initial* endowment of the mind, whose permanent characteristic is a restorative and predictive capacity. Likewise, Locke's definitive statement of the theory of sensation and reflection (derided by Harris as 'a kind of logical Chemistry'), is silently ignored and replaced by what Harris calls 'the *intellectual* Scheme, which never forgets Deity, postpones every thing corporeal to the *primary mental Cause* . . . the origin of intelligible Ideas' (*Hermes*, p. 393). In this, the most important chapter that Harris wrote, the mind is born with the actual power and the latent capacity to comprehend and analyse a world of providential design. At this point not even Harris can ignore the implications for modern philosophy:

We must now say—*Nil est in* SENSU, *quod non prius fuit in* INTELLECTU. For tho' the contrary may be true with respect to Knowledge merely human, yet never can it be true with respect to Knowledge universally, unless we give Precedence to ATOMS and LIFELESS BODY, making MIND, among other things, to be struck out by a lucky Concourse.

'Tis far from the design of this Treatise, to insinuate that Atheism is the Hypothesis of our later Metaphysicians. But yet 'tis somewhat remarkable, in their several Systems, how readily they admit of the above *Precedence*.

For mark the Order of things, according to *their* account of them. First comes that huge Body, *the sensible World*. Then this and its Attributes beget *sensible Ideas*. Then out of sensible Ideas, by a kind of lopping and pruning, are made *Ideas intelligible, whether specific or general*. Thus, should they admit that MIND was coeval with BODY, yet till BODY gave it IDEAS, and awakened its dormant Powers, it could at best have been nothing more, than a sort of dead Capacity; for INNATE IDEAS it could not possibly have any. (*Hermes*, pp. 391–3)

Similarly, in *Philosophical Arrangements*, Harris again ridicules the Epicurean and Lucretian doctrines of atoms and void, and the idea of a wholly material universe created by chance. Epicureanism was as large and vaguely defined a target as materialism itself, but its local and modern reference was first to Hobbes (*Leviathan* and *De Corpore*, 1655), and then to Locke's sensational theories of knowledge. The latter had been updated and narrowed since the publication of *Hermes* by Locke's French disciple Condillac, whose *Essai sur l'origine des connaissances humaines* (1746 and 1771) had been translated by Thomas Nugent and published by Harris's own bookseller John Nourse in 1756. For Condillac, *all* human faculties originated in the single phenomenon of sensation, and only before the Fall had it been possible for the soul to have 'ideas precedent to the use of its senses'. As fallen beings, 'we have no ideas but what come from the senses'.[13]

Harris did not need to have read Condillac to find such an idea philosophically preposterous and morally degrading, but if he had looked into the work of his exact contemporary in France, La Mettrie, he would have found much worse. La Mettrie's provocative and perhaps not entirely serious *L'Homme machine* (1748) reduced Descartes's dualism of body and soul to a single mechanical theorem: 'The human body is a machine which winds its own springs . . . I believe that thought is so little compatible with organized matter, that it seems to be one of its properties on a par with electricity, the faculty of motion, impenetrability, extension, etc.'[14] Currents of materialist thought similar to those in La Mettrie and Condillac could also have been found in Toland's Deist *Letters to Serena* (1704), in Holbach's *Système de la nature* (1770), and in Helvetius's sensationalist theories in *De l'homme, de ces facultés, et de son éducation* (1772).[15] The real purpose of *Philosophical Arrangements*, therefore, was threefold: to continue the argument for the priority of mind rather than sense already developed in *Three Treatises* and *Hermes*; to demonstrate through the Aristotelian doctrine of categories that a *method*

---

[13] T. Nugent (trans.), *An Essay on the Origin of Human Knowledge, being a Supplement to Mr. Locke's Essay on Human Understanding. Translated from the French of the Abbé de Condillac* (1756), 17.

[14] J. O. de La Mettrie, *Man a Machine*, cited from the French and English text, with notes by G. C. Bussey (La Salle, Ill., 1912; 1961 edn.), 93, 143–4. See also Chomsky, *Cartesian Linguistics*, pp. 9-10, 81 f., and I. Robinson, *The New Grammarians's Funeral: A Critique of Noam Chomsky's Linguistics* (Cambridge, 1975), 58 f. (against the mechanical or 'scientific' linguistic model).

[15] La Mettrie, *Man a Machine*, pp. 165–74.

already existed for such a theory; and to insert an active, participatory, and shaping function for the mind above that of a merely passive reception of inert sensory data. Properly understood, the intellect was a human function of divine processes, a dynamic energy in Nature itself.

To achieve his purposes, Harris sought ammunition and authority for his counter-attack in the 'pre-scientific' sources of Greek philosophy. His footnotes make this obvious. But apart from the Greeks, and of course Berkeley and Shaftesbury's writings, there were also English sources available to him among the Idealist group of Salisbury metaphysicians— most notably in Edward Herbert's theory of 'Common Notions' and *Zetetica* (his term for the predicaments or categories). Herbert's *De Veritate* (1624) was the first English book to establish the Ciceronian idea of grounding certainty upon 'internal intuitions presented to the mind independently of experience'. For someone determined to knock down Hobbes and Locke, Herbert provided perfect support, since he had also argued that 'our mind clearly corresponds to God and our body to the world . . . Mind, since it is spiritual cannot be passive'.[16] As we have seen, Harris only needed to step outside his Salisbury front door, as it were, to find that John Norris and Arthur Collier had been saying all this from the 1680s up to 1713. Their neo-Platonist Idealism, and particuarly their arguments for the priority of mind as the foundation of knowledge, may have influenced him. Herbert is the *only* English metaphysician mentioned in the opening chapter of Harris's book.

Philosophical Arrangements was particularly designed to show the unflatteringly derivative quality of modern philosophy, or at least the limits of its claims to novelty. The notion of the mind at birth as a passive receptor, a blank sheet on which experience and ideas are inscribed by experience, had been

---

[16] Carré (trans.), *De Veritate* (1624; Bristol, 1937), pp. 42, 169, 175. Carré points out that Herbert's influence on 17th- and 18th-cent. philosophers 'is not to be supposed', and that 'similarities of ideas are as likely to be due to common sources in Greek thought' (p. 42). Nevertheless, Carré himself observes that the parallels between Herbert and Descartes, Leibniz, the Cambridge Platonists, and the works of Reid and Hamilton, are 'remarkable' (p. 43). Herbert's sources, as with Harris, are the Stoics, mediated through Cicero's *De Natura Deorum*. For unexamined parallels between Harris and Herbert, see Chomsky, *Cartesian Linguistics*, pp. 60–3.

unambiguously associated for more than eighty years with John Locke's *Essay concerning Human Understanding*. Harris might have gone to Platonic Forms for a counter-argument, or to an intermediate source such as Herbert. But he does neither. If the mind at birth was not a *tabula rasa* but was already stored with an infinite *capacity*, the most authoritative argument against Locke was found in Aristotle. A single quotation from Aristotle's *De Anima* (3. 4. 11) did the job, and put before the attentive reader the clear implication that in this respect, as in many others, the moderns had grown tall by climbing on the backs of the ancients.[17]

Harris's silence about his non-classical sources is remarkable, but it is also explicable. As the declining reputation of Shaftesbury had indicated in Harris's own time, there were both problems and dangers in footnoting one's allegiance to English discussions of such contentious ideas as an intuitive moral sense, or to the belief that Religion and Reason were the same thing. By the second half of the eighteenth century Herbert's idea of Common Notions had been appropriated by the Deists, especially by Toland, Collins, and Tindal. In 1680 Herbert had been classed with Spinoza; from 1754 to the end of the century a standard work on the Deists described the latter as the most eminent among the heretics.[18] Harris neither wanted nor could afford to be associated with this sort of controversy. After all, his well-known kinship with Shaftesbury and his descendants was burden enough. There is one solitary reference to Shaftesbury (as the source of a translation) in *Philosophical Arrangements*. Nevertheless, this was enough for

[17] *Philosophical Arrangements*, p. 120. Harris's patrician footnote is almost comically dismissive: 'The Human Intellect was supposed by the Peripatetics to be pure and absolute Capacity; to be no particular thing, till it began to comprehend things . . . a Writing Tablet, where nothing as yet is in Actuality written . . . But this in the way of disgression—'Tis only the short Specimen of an ancient Speculation, which gives us reasons, why the human Intellect can have no *Innate Ideas*.' For a more detailed discussion of innate ideas, see pp. 413–17 of *Philosophical Arrangements* (ch. 17).

[18] J. Leland, *A View of the Principal Deistical Writers that have appeared in England in the last and present century*, 3 vols. (1754–6); 2 vols. (1757, 1764, 1766): see 5th edn., 2 vols. (1798), i. 3. Leland's list includes Hobbes, Toland, Shaftesbury, Collins, Tindal, Bolingbroke, Hume, and Thomas Chubb (i. 214–83). Herbert is called 'one of the first that formed deism into a system, and asserted the sufficiency, universality, and absolute perfection, of natural religion, with a view to discard all extraordinary revelation as useless and needless' (i. 3).

Johnson to dub Harris an 'Infidel' for writing it, and as 'foolish' for succumbing to Shaftesbury's influence.[19]

Either through ingratitude or ignorance, the moderns had rejected their intellectual inheritance, and Harris's purpose was to re-call, re-cognize, and re-mind the modern memory. An awareness of one's intellectual indebtedness was not a cause for anxiety but a reason for confidence. Even so, Harris's book is neither a slavish defence of Aristotle nor is it a blinkered rejection of everything in Locke. In fact, he agreed with Locke's *tabula rasa* theory, but only in so far as it denied the possibility of innate propositions. He preferred Aristotle's notion of an *instinctive* sense of self-existence and, more importantly, the idea that the mind naturally possessed an active *potency*, a capacity for propositional knowledge which was directed *on to* the material world, not simply acted upon by the latter's chaos of Protean particulars. Like Herbert before him and Kant after him, Harris argued for an organic relationship between Nature and human thought, a reciprocity between objects and understanding.

In letter 8 of *Several Letters written by a Noble Lord to a Young Man at the University* (written 1706–10, published 1732), Shaftesbury argued that Locke's liberation of the mind from scholasticism and Cartesianism had gone too far, and that the individual's moral sense not only pre-existed ratiocination but was the source of innate ideas of 'Order, Administration, and a GOD'. Such ideas 'infallibly, inevitably, necessarily spring up' in every man.[20] Harris would have agreed with this, but again, the 1770s were a long way from the first decade of the eighteenth century. Shaftesbury's problem lay in propagating the valuable skills of logical procedure without appearing to clog the reader's mind with pedantry and a specialized dialect. The title of *The Moralists: A Philosophical Rhapsody* hints at the solution, i.e. scientific knowledge rigorously sought through an informal discourse of self-examination. This

---

[19] *Thraliana: The Diary of Mrs Hester Lynch Piozzi, 1776–1809*, ed. Balderston, i. 35 (28 May 1977).

[20] *Several Letters* (1732), p. 28: dated 3 June 1709. Hobbes, said Shaftesbury, was bad enough, but it was 'Mr. LOCKE that struck out all Fundamentals, threw all *Order* and *Virtue* out of the World, and made the very *Ideas* of these (which are the same as those of God) *unnatural*, and without Foundation in our Minds. *Innate* is a Word he poorly plays upon' (p. 28).

device could release philosophy from its popular discredit and its imprisonment in 'colleges and cells', where its chief pupils were 'Empirics and pedantic sophists', and restore its value as a means of educating the 'statesmen' for civic employment. This was Harris's cue. He begins his *Philosophical Arrangements* where Shaftesbury left off, with a demonstration that philosophy is a matter of practical civic value, and his English exemplars include men who combined outstanding and active civic careers with the contemplative virtues: Sir Thomas More, Sir Philip Sidney, Raleigh, Herbert of Cherbury, Milton, Algernon Sidney, Sir William Temple, and his dedicatee, Lord Hyde. Again, John Locke is significantly omitted; but so is his mentor Shaftesbury.

Philosophical Arrangements thus replaced abhorrent materialist notions by the idea of Divine Providence, the priority of mind over body, and the perception of the Genus *before* the Species. Inverting the ancient scholastic maxim, and in so doing reversing materialist psychology, Harris stated that nothing existed in Nature which did not first exist in mind (*nihil est in sensu quod non fuit prius in intellectu*), and as if to create in style what he could not demonstrate in logical argument, he regularly punctuates his explanatory mode of discourse with a series of soaring and romantic prose lyrics on the creative power and astonishing energy of the mind itself:

as there are no FORMS of art, which did not pre-exist in the Mind of Man, so are there no FORMS of Nature, which did not pre-exist in the mind of GOD. 'Tis through this we comprehend, how MIND or INTELLECT IS THE REGION OF FORMS, in a far more noble and exalted sense, than by being the *passive* Receptacle through Impressions from Objects without. It is their Region, not by being the Spot into which they migrate as strangers, but in which they dwell as *auchthons*, the *original Natives* of the Country. 'Tis in Mind they first exist, before Matter can receive them; 'tis from Mind, when they adorn Matter, that they primarily proceed: so that, whether we contemplate the works of Art, or the more excellent Works of Nature, all that we look at, as beautiful, or listen to, as harmonious, is the genuine EFFLUENCE OR EMANATION OF MIND. (pp. 112–14)

So to some extent we must interpret the overt sources, reconstruct the suppressed targets, and recover the critical moment in order to appreciate the full significance of Harris's argument. What looks like an illustrative discourse is in fact a reactive and corrective redirection, an argument by inference and implication. Put bluntly, each time Harris mentions Aristotle and his commentators (as well as his intellectual ally Lord Monboddo), he has chosen *not* to acknowledge the existence of Descartes, Hobbes, Locke, and their materialistic progeny. Praise of the former group implies inadequacy in the latter, and the reader is left to construct the bridge between the inspiring universals of ancient philosophy and the purblind parochialism of the materialist moderns. Unless this is seen from the outset, the controversial reception of Harris's book seems unintelligible. Its style is a strategy constructed in advance to deflect hostility. Its author's constitutional dislike of public wrangling and controversy, as well as his determination to speak ill of nobody, means that we need to read between the lines, otherwise his book may be dismissed, as it has often been, as a lofty, harmless, even pedantic piece of antiquarianism, with all its talk about the ten Predicaments and its invocation of the Stagyrite's authority.

But having said that, we might recall that it was just as much the manner as the matter of *Hermes* that was to cause such frenzied rage in the mind of Horne Tooke in 1778. The *manner* of Harris's argument contained the message, and in the case of *Hermes* this was deconstructed by Tooke as a sign of a tyrannical Establishment, an authoritarian scholarship, a doctrine of legal precedent, a linguistic manifestation of power politics. Contemporary reviewers of *Philosophical Arrangements* noted time and time again that its *raison d'être* was the defeat of modern materialism.[21]

IV

Correctly anticipating that his discerning readership would be small and select, Harris nevertheless hoped that a significant number would come from Scotland. Monboddo, perhaps with David Hume in mind, was not so confident. He wrote to caution his friend on 1 June 1774:

You seem to have a very good Opinion of the Learning of my Countrymen. The Truth is that we have more smatterers in Greek &

[21] See e.g. *The Critical Review*, 50 (July 1775), 8, *and* NS 35 (1802), 261.

Latin, and Matters of Taste, than you have, in proportion to our Numbers; but you have more real Scholars, and the antient Learning has I believe still more friends left with you than with us; and there are some Men among us, one particularly whom I think you know, who, not being Antient Scholars themselves, would willingly explode that Learning altogether, without putting anything in place of it except their own writings. From these your Metaphysics, so different from their own, cannot expect a good Reception. But there are some few in this Country beside myself who, having got a Taste of it in your Hermes, will read your Categories with great delight.[22]

Sensing this kind of reception in advance, Harris made some attempt to soften the sometimes stark opposition between the ancients and the moderns and to seduce the English reader by an appeal to literary patriotism. Monboddo had taken his cue from Harris's earlier letter, written in the middle of his 'Parliamentary time' (23 March 1774):

my Categories are going on, and many sheets are printed off. As the Names, Categories and Predicaments are rather scholastic, though I have not rejected them, yet I have chosen to make the running title of my book *Philosophical Arrangements*; a name more intelligible, and yet I think by no means improper, as the scope of the work is to *class* our Ideas, and *arrange* them after the most clear and comprehensive method.[23]

Accordingly, Shakespeare and Milton are now used to illustrate in literature not only the technical logical teachings of Aristotle, but also to display the inherent or 'Natural' logic in all forms of verbal discourse. Materialists and atheists who would put forward empiricism and mathematical science as the keys to understanding both man and his world would not like this third instalment of Harris's lifelong assertion of the

---

[22] Harris Papers, 'Private Correspondence', vol. 31, part 3.

[23] Knight, *Lord Monboddo*, pp. 88–9. In the National Library of Scotland, MS 24501, there are 5 letters from Monboddo to Harris and 9 from Harris (1768–79). All except 5 from Harris were published in the appendix to the *Sixth Report of the Royal Commission on Historical Manuscripts* (pp. 673–81). The unpublished letters are chiefly about the superiority of the 'Republic of Letters' over 'the Din of Politicks', the progress of each other's writing, with some anecdotes about the difficulty of acquiring good books: thus, on 18 Mar. 1773, Harris remarks that 'Voltaire seems to sleep, & Rousseau to be forgot. I heard it in our House lately asserted that Voltaire, Rousseau, Bolingbroke, and Davd. Hume were the four Evangelists of modern Free-thinkers.'

priority of mind. But at least they could not also charge him with neglect of his own literary culture.

Logic inhabits the attics and the cellars of the Augustan mind; it is both a place of airy speculation and a breeding-ground for diseases of the intellect, eliciting enthusiasm and fear, obsession and loathing. On the one side there is what Johnson called 'lofty towers of serene learning' (*Rambler* 117), and on the other that aspect of puerile education in mental gymnastics alluded to in Stephen Dedalus's words in *Ulysses*: 'The Schoolmen were schoolboys first . . . Aristotle was once Plato's schoolboy' (*Ulysses*, p. 237). Harris's aim (shared by Shaftesbury and many others before and since) was to do for logic what Addison had done for philosophy, to bring it out of 'Closets and Libraries, Schools and Colleges' into the world of common discourse. He was acutely aware that what he was attempting in his penultimate book would meet a partisan reception, and that his subject would be off-putting. His strategies therefore avoided any mention of either logic or Aristotle in his title. Avoiding the use of traditional terms (universal genera, categories, predicaments), and all esoteric variants (such as Herbert's *Zetetica*), he chose the neutral word 'Arrangements'. He also cut out those all too familiar textbook examples of logical terms and distinctions such as man and horse, Porphyry's tree, and so on, replacing them with a wide range of classical and modern literary examples. In place of that traditional emblem of the clenched fist of logic, combative, esoteric, exclusive, and schematic, he put an openness and an accessibility into its fundamental processes through the avenues of the best literature. As Johnson illustrated lexicographical definition by the best literary usage, Harris defined Aristotelian categories by their application and function in *Macbeth* and *Paradise Lost*. In the opening lines of Shakespeare's tragedy ('When shall we three meet again | . . . When the battle's lost and won') he finds the predicaments of *where*, and *when*. The point is not to prove that Shakespeare knew Aristotle's categorical logic, or that in some way everything had been written in advance by Aristotle, but to show that there was a universal system of logical categories on which all human discourse rests. Illustrating his logical theories with literature certainly alleviated what might otherwise appear 'dry and severe', but there was also his conviction that in Art reposed the finest manifestation of a universal and ultimately divine logos. In his letter (17 April 1775) accompanying a presentation copy of the book to Elizabeth Montagu, the author of *An Essay on the Writings of Shakespeare* (1769 and 1777), Harris put it this way:

what could I do?—I was no Poet. Had I attempted to versify I might do more mischief than good—I adopted therefore what I thought ye wiser measure, & instead of my own bad Poetry enrich my Work with the Poetry of ye others & those ye most admired for their superior genius.

Shakespear stands high in this list, & tho I dont apprehend he ever heard a Syllable of the Peripatetic Categories (or of Pythagoras & Archytas from whom they were derived), yet his Sentiments contribute not a little to their illustration. A good Argument in favour of these Arrangements, because if they were not founded in Nature, he, who had no other Guide than Nature, could never have furnished such examples . . . pure Nature & Genius & Art co-incide, like Radii of the same circle; they have a common Centre, which Centre is Truth.

I have said nothing on this occasion about Shakespear which I do not firmly believe.[24]

Two volumes of Monboddo's *Of the Origin and Progress of Language* came out before Harris's *Philosophical Arrangements*, citing *Hermes* many times and giving some advance publicity for Harris's next book: 'the best book of metaphysics in the

---

[24] Harris Papers, vol. 1, part 1, 'Copies of my 3d Volume', pp. 4–5. Harris was recalling William Petvin's remark that the 'Universal Mind' is like 'a Circle or Sphere whose Centre is every where and circumference no where' (Harris Papers, Letter Book, vol. 31, part 1, p. 84, 16 July 1746). Harris sent copies and accompanying letters to Lords Marchmont, North, Monboddo, the Bishop of Chester, the Earl of Bute, his godson Cropley Ashley, Lady Browne, and the king. Harris's memorandum for 12 Apr. 1775 is as follows: 'I went to the Queen's house, and left a Present of my 3d. Volume to his Majesty on the Library Table—at the same time I presented a Copy of my three Volumes to her Majesty, with a Letter accompanying—Her Majesty was graciously pleased to see me, and thank me for my Present—politely said, she should read them, & what she did not understand she shd. apply to me to explain—she was so good to converse with me with her usual good sense & affability for about half an hour. The same day in the afternoon I sent two Copies of my 3d. Volume to Mr Batt, to be shipt for Berlin, one to his Prussian Majesty, another to my Son.' Other presentation copies went to Dr. Heberden, the Bishops of Oxford and Salisbury, Lord Hyde, his brothers Thomas and George Harris, Edward Hooper, Elizabeth Carter, 'Athenian' Stuart, Richard Owen Cambridge, the British Museum and Bodleian libraries, Dr Warton, and others.

English language' (i. 76). Monboddo had claimed that *Hermes* had had a seminal influence upon him. It 'first introduced me to the Greek philosophy'; and he went on to say that his own *magnum opus* on language could be seen as 'not a bad second part' to *Hermes*, as well as an imitation of its style and 'philosophical accuracy'.[25] Monboddo may have been Harris's greatest acolyte, but his mentor expressed only diffidence. After the *Arrangements* had appeared in print, Harris wrote to Monboddo on 11 February 1775:

Greek and Latin, having little connection with the Philosophy in fashion *here*, are I am afraid not much cultivated. This is rather against one, who has filled his pages with a great deal of that unfashionable lore. I hope however to find a few, scattered up and down in corners out of sight, who may think my endeavours laudable, and worthy of their attention. I have great hopes from the scholars of North Britain. They have not yet transferred the whole of Philosophy from the head to the hands; that is to say, from Syllogism and Theory, to Air-pumps and the Electric Apparatus. I have great hopes to find advocates in them . . . to support the cause of Metaphysics, and the first Philosophy, which—because they have nothing to do with experiment—are ·therefore boldly called 'nonsense.' I can only add, that, if this be true, then is Euclid 'nonsense' also, for I never heard that his theorems depended upon experiment.[26]

Harris certainly gained ground in Scotland, particularly among the members of the Scottish common-sense school. Thomas Reid, Professor of Moral Philosophy in the University of Glasgow, dealt savagely with Descartes, Locke, Malebranche, Norris, Collier, Berkeley, and others, in 1785, but he singled out Harris and Monboddo for particular praise on a matter of crucial importance: they were the *only* modern writers who properly understood the term *idea*![27]

---

[25] Knight, *Lord Monboddo*, pp. 48–9. [26] Ibid. 90–1.

[27] T. Reid, *Essays on the Intellectual Powers of Man* (Edinburgh, 1785), 474. Reid wishes to return the use of the term *idea* to its Platonic original ('the essence of a species'). Thus, 'If the word idea be restricted to the meaning it bore among the Platonists and Pythagoreans, many things which Mr LOCKE has said with regard to ideas will be just and true, and others will not . . . It will be so far from the truth, that all our simple ideas are got immediately, either from sensation or reflection, or from consciousness; that no simple idea is got by either, without the co-operation of other powers.' It is often forgotten that when Johnson kicked a large stone 'with mighty

V

Harris's last book, *Philological Inquiries* was completed in three parts and published one year after his death. Its first part (chapters 1–7) was an expansion and rewriting of an unsigned and undated pamphlet separately published in 1752 entitled *Upon the Rise and Progress of Criticism*. Its third part, a history of the 'Literature of the Middle Ages', was in press by 18 June 1776, and Monboddo was the first to register the literary world's high expectations of Harris's final venture, pointing out the absence of such a study in critical histories to date. For Samuel Johnson, this was the only work by Harris to capture and retain his unqualified attention: as Thomas Tyers remarked, *Philological Inquiries* 'had attractions that engaged him to the end'.[28]

Harris's purpose was to synthesize the study of literature with certain structural philosophic ideas, using the latter as presuppositions to 'explain' the causes of the former. An early reviewer recommended it to young men in the universities, noting that it was written in a more popular style and went less deeply into logic and metaphysics than Harris's earlier works, and later critical opinion in England and France noted both its promise and also its disappointment. For the first time in English Harris offered, in George Saintsbury's words, 'not merely a philosophic-literary view of criticism' but also 'an inquiry into those regions of literature on which his predecessors have turned a blind eye'. Saintsbury also expressed the characteristic conclusion: 'As is the exaltation of the promise, so is the aggravation of the disappointment', and, Lawrence Lipking adds, 'very genially and politely, Harris had turned his back on his chance for historical importance'.[29]

force' (Boswell's *Life of Johnson*, 6 Aug. 1763), to disprove Berkeley's idealism, he did so in order to *assert* his belief in '*the first* truths of *Père Bouffier*, or the *original principles* of Reid and of Beattie; without admitting which, we can no more argue in metaphysicks, than we can argue in mathematicks without axioms'.

[28] T. Tyers, *A Biographical Sketch of Dr. Johnson* (1784), in *Johnsonian Miscellanies*, ed. G. B. Hill, 2 vols. (Oxford, 1897). ii. 344: 'The Posthumous volumes of Mr. Harris of Salisbury (which treated of subjects that were congenial with his own professional studies) had attractions that engaged him to the end.' *Philological Inquiries* was published in 2 vols.

[29] Lipking, *The Ordering of the Arts in Eighteenth-Century England*, pp. 98–9, citing G. Saintsbury's *History of English Criticism* (Edinburgh, 1930), 207–8.

Philosophical Inquiries is not, of course, the book on which Harris's reputation rests: this had already been firmly established by his previous three books. *Hermes* alone was sufficient to guarantee his continuing importance. But, it is generally agreed, something went wrong in this last one. Harris's attempt to look at literary history with the eye of a philosopher, a challenging and unprecedented objective, was to reveal the perils of a method which, in his previous books, had too comfortably inscribed the present in the margin of the past. What he needed for this project was not more learning but a sophisticated theory of literary causality and a definable concept of intellectual history. In the event, we are presented with a series of miscellaneous chronological observations on literary phenomena without a theoretical framework in which to interpret their significance. There are many gaps, the widest being Harris's indecision about the relationship between *author* and *work*, and the inter-relationships between both of these and any notion of cultural pattern.

Even so, the problem which Harris defined is actually more important than his ambiguous solution of it. The contemporary sources now needed by Harris to complete a historical survey turn out to be those generated by close *personal* loyalties which, considerable as they are, nevertheless drove him away from philosophical detachment towards a very personal com-memoration. Harris privileges the notion of 'author' rather than 'authority' for the first time, and yet (in the case of Fielding, or Lillo, among others) he also continues to allocate value in contemporary literature in so far as it may be slotted into an authoritative classical tradition. The problem is one of critical perspective contaminated by personal loyalties, a shortage of appropriate contemporary material for an a priori method, and a barely conscious shifting sensibility—both Harris's own, and that of his time. Individual works by individual authors known to Harris as friends, allies, and clients begin to complicate a previously stable pantheon of authoritative referents at a time when the personality of the 'author' is becoming a central critical issue.

Thus, Harris's last book inscribes modern literary practice in the margins of an autobiographical record of his own life of classical scholarship. The critical and methodological rigour characteristic of *Hermes* is here skewed, albeit attractively, by deep affections and personal loyalties. Having marginalized modern philosophers in *Philosophical Arrangements*, defining their presence by erasure, his last book privileges people he knew and long-dead authors he had come to know (both groups becoming, as it were, personal friends), whose cultural loyalties he shared. Paradoxically, then, this attempt at a systematic and 'philosophical' literary theory is at the same time Harris's most personal book, a retrospect on his whole literary career, 'a monument of affection to his numerous friends'; and although it continues Harris's lifelong campaign to establish 'the dignity of mind and its objects in opposition to the doctrines of chance, fatality, and materialism', it also puts forward a highly personal view of culture as the instrument of human progress, and an acknowledgement that literary tradition is not solely determined by ancient and venerated authority figures or long-established social and linguistic structures, but also by modern, individual artistic sensibilities.

Creating a monument to friendship was important for Harris in several ways: it established his sources, allies, and loyalties; it reassured him that his life of writing had not been solitary and without influence; and it placed his own 'philosophical criticism' in a continuing tradition, confirming that the theoretical position which he himself represented could be confirmed by the lives and works of others. The many retrospective friendships celebrated here required a long and busy life, thus Harris claims the privilege of age for his commemoration. But if the idea of a vigorous continuity was to be shown, Harris could not afford an elegiac, resigned, or patronizing tone. The greatest weakness of the book is undoubtedly its annoyingly miscellaneous method and its theoretical shallowness; yet, in the Socratic sense, this was to be an examination of his own life, and the examination itself was to be of primary value, as well as a key element in his mapping of European intellectual continuities.

Its dedicatee was his cousin Edward Hooper, an intimate friend for more than fifty years, but with no known literary interests. Harris's son has a better qualification, for he had acquired copies of the Livy manuscripts in the Escurial library during the time he was Minister Plenipotentiary to the court of Madrid, and he was instrumental in having gifts of books sent to his father from Prince Potemkin at the court of Catherine the Great. With these two exceptions, every one of the friends mentioned in the book is either a critic or a scholar, and every book cited in part 1 had a personal significance for him, either because he had helped in its composition, or because it had been dedicated or presented to him by its author, or because its author had contributed to Harris's own publications, or simply because he had known its author on reasonably intimate terms. Part 1 is a history of Harris's own life measured out in his friend's books, and part 3 signals Harris's desire to connect provincial Salisbury with European culture at large: his twelfth-century 'countryman' John of Salisbury is given more space than Aristotle.

The fourth chapter of part 1 is devoted to modern writers of the 'explanatory' kind, and Harris cites Joshua Reynolds for his *Discourses on Art*, Thomas Warton for his edition of Theocritus and the *History of English Poetry*, Joseph Warton for his *Essay on Pope*, John Upton for his edition of *The Faerie Queene* and of Arrian's Epictetus, Elizabeth Montagu for her *Essay on Shakespeare*, Samuel Johnson for the *Dictionary*, Robert Lowth for his *Short Introduction to English Grammar*, Floyer Sydenham for his edition of Plato's *Dialogues*, Elizabeth Carter for her translation of Epictetus, his deceased uncle Maurice Ashley for his Xenophon, Jonathan Toup for his Emendations on Suidas and his Longinus, and Dr John Taylor for his Demosthenes, Lysias, and his book on Greek epigraphy. Samuel Clarke, the brother of the Dean of Salisbury, an anti-Lockian correspondent of Kames, Collier, and Leibniz, is cited for his edition of the *Odyssey*, 1740 (not, we should note, for his best-known work of intellectual deism directed against Hobbes and Spinoza, *A Discourse concerning the Being and Attributes of God* 1705–6). Fielding's *Joseph Andrews* and *Tom Jones* are listed as 'master-pieces of the comic epopee' (chapter 7), and among others there is James 'Athenian' Stuart (Harris's illustrator and the co-author of *Antiquities of Athens*), Lord Lyttelton (for his *Life of Henry II*, for which Harris and Warton had personally surveyed the ruins of Clarendon

Palace),[30] Richard Owen Cambridge (for *The Scribleriad*), Lord Monboddo (for *Ancient Metaphysics*), and Daines Barrington, vice-president of the Society of Antiquaries (for his *Observations on the Statutes*, 1766). The most self-indulgent insertion in this generally self-conscious gallery of friends is in chapter 5, which includes a stanza from one of the Harris family's own private theatrical performances in Salisbury, an unpublished pastoral drama entitled 'Perdita to Florizel', probably written by Louisa Harris.

Scholarship, friendship, and paternity are somewhat awkward parameters for a taxonomy of literary theory, textual editing, cultural history, prosodic theory, modern fiction, poetry, history, and the law. For the biographer, however, this is more than adequate compensation for the dispassionate style and shadowy presences lurking beneath *Philosophical Arrangements*. Moreover, *Philological Inquiries* also celebrates some of the *places* of special significance to Harris: apart from Salisbury itself, there is Henry Hoare's Stourhead, Lyttelton's Hagley, Cobham's Stowe, and Bubb Dodington's Mount Edgcumb, each of them (like Fielding's celebration of Hagley Park and Ralph Allen's Prior Park in the composite image of Paradise Hall in *Tom Jones*) the physical manifestations of a style of living and learning of key importance to an exponent of Harmony. Throughout the book Harris is alert to the *locus amoenus*, the environment of productive learning, and to the realization of harmonic relationships between theory and practice. If he loses sight of his philosophical aim to discover the causes of literature as such, compensation is found in the discovery of individual men and communities which, like his own Salisbury set, kept alive certain cultural continuities.

Part 1 of *Philological Inquiries* explains the rationale for the whole book, and Harris begins well (though this may be because this part had been written thirty years earlier).

---

[30] BM Add. MSS 18729, 'James Harris Autograph Papers 1768–9', draft of a letter to Lyttelton, 28 Dec. 1769: contemplating the ruins, Harris thought of 'Henry on his Throne, determined with a kind of Indignation to support Civil Power against Papal Usurpation; his Earls & sturdy Barons on one side of him, with all yt Ferocity & Pride, so peculiar to feudal chiefs; on ye other side the great Becket, with the Prelates his partisans breathing forth ye Spirit of Sacerdotal Empire . . . If I have omitted any thing in ye above Narrative, my worthy & learned Friend Dr Warton, who was my Companion in ye Survey of these Ruins, will be able to supply the defect.'

Philology (or criticism) is to the literary world what natural philosophy (or science) is to the material world: both are concerned with questions of causation. The subject is then subdivided into three parts: an investigation into the rise of criticism and critics; an exemplification of critical doctrines in selected ancient and modern authors; and an essay on the taste and literature of the medieval period, this being the period in which such things were most precarious. The first part is Harris's version of the Art of Poetry, and his guides are Aristotle and Longinus on poetry and rhetoric, and, among the Romans, Cicero's *De Oratore*, Horace's *Art of Poetry*, and Quintilian. Broadly speaking, these are examples of what Harris calls 'philosophical criticism', followed by the dependent genre of 'historical criticism', which includes Proclus's commentary on Plato and Suidas on lexicography. After the Dark Ages, to which Harris will return, philosophical criticism concerning the causes and principles of writing in general is further exemplified by Vida, Scaliger, Rapin, Bouhours, Boileau, Bossu, Roscommon, Buckingham, Shaftesbury, and Pope. In painting, there is Reynolds. Among the modern critics of the 'Explanatory Kind', Harris includes both Wartons, Tyrwhitt, Upton, Addison, Elizabeth Montagu, Johnson, and Lowth (in lexicography and grammar), and then a group of translators (Casaubon, Elizabeth Carter, Sydenham, and Maurice Ashley).

Part 2 lists the 'corrective' critics, or textual editors such as the Scaligers, Casaubons, Burmann, Wasser, Bentley, Toup, Taylor, and Upton. Here, Bentley is singled out for his 'rage of conjecture' in his edition of Milton. A theory of contraries is outlined, reminiscent of Fielding's 'Principle of Contrast', and illustrated by examples drawn from Virgil, Homer, Shakespeare, Milton, and Salvator Rosa (see further Appendix IV below). Chapter 2 is on the syllabic metre of Greek and Latin, and the accentual metrics of English, with further parallels between metre and musical theory. In prose style, the periodic sentence is praised, with the opening of *Hermes* and *Philosophical Arrangements* given as examples. In general, a good writing style is like Garrick's acting, its latent technical mastery is concealed by an appearance of elegant ease.

The central section in Harris's discussion of drama is

chapter 6, and an analysis of the peripeteia or 'tragic revolutions' of Sophocles' *Oedipus Tyrranus*, Milton's *Samson Agonistes*, *King Lear* (Shakespeare's version, not Nahum Tate's), *Othello*, and George Lillo's sentimental and didactic Christian tragedy, *Guilt its own Punishment; or, Fatal Curiosity.* No doubt Harris was well aware of Dryden's comparison between Jonson's *The Silent Woman* and Shakespeare's dramatic methods, and that between Addison's *Cato* and Shakespeare's *Othello* in Johnson's 'Preface to Shakespeare', but Harris may well have had a particular reason for selecting Lillo. Whereas most modern comedies fell back on a miraculous reformation in the fifth act, and most tragedies on a terminal slaughter, Lillo's was a rare contemporary example of a well-made and classically shaped tragedy which resisted such trite resolutions to the notorious problem of the fifth act.

Harris's choice was almost certainly an act of deference to 'a witty friend of mine, who was himself a dramatic writer, [who] used pleasantly, though perhaps rather freely, to damn the man who invented fifth acts'.[31] This was Fielding. Lillo's tragedy had been first performed under Fielding's management and direction by the 'Great Mogul's Company of Comedians' at the Little Theatre in the Haymarket on 27 May 1736. When it was performed again in March 1737 it was preceded by Fielding's commendatory prologue and ran for eleven nights before the Stage Licensing Act (21 June) put Fielding out of business as a playwright and producer (a disgraceful episode described acerbically in Harris's Memoir of Fielding). Understandably, Fielding admired Lillo's tragedy. In *The Champion* for 26 February 1740 he described it as 'a Master-Piece of its kind and inferior only to Shakespeare's best pieces', and Lillo himself 'the best Tragic Poet of his Age'. After years of neglect, Harris's analysis of the play in *Philological Inquiries* brought it back to life, and at least ten printed versions appeared between 1780 and 1830. Harris saw it as an excellent example of classical tragic form, not only for its successful observation of the unities, but also for its

---

[31] *Philological Inquiries* is the least accessible of Harris's books, therefore all quotations from it are from the single-vol. *Works of James Harris*, ed. by his son: see p. 433. All subsequent page references are to this modern-spelling edn. and are given parenthetically.

psychologically coherent exorcizing of pity and terror, its probability of plot and character and (for him) its excellent and 'striking Revolution', comparable with Milton's *Samson Agonistes*. Henry Mackenzie, though prompted to read and amend Lillo's play for production in 1782 by perusing Harris's book, turned this 'classical' elegance into a 5-act sentimentalization, not to say melodrama, which would hardly have pleased Harris.[32]

Unlike Johnson, Harris defends the doctrine of unities on the grounds that 'geniuses, though prior to systems, were [not] prior to rules, because rules from the beginning existed in their own minds, and were part of that immutable truth, which is eternal and everywhere' (p. 452). On the other hand, Harris's defence of the imagination of the dramatic poet and his audience is an exact analogy of Johnson's argument against the prescriptive critics in the *Preface to Shakespeare*. Harris argues:

The real place of every drama is a stage; that is, a space of a few fathoms deep, and a few fathoms broad. Its real time is the time it takes in acting, a limited duration, seldom exceeding a few hours.

Now imagination, by the help of scenes, can enlarge this stage into a dwelling, a palace, a city, &c; and it is a decent regard to this which constitutes probable space . . . the mind . . . can enlarge without violence a few hours into a day or two; and it is in a decent regard to this, we may perceive the rise of probable time . . .

. . . as to time, we may suppose a play, where lady Desmond, in the first act, shall dance at the court of Richard the Third, and be alive, in the last act, during the reign of James the First . . .

. . . Poets, though bound by the laws of common sense, are not bound to the rigours of historical fact. (pp. 448–9)

Part 3 of *Philological Inquiries* promises much, if only because it had few competitors or precedents; and it was for this reason that part 3 was rapidly translated into French by Antoine Boulard in 1785, with the comment that it was 'une agréable esquisse d'un ouvrage qui manque à notre littérature'.[33] As an exponent of Greek language and culture, the subject of

---

[32] See G. Lillo, *Fatal Curiosity*, ed. W. M. McBurney (1967), pp. ix–xvi.

[33] A. M. H. Boulard. *Histoire littéraire du moyen age* (Paris, 1785), p. vi. For a review, see *L'Année littéraire*, 32 (1785), 145–68.

medieval literature was of clear interest to Harris, and although we might expect an unmitigated lament for the world's subsequent degeneration after the sack of Rome by the Goths in the fifth century and the toppling of Byzantium by the Turks in the fifteenth, the purpose here was not to bemoan the gloom of the Dark Ages but to celebrate its pinpricks of continuing illumination. Certainly, Harris had little sympathy for the perpetrators of religious wars in an age of 'monkery and legends . . . of crusades to conquer infidels and extirpate heretics'. This was, after all, only the 'twilight of a summer's night' in which commentators such as Ammonius and Simplicius worked by the light of guttering candles. The destruction of the Alexandrian library was indisputably the disaster which inaugurated an 'age of barbarity and ignorance' almost as horrific as the classical Epicurean and Democritean notions of atoms and void.

Harris's literary judgement is at odds here with his belief in a providential world order. His optimistic critical position battles awkwardly with an historical and moral conviction that there is in human history an inevitable tendency towards degeneration. His unoriginal solution is to regard the Dark Ages as a misguided interlude of wilful ignorance. His recuperative instincts therefore search out pinpricks of light and evidence of precarious continuities among the ruins, linking Epicurus' garden, Plato's Academy, Aristotle's Lycaeum, Zeno's Stoa, and the University of Oxford as places of contemplation and philosophical seriousness. He happily related Athenian Stuart's anecdote that modern-day Athenians still play the lyre, that Mount Hymettus is still famous for its honey, and that an old Greek pilot showed Commodore George Anson the very spot where the Grecian fleet lay at the siege of Troy (p. 476). He finds resemblances of thought and expression between the ancient Greeks, the medieval Arabs, and Shakespeare's *Julius Caesar* (p. 487), and argues that the same taste created Petrarch's retreat at Vaucluse and Valentine Morris's exquisitely landscaped gardens at Piercefield (p. 528).[34] The rationale for this phenomenon is that

---

[34] Piercefield: the estate and gardens near Chepstow, Monmouthshire, owned and laid out by Valentine Morris (d. 1788) who became Lieutenant-Governor of St Vincent's. Described in *The Gentleman's Magazine*, 69 (1799), 1037, as 'this paradise of

'Human institutions perish, but nature is permanent' (p. 467). Harris nevertheless remarks that some things done in the name of historical convention are intolerable: thus the syllabus provided for the education of a medieval historian at Westminster and Oxford is inappropriate for the student of 1780 (p. 501). Harris uses his scholarship to stitch together a coat made of many colours, but the result is a garment with as many holes as patches. Loss, distance, absence, and obscurity find a modicum of compensation in the accounts brought back by travellers to modern Greece, including those of Stuart and Revett, and of Lyttelton, but, like the *Greek Anthology* itself, these are fragments shored against the ruins. Little wonder, therefore, that he points out with great relief that it was the 'dark and barbarous' medieval period which produced the 'lightness' and 'uniformity' of 'the completest gothic building now extant', Salisbury cathedral (p. 513).

In his final chapter Harris confronts the question of moral and cultural decline directly. He rejects those who bolster it, like Swift, with the 'filth' of misanthropic satire, and evokes Fielding's argument that those who condemn their own time forget to include themselves in the general condemnation:

Bad opinions of Mankind naturally lead us to misanthropy. If these bad opinions go further, and are applied to the universe, then they lead to something worse, for they lead to atheism.[35]

If human history is cyclical and repetitive, the memory of an historian or a scholar and the procedures of a philosopher have a profoundly social function, to remind the uninstructed and weak minds that tend to negative ways of thinking that

---

the fairies', which 'no language can describe', the gardens were always kept open to the public, and featured rocks, an echoing cave, woods, serpentine walks, and a temple, by comparison with which the romantic setting of Tintern Abbey was but 'the porter's lodge'. For an account of Morris's emblematic career, his almost legendary benevolence, generosity, and misfortunes, see *The Gentleman's Magazine*, 59 (1789), 862–4, and C. Shepherd, *A Tour through Wales and the Central Parts of England* (1799).

[35] Harris's own annotated copy of Swift's *History of the Four Last Years of the Queen* (1758) provides an additional elaboration of his dislike, as well as a startling contrast between their radically different sensibilities and literary methods: 'See here & in the following pages the Author of the Tale of a Tub, of John Bull [*sic*], of ye Examiners, of the Drapers Letters, of Gulliver, of a thousand other abusive pieces both in prose & verse, see I say this worthy Author, pleading against the Liberty of the Press; the Man, whose whole Life was a Libel on his Country-men & Country, on Religion and Human Nature.' This book is now in the Osborn Collection at Yale University.

every man is a microcosm of the race as a whole and that there is a 'providential circulation, which never ceases for a moment through every part of the creation'. Harris's last word is Tiresias-like, and expressed using the words of Aeneas to the Cumean prophetess:

> Virgin, no scenes of ill
> To me or new, or unexpected rise;
> I've seen 'em all; have seen, and long before
> Within myself revolv'd 'em in my mind.

Almost everything in Harris's last book points to an ending. Those sibylline words just cited above might be read as a final act of self-justification, an arrogant exercise in solipsis, an attempt at inclusive human consciousness as if to say that nobody who had not surveyed the vast reaches of classical literature and philosophy could gainsay the metaphysic of universals and benevolent optimism. Harris chose to work in isolation from the critical discourse of his own time, and often looks like a philosophical Allworthy figure inhabiting an intellectual Paradise Hall surrounded by materialist neighbours. But his discernment of a vital continuity should not be underestimated: the study of classical culture and the modern writings of his friends and contemporaries were part of a continuum, linked by a common obligation of the human mind to acquaint itself with Ideas as the rationale for one's social being. His abiding ambition was to remove that fear of speculation which prevented the thinking mind from rising above its immediate practical interests. As Schelling was to say again in 1802, 'only Ideas provide action with energy and ethical significance'.[36]

Harris's cultural role has always been too narrowly defined as that of a mediator of classical paradigms for modern artistic and intellectual writing at a time when such ideas and models were being rapidly dismantled, even ridiculed. For this reason, the flaws and ambiguities in his last book are symptomatic of its time, caught between an aesthetic which was hierarchical and highly formalized, and a new aesthetic which by comparison seemed subjective and idiosyncratic.

---

[36] Cited in J. Habermas, *Knowledge and Human Interests*, trans. J. J. Shapiro (Boston, 1968), 301.

Roland Barthes may be right in attributing to English empiricism and French rationalism the production of the modern figure of the 'author', the 'prestige of the individual . . . the human person'.[37] For Harris, however, any theory of modern writing which omitted the equally 'human persons' of Horace, Virgil, Cicero, Aristotle, Shakespeare, Milton, and Dryden, would seem the perfect example of shallow modernist illusions. Of course, the idea of writing liberating what Barthes calls an 'anti-theological activity', denying a fixed meaning to 'God and his hypostases—reason, science, law', would have been no less abhorrent to Harris than to the age in which he lived. Yet his last book, whether we regard it as 'pre-Romantic' or not, implicitly recognizes a more generous and ampler concept of 'author' (including the critic, editor, lexicographer, and novelist) in precisely Barthes's modern sense, neither as a classic nor as a 'personality', but as a medium through which timeless texts may continue to be transmitted and interpreted.

[37] R. Barthes, 'The Death of the Author', in *Image Music Text*, trans. S. Heath (1977), 142–3.

9

The Very Soul of Harmony

On 20 April 1776 Elizabeth Harris wrote to her son in St Petersburg from the Twickenham home of Richard Owen Cambridge about a dinner party that had just been held there. The place had been a favourite one for the Harris family for many years. It was a convenient retreat from London, or, in Elizabeth's words, its fine walks, situation, and weather were 'a prodigious treat to us cockneys, who are accustomed to see nothing from our windows but a fishmonger's shop and the lame, idle shoeblack'. The dinner party of 18 April, which Gibbon also attended, is described as follows:

Tuesday Dr Johnson, his fellow-traveller through the Scotch Western Isles, Mr Boswell, and Sir Joshua Reynolds dined here. I have long wished to be in company with this said Johnson; his conversation is the same as his writing, but a dreadful voice and manner. He is certainly amusing as a novelty, but seems not possessed of any benevolence, is beyond all description awkward, and more beastly in his dress and person than anything I ever beheld. He feeds nastily and ferociously, and eats quantities most unthankfully. As to Boswell, he appears a low-bred kind of being.[1]

If Johnson, the tone-deaf moralist, was a socially ungraceful house guest, James Harris had his own problems, quite apart from a wife who allowed a deficiency of social sophistication to mask intellectual genius. He was, in this very month, and in the words of Elizabeth, 'a poor lame soul', plagued by the gout which had afflicted him from at least the winter of 1767. Harris was able to joke about his affliction, and this was perhaps just as well, for in George Huddesford's satire on the military reviews in Essex, 'Warley' (1778), addressed to Reynolds as 'the First Artist in Europe', occurs this vicious couplet: 'View yon Weavers of dull Philosophical Prose, | Led by club-footed Hermes from Sal'sbury Close.'[2]

[1] *Series of Letters*, i. 302–3 (20 Apr. 1775).
[2] Quoted in F. W. Hilles, *The Literary Career of Joshua Reynolds* (Cambridge, 1936), 96. Hilles notes (p. 117) that 'there can be little doubt that the bulk of Sir Joshua's

10. James Harris, by George Romney.

I

For the last five years of his life there were, however, many
more plaudits than brickbats for his literary reputation. His
fame made him an honoured guest at the private concerts at
the Burneys' London home, and there was a trickle of good
news from further afield. Harris had sent copies of his books to
Revd Dr Jeans, chaplain to the British embassy in Paris, and
to its secretary, Colonel St Paul. The latter had sent his copy
of *Philosophical Arrangements* to D'Alembert, perpetual secretary
of the Académie française 'who was very anxious to see it'.[3]
On 18 November 1776 Harris was taken off to dine at
Twickenham by Edward Gibbon, who was to use and
comment favourably on *Philological Inquiries* and *Philosophical
Arrangements* in volume vii of *The Decline and Fall of the Roman
Empire*. And there was, of course, the continuous praise of his
disciple Monboddo. Horne Tooke's vinegarish attack on
Harris in the *Letter to Mr Dunning* might have spoiled May
1778, but the most noticeable change in these years was his
frequent absence from parliamentary duties. He spent time in
the company of John Wilkes, whom he described as 'a
pleasant social companion [and] a lover of letters',[4] and he
continued, tirelessly, his search for new discoveries. On 23
May 1779 he wrote to his son in St Petersburg: 'I must repeat
old questions. What have you heard about the Hymn of
Homer? What about the MS of Strabo wanted by the Dean of
Christchurch, Oxon, for a new edition of that author? If there
be such a MS as the last, and it may be borrowed, I believe
the University would find people to be security for its safe
return.'[5] Harris did not receive those manuscripts, but in
March 1780 he wrote to acknowledge a gift from Catherine
the Great's first minister, Prince Potemkin, who had sent him
a Greek translation of Virgil's *Georgics*. Catherine herself had
discussed Harris's books with his son, the British envoy to St
Petersburg, and while he was in London a 'coffee-house' was
kept each morning at the Charles Street house, where, Mrs

reading could be characterized as philosophical. The writings of Harris and Beattie
he preferred to those of Richardson and Sterne.'

[3] *Series of Letters*, i. 308 (13 July 1775).
[4] Ibid. i. 384–5 (1 Apr. 1778).   [5] *Series of Letters*, ii. 408.

Harris remarked, 'many clever people resort'. In Salisbury, the *Salisbury Journal* reported that the town's theatrical performances had been 'exceeded by none but those of the metropolis' (27 February 1775), and the assemblies in the winter of 1777 had 'beat Bath hollow . . . it seem'd as if Bath had come to Salisbury', Elizabeth Harris observed. She also informed her son that Harris had written to Lord North for permission to absent himself from parliamentary duties (12 November 1777).

As in every previous year since 1761, Harris transported his family from Salisbury to their London home (for many years now in Charles Street, off the south-western corner of Berkeley Square) in mid-January for four months. Apart from parliament, there were concerts to attend and some new things to see. Having initially declined an invitation to view Mrs Elizabeth Montagu's splendid new house designed and decorated by his friend James 'Athenian' Stuart in Portman Square because of gout—'It is the only Instance of his Life, when, by waiting on Mrs. Montagu, he thought there was a possibility of making a false step',[6] he wrote—he eventually complied in 1779. The house 'made me imagine I was at Athens, in a House of Pericles, built by Phidias . . . I felt a more solid satisfaction of reflecting, that, in my own Country, the Genius of Phidias could still produce an Architect, and the Genius of Pericles still produce a Patroness.'[7] As Harris's reputation increased, the family's London activities gradually contracted. When close friends dropped in for breakfast, they generally did so in Salisbury or in Bath; and when they all arrived in Bath on 30 September 1777, James's status as a member of the Queen's establishment entitled him to a 'most violent Peal' of bells from Bath abbey. They enjoyed the accolade, for which they felt obliged to pay the customary garnish, twice, but their reaction was to make for a music shop as quickly as propriety allowed, rent a pianoforte, and retire to their lodgings.[8]

[6] Quoted in R. Blunt, *Mrs Montagu: 'Queen of the Blues' Her Letters and Friendships from 1762 to 1800*, 2 vols. (n.d.), i. 309 (31 Mar. 1776).

[7] Ibid. ii. 100 (undated).

[8] Elizabeth Harris to Gertrude (30 Sept. 1777), Lowry Cole Papers, PRO 30/43/2, adding, 'Louisa and I are like Squire *Blunderhead* much pleased with the civilities shown us.'

The only place that could provide peace and quiet was Durnford, but there was to be no quiet exit from national politics. Harris's parliamentary career closed with a bang rather than a whimper, with the national crisis and 'great terror' of the Gordon riots. At one point seven fires raged at the same time in different parts of the city, his cousin Edward Hooper went about armed, the City was put under martial law (8 June), Newgate prison was opened and fired, and an armed guard was posted at the British Museum. In Salisbury itself there was intense military activity in preparation for an expected French and Spanish invasion (19 August 1780). The military camp near Salisbury which Harris had helped to establish ensured that the vibration of the metropolitan drumbeat was felt on their own doorstep. While Harris sat painfully through the debates with increasing distaste at 'Patriotic' rhetoric, Mrs Harris in Salisbury observed (16 September 1780) that 'a great number of prisoners, French and Spanish, have come through here from Plymouth . . . French prisoners pass my dressing room every day; they are very lively'.[9]

Having been a constant attender in the House of Commons up to 1777, Harris's fourth parliament became irksome, sometimes alarming, and always physically difficult. Acknowledging, perhaps diplomatically, that Harris's absence from the House was caused by ill-health rather than political disaffection, Lord North wrote solicitously to him on 12 June 1779 to ask for his assistance in getting Lord Hyde (who had lost the seat for the University of Cambridge) re-elected to Christchurch. In private, however, Harris had been thoroughly disillusioned with politics for some months. On 18 February 1779 he recorded his disdain for continued parliamentary activity: 'Tired with Tautology and futile Patiotism, I have not deemed our trifling debates worth recording.'[10] The parliamentary sittings of 9–12 February were 'the nights of

---

[9] *Series of Letters*, i. 434 (16 Sept. 1779).

[10] Malmesbury Papers, 'Parliamentary Matters from January 1779 to May 9 when I left London' (18 Feb. 1779), 476B–482B. On 24 Dec. 1779, Harris's son wrote to Gertrude: 'I am sorry any infirmity should prevent his going to Town, but think his conscience may tell him, he has done enough on Parliament, and if the account between him and ministry was to be balanc'd the remainder would be in his favour': 'Letters of James, First Earl of Malmesbury', Merton College, Oxford: F33. (a).

those insolent Insults [on Charles Fox] of the Canaille Patriots, headed by their Leaders and their Servants, when decent Families were kept up then each night, Windows broke, Houses pillaged, and Fire arms fired into them charged with Bullets'. Harris stayed away from the House, until his colleague John Robinson urged his attendance, for at least another ten days until the crisis might be over.

Characteristically, neither physical illness nor political strife distracted Harris from pursuing selected and more amenable literary and social duties. Part of his time was devoted to seeing his last book through the press. Much of it had been composed while walking in his garden at the Manor House in Great Durnford:

> where having arranged his Ideas he used to come in & indite to one of his daughters (both of whom occasionally wrote for him) and as she was assisting he would frequently make her read over a paragraph and ask her if she understood it?—And if any passage did not appear clear and intelligible to her he would alter it, saying it must be obscurely expressed if the meaning were not obvious at first sight to an attentive reader.[11]

He was able to correct and return the proof sheets of *Philological Inquiries* to Nourse by 15 October 1779, twelve days after receiving the news that his daughter Gertrude had survived the shipwreck of the Russian frigate *Natalia* on 3 October, returning from visiting her brother in St Petersburg.[12] On 30 November Mrs Harris heard that Lyttelton, James's exact contemporary and by now a close friend, had died suddenly.

The year 1780 began just like any other year since Harris's election to Parliament back in 1761. After moving to London in mid-January, the combination of parliamentary duties and social activities filled his time. But this year there were longer periods spent in Bath, and the reason for this would soon become clear. In the meantime, John Henry Jacob, the son of his Salisbury physician, was on a European tour, and Harris wrote several enthusiastic letters to him from his London house in Charles Street, or from the House of Commons (December 1779 to November 1780), relaying news of new

[11] Gertrude Harris, 'Memoir'.
[12] Harris to Gertrude from Salisbury, PRO Kew 30/43/2.

books and authors he had met, such as Patrick Brydone, the author of *A Tour Through Sicily and Malta* (1773), and the forthcoming account of Captain Cook's circumnavigation of the globe.[13] He attended assemblies and plays in Salisbury, recitals by Bach in London, and private concerts in Bath given by Rauzzini with the help of Louisa. In London there had been an exhibition at the Royal Academy in the Strand in May where he admired the recent works of Reynolds, Gainsborough, Loutherbourg (Garrick's scene designer), and Zoffany. Of the latter's *tour de force, The Tribuna of the Uffizi*, painted for Queen Charlotte, Harris remarked that 'a more complicated and comprehensive Piece was never seen' (9 May 1780). Gout eventually prevented him from regularly attending the fortnightly Salisbury Assemblies, and walking became painful (29 May). Even so, on 26 September (in Bath) he attended an exhibition of landscape paintings by the locally born landscape painter John Taylor, a student of Hayman, 'a Gentleman not a Professor . . . His pictures are enriched, like the works of Claude, with Architecture, and putting one in mind of him, the Poussins and Salvator Rosa. This Gentleman for Pictures, and Lady Beauclerc for Drawings, tho no Professors are capital in their way, and yet, which is extraordinary, were never out of England' (26 September 1780).[14]

Harris had always been quick to identify the points at which English culture might be compared favourably with continental achievements; but he also noted signs of dissolution. With dismay, he observed that Lord Temple's house at Eastbury (once the home of Bubb Dodington), than which 'there was not a more stately Pile in the West of England' and which 'might have kept pace with the Pyramids', had recently been demolished and its materials sold off.[15] However, the

---

[13] These letters are still in private hands, and copies were kindly made available to me by Mr John Jacob of Durrington Manor, Salisbury.

[14] Harris to Henry Jacob (26 Sept. 1780), pp. 1–2. Taylor, it seems, never achieved the status promised by his early distinction. Smollett, in the manner of Matthew Bramble, had reacted to Taylor's landscapes in a very similar way between 1766–70: 'If there is any taste for ingenuity left in a degenerate age, fast sinking into barbarism, this artist, I apprehend, will make a capital figure, as soon as his works are known', *The Expedition of Humphry Clinker* [1771], ed. L. M. Knapp (Oxford, 1984), 76.

[15] Harris to Jacob (26 Sept. 1788), and cf. Harris to his son, *Series of letters*, i. 471.

health of Mrs Harris, perhaps the main reason for being in Bath, was apparently responding well to hydrotherapy.

But some time towards the end of 1780 Harris himself returned unexpectedly to the Close. This was to be his last illness. His Salisbury physician Dr Jacob advised him to return to Bath, and a house was eventually engaged for him on North Parade. Gertrude reported Harris's prescient response to these arrangements: ' "You have taken a House for me at Bath, but you must now do me another Service . . . and put off that House—I shall very soon want no House at all"—this he said with emotion and then quitted the room.'[16] Having forbidden music for the last three or four weeks of his life because of its emotional effect on him, his preferred consolation in his last days was remembrance of his great friendship with Fielding almost thirty years before. He savoured yet again his friend's portrait of their mutual friend William Young, and through Parson Adams's words recalled the Christian Stoicism which Fielding said he had first learnt from Harris:

One evening & one of the last of his existence he desired to have Joseph Andrews read aloud, and he himself selected the different chapters from which Louisa read several passages, to which he listened with attention, expecially to that part of the 3d Chapter in the second book wherein Parson Adams discourses his Host on religion, and the Immortality of the Soul.[17]

On 16 January 1781 the younger James wrote to his sister Gertrude one of those letters dreaded by all expatriates for whom the tyranny of distance carries a permanent sense of dread. In transmitting his new year wishes he also communicated his anxiety about his parents' declining health, urging Gertrude to move the family to Bath:

I trust in God you will go to Bath, remain at Bath, settle at Bath, never leave Bath, it is the very place suited both to my father & mother, society, amusements, no trouble, no dull visits, no parliament—& as for the expense, surely it cannot enter into the

[16] Gertrude Harris, 'Memoir' p. 2.
[17] Gertrude Harris, 'Portrait of . . . my Mother' (1806), PRO Kew 30/43/1/4, p. 18.

account, when so important a point as health is in the respect, every other consideration is secondary.[18]

But it was too late. James Harris had died at home in the Close at a quarter to four on Friday, 22 December, in the arms of his friend, the physician Dr Jacob, more than three weeks before this letter was even written.[19] Gertrude's letter of 25 December informing the younger James of their father's death was still on its way to St Petersburg. Mrs Harris, who by now was unable to write a legible hand due to arthritis, survived her husband by only ten months, dying of a syncope on 16 October 1781 in Gay Street, Bath. Sir James Harris, his wife, and 3-year-old son arrived from St Petersburg just two days before Elizabeth died. After the death of both parents, the house in the Close, for so long the heart of the family's corporate life, as well as the nerve-centre of Salisbury's musical, theatrical, and social energy, fell silent. Gertrude added a poignant note to her manuscript memoir: 'As none of us liked returning to the old House in the Close, we very soon returned to Durnford.'[20]

Harris's death signalled an immediate cultural crisis in Salisbury. The town had lost its most distinguished literary figure, the festival's maestro, and the subscription concert director. Moreover, the master of the lyre was to be buried only nine days after the funeral of his closest festival collaborator, the cathedral organist Dr John Stephens, on 28 December. The six pall-bearers (perhaps Dr Jacob, Canon Bowles, Richard Owen Cambridge, Edward Hooper, and Charles Batt) were met at the west door of the cathedral by the whole choir, and on the following Sunday the Revd Chafy preached a sermon which served as a commemoration among Harris's immediate friends and neighbours. Some years later,

---

[18] 'Letters of James, First Earl of Malmesbury', deposited in Merton College, Oxford: F33. (a). Harris had complained to the Earl of Rochford, Principal Secretary of State, on 21 Apr. 1771, that he neither could nor should subsidize his son's ambassadorial expenses out of his own pocket: see R. A. Roberts (ed.), *Calendar of Home Office Papers of the Reign of George III 1770–72* (1881), 246.

[19] Lord Pembroke to Lord Herbert (24 Dec. 1780), in Lord Herbert (ed.), *Pembroke Papers (1780–1794): Letters and Diaries of Henry, Tenth Earl of Pembroke and his Circle* (1950), 75. Lady Pembroke had seen Dr Jacob 'at the Play; very melancholy' in the evening.

[20] Gertrude Harris, 'Portrait . . . of my Mother' (1806), p. 18. The family spent most of its time in Bath before removing to Durnford.

John Bacon, the Royal Academy's first Gold Medallist and the sculptor of Johnson's bust at Pembroke College, designed a permanent monument to Harris which was erected in the north transept of Salisbury cathedral, a female figure of Philosophy holding a cameo portrait of Harris in profile, with a Latin inscription celebrating his public career and his scholarship in the fields of grammar, logic, and ethics.[21]

II

Harris had travelled a long way; but however great the universal focus he maintained in his work on aesthetic and philosophical theories, those interests had been settled at an early age, and they had been shaped by the passionate single-mindedness, and perhaps the isolation, of the autodidact. By fortune born into a cultured family in a musical city, he had bought Handel's *Suites de pieces pour le clavecin* (1729) when he was 20 and supposed to be practising the law rather than the harpsichord.[22] Philosophical interests were virtually guaranteed by his kinship with the Shaftesbury family and a strong local tradition of Idealist thinkers virtually unique to Salisbury at the time. Harris converted a provincial ethos of competence into public performance subject to the full and rigorous scrutiny of a Johnson, a Tooke, and two subsequent centuries of expert linguistic scrutiny. His most important book had to await the 1960s for its full appreciation. His entry into the public arena of national politics was, if compared with his father's repeated failure, almost effortless; promotion came

---

[21] Malmesbury 'Memoirs of the Life and Character of the Author', *The Works of James Harris, Esq.*, 2 vols. (1801), i. p. xxi. The *Salisbury Journal* account (Monday, 1 Jan. 1781) of the funeral on the previous Thursday speaks of the choir's 'last melancholy tribute of piety over those ashes, once animated with the very soul of harmony', and announced that part of the *Messiah* and Handel's 'Funeral Anthem' would be performed on the following Tuesday's subscription concert. Parry replaced Stephens as cathedral organist, and, as a measure of Harris's importance in the city's musical life, the subscribers met on 15 Jan. to determine the future management of the Salisbury subscription concert: Corfe became the new director of the band, and Harris's protégé William Benson Earle took over the festival's management until his death in 1796: see R. Benson and H. Hatcher (eds.), *The History of Modern Wiltshire by Sir Richard Colt Hoare: Old and New Sarum* (1843), 586.

[22] Gerald Coke Handel Collection: see J. Simon (ed.), *Handel: A Celebration of his Life and Times, 1685–1759* (1986), 186.

very quickly, and it was clearly prompted by the kind of man he was, and the kind of talents he possessed. He was a philosopher in his study, but a man of the world when he needed to be.

In every sense Harris was a man of his own time. His Idealism arose from his opposition to the prevailing materialism; thus his sometimes severely abstract style and his subject-matter were to meet with less informed sympathy than he might have wished. Generally, one discerns among Harris's contemporary readers more *respect* than understanding or enthusiasm, a perception that his impressive scholarship was admirable, generally unanswerable, but yet could be safely filed away as being too hard for immediate attention. Yet understanding, respect, and enthusiasm existed to a remarkable degree among those who, like Fielding, Upton, Boswell, Lowth, Monboddo, and Gibbon, knew the man more than casually. There was, after all, little natural enthusiasm for abstract aesthetic and philosophical theory in England, and, since the Deists had queered the pitch, Harris's brand of pantheistic rationalism (what he called a 'Providential Circulation'), not to mention his actual and intellectual kinship with Shaftesbury, was ahead of its time. It was left to later, and German, commentators to discern that there was, as early as *Three Treatises*, a strain of 'pre-Romanticism' in his writing.[23]

'Harris's England', like 'Pope's England' in *The Dunciad*, had been swept by the materialist ethic of Cartesian–Newtonian mechanism, and a mania for sectarian subdivisions in the worlds of learning, politics, religion, and the arts. In reconstituting the classical tradition, in which the best had been thought by the best of men, Harris (like Reynolds) could justifiably feel that he had inscribed English culture in its proper classical and post-Renaissance context. Unlike Reynolds, whose aesthetic theory is essentially tragic, Harris's epistemology maintained a boundless optimism: there was nothing which the human mind was incapable of achieving by means of the language of the arts. The paradox is that his world-view, based as it is on harmonic and analogical

---

[23] See e.g. Funke, *Englische Sprachphilosophie, passim.*

presuppositions, is in the end more rather than less geo-
metrical than even La Mettrie's *Man a Machine*, with the
crucial proviso that human systems were always to be seen as
functions of a divine order. Ideologically, therefore, Harris's
writings are simultaneously perfectly Augustan and self-
consciously oppositional: his works see the individual in a
tradition, a work in a genre, a genre in a hierarchy, a *parole* in
a *langue*, Art itself in an overarching metaphysical structure.
He asserts concord, pattern, analogy, harmonic relationships,
and intelligible causes in art, language, literature, and
criticism, but for a culture which seemed increasingly to prefer
the narrowly nationalistic, parochial, materialistic, experi-
ential, and egocentric, and which also seemed perversely to
ignore the source of its own power in what he variously termed
the 'first philosophy' and 'the dignity of Mind'.

Harris's religious views, to readers like Johnson and Mrs
Thrale at least, were too much a matter of inference, too easily
smeared with the reputation of Shaftesbury. It is not difficult
to see how enthusiasm for correct social conduct and classical
philosophy could be seen to downgrade Christian duties. But
their imputation about Harris's impiety was quite obviously a
misreading of both the man and his work. Nobody who had
listened to his music or discerned the metaphysical basis of his
epistemology could be in any doubt as to the reality of his
spiritual convictions. One might raise the same suspicions
with greater plausibility about Johnson's own *Rasselas*. Harris's
rationalism is more clearly based on metaphysical conviction
than anything Johnson ever wrote: his theories of logic,
language, and art rest on a fundamental belief that human
forms of artistic expression are links and pathways to
inspirational ends. That is why he admired Handel and, to a
lesser extent, Fielding; and that is why he scorned Locke and
feared the effects of Swift's satire.

III

Harris did not live long enough to learn that his posthumous
volumes were to become the favoured reading of his intellectual antagonist, Johnson, or that his Idealism was to be
carried on in Reynolds's *Discourses on Art*: that his linguistic

theories were to be developed by Monboddo, and were to influence Adam Smith, Hamann, Herder, von Humboldt, Jeremy Bentham, Dugald Stewart, James Beattie, Hugh Blair, and Joseph Priestley.[24] He could not have known that his Universal Grammar was to liberate the teaching of English from the tyranny of Latin. Nor could he have foreseen that the idea of a linear progress through a series of triumphs and progresses was, as he always believed it to be, a modernist illusion, nor anticipated that William Blake's violent opposition to Reynolds's aesthetic theories would stimulate a view of the mind precisely akin to his own:

Man brings All that he has or can have Into the World with him. Man is Born like a Garden ready planted and Sown . . . I always thought that the Human Mind was the most Prolific of All Things & inexhaustible.[25]

In commemorating his friendships in his last book Harris recognized for the first time that the causes of art and knowledge may not, after all, lie in any abstract metaphysic, but in a none the less mysterious process in individual men, himself included. Since he had known nearly all the leading English figures of his day in painting, literature, philosophy, music, literary criticism, politics, and the natural sciences, this is both an unsurprising and a fitting conclusion, particularly for a life devoted to the Shaftesburian idea that the contemplative life need not and should not be opposed to a civic career.

Harris is remembered as a musical theorist by musicologists, as a literary theorist by historians of criticism, as a universal grammarian by linguists. Salisbury citizens without the remotest interest in universal grammar remembered him primarily for his role as maestro. In a memorable phrase, the *Salisbury Journal* noted that he had been 'animated with the very soul of harmony'. The purpose of this book has been to

[24] For a discussion of Harris's influence, see Joly's edn. of Thurot's trans. of *Hérmès*, pp. 8–13, 24–57.
[25] See Sir Joshua Reynolds, *Discourses on Art* (1769–91), ed. R. Wark (New Haven and London, 1959), 310 (Blake's annotation in his 1798 copy of Reynolds). For a possible connection between Blake and Harris's *Daphnis and Amaryllis*, see M. W. England, 'The Satiric Blake: Apprenticeship at the Haymarket?', *Bulletin of the New York Public Library*, 73 (1969), 536–8.

reveal not only the variety of Harris's interest and achievements but also their emanation from a particular and highly 'associated' sensibility characterized by a tenacious artistic devotion in which each interest fed the other. His life reveals everywhere a desire to propagate and share his cultural experience, to involve others and be involved. He poured his energy into a long-running and popular musical festival which he himself made no effort to memorialize. Literary friendships were his life-blood; and language was his tool of knowledge. In other words, Harris sustained a deep-seated belief in the necessary integration of things, and this integration originated with the individual self. Not the least important example of this was his attitude to the practical education of his family. He was refreshingly free of snobbery and prejudices about gender-roles, freely extending the benefits of his learning to women writers, such as Sarah Fielding, Jane Collier, Elizabeth Carter, Mrs Thrale, Elizabeth Montagu, and others, and yet demanding the highest standards from them in exchange for his assistance. With his own teenage daughters the same uninhibited motives are revealed. He set them to work on exactly those topics which exercised his own philosophical imagination. Gertrude's childhood exercise book for 1766 still survives, and it contains 'little Essays . . . written by my father for my sister and myself who used to write them in his room as a kind of Exercise generally before breakfast'. The subjects were set weekly and included 'On Happiness', 'On Language' (beginning 'A Language is a collection of sounds articulate by which a number of Persons communicate their sentiments one to another'), 'On the Capital Parts of Speech', 'On the Terrestrial Globe', 'On Poetry', 'Upon Deceit and Fraud'. One essay in particular, 'Upon Virtue', reflects that holistic, harmonic ideal of inner and outer conduct which many of his contemporaries signified when they called him 'the amiable Mr Harris':

In all our actions, in all our dealings, we ought to have a proper regard both to others, and to ourselves. A proper conduct, in our dealings with respect to others, constitutes the Virtue of Justice; with respect to ourselves, the Virtue of Prudence.[26]

---

[26] Gertrude Harris, Exercise Book, PRO Kew GD 30/43/1, 18 June: the essays date from 21 May to 14 Aug. ('On Gardening'), 1766. Her sister Louisa was 13 years

Fielding had fashioned his greatest novel out of such material. Harris himself had fashioned a life. The fact that such a maxim comes from Aristotle simply reinforces the historical continuities that Harris had spent his life attempting to retrieve and transmit.

Harris's published works offer a stiff challenge to the intellect, and sometimes the patience, of his readers. No one could claim that aesthetic, linguistic, and epistemological theories could be taken as light reading to fill a moment's idleness. Even though his friendship brought pleasure, encouragement, and in some cases consolation, to those who knew him well, the qualification for that friendship was nevertheless a demanding consanguinity of mind. The sinner in Boswell shrewdly identified both the attraction and the embarrassment of having Harris as a friend: the exemplary life and the inspirational ideals in his writing were no less admirable for being beyond the reach of ordinary men.

old at the time, and Gertrude was 16. In her 'Portrait . . . of my Mother' (1806) Gertrude remarks, revealingly: 'it was not till after my mother's death, that I was informed by my sister (to whom she had told it long after we were all grown up) that I had been her favorite child; whilst I had myself always supposed it had been my Brother', PRO Kew 30/43/1/4, p. 4..

Appendix I

Concord

HARRIS's only lengthy published poem was in print by the end of November 1751. It was privately printed in a very small London edition, and although no printer's name appears on the title-page, it was probably done for the printer of *Three Treatises* and *Hermes*, John Nourse. In his letter of 26 November to James, George William Harris promised to bring four copies of the printed poem to Salisbury, together with news of the reception of *Hermes* by some of Harris's friends. Its only 'public' appearance was in an inaccurate and incomplete reprinting in volume xii (pp. 53–9) of Frances Fawkes and William Woty's *The Poetical Calendar* (1763), where the dedication to the Earl of Radnor is included, but the prefatory 'Argument' is omitted. Harris's holograph and three of the original four printed copies survive in the Harris papers, and the poem is here printed from Mrs Elizabeth Harris's signed copy, which identifies the dedicatee as Sir John Robartes, created fourth Earl of Radnor in 1741.

Sir John Robartes, FRS (1686–1764) had been a family friend since, at the very latest, February 1747, and although art and science had been the pastimes of his family since the time of the first earl (1650–1718), more precise information about his friendship with the Harris family is yet to be established. His Twickenham home, Radnor House, which he inhabited from 1741, lay about 400 yards from Walpole's Strawberry Hill to the south-west and at the same distance from Pope's villa to the north. He was therefore also a near neighbour of Richard Owen Cambridge, whose home was a frequent holiday retreat for the family of James Harris. Radnor drew a secret service pension of £2,000 from the Duke of Newcastle between 1754–6, but was better known as a collector of paintings, a dilettante with a reputation for a restless and somewhat whimsical interest in landscape design. Horace Walpole nicknamed Radnor's Twickenham house 'Mabland', and acidly remarked to Conway, 8 November 1752: 'Have you any Lord Radnor [near you] that plants trees to intercept his own prospect, that he may cut them down again to make an alteration?'

Even so, it was Radnor who performed the role of Harris's aesthetic mentor. He helped Harris himself to acquire an art collection. In June 1762, assisted by Thomas Harris, Radnor purchased Canaletto's painting of the Horse Guards and St James's

Park, together with a Hobbema and 'a great Ruysdael . . . the best picture of the kind I ever saw'. Each was intended for James, and the Canaletto and Ruysdael were left to Harris in Radnor's will (*Series of Letters*, i. 85). Harris went on to acquire paintings by Luca Giordano from his son in Berlin, on the recommendation of Raphael Mengs (including the 'Apotheosis of Trajan', the 'Flight into Egypt', and Mengs's miniature self-portrait). George Romney was chosen to paint five Harris portraits and was at work on the commission in 1778.

When Radnor died unmarried in 1764 the Radnor line died with him. There appears to be no connection between Harris's dedicatee of 1751 and Harris's Salisbury neighbour at Longford Castle, William Bouverie, Viscount Folkestone, a much more impressive art collector, created Earl of Radnor on 31 October 1765. As has been indicated in Chapter 3 above, *Concord* is primarily a versification of the third Earl of Shaftesbury's benevolist aesthetic theories in the *Characteristicks*, although there are clear echoes also of Pope's *Windsor Forest*, *Essay on Man*, and Milton's *Paradise Lost*, as well as a general indebtedness to the Platonic notion of Ideal Forms. Most of the ideas of cosmic order in Harris's poem had become commonplace by 1751. For a useful survey, see M. Battestin, *The Providence of Wit: Aspects of Form in Augustan Literature and the Arts* (Oxford, 1974), 1–57.

THE ARGUMENT

Proposition. General Sympathy of all Things congenial: Exemplified in the Elements, in Vegetables, in Brutes, in Man. Cause of this; One Principle or Source of all Things, which preserves them all by different manners of Sympathy; Things inanimate, by Cohesion; Plants, by Vegetation; Animals by Sensation; Man by all these, and by Reason super-added. Hence the general Sympathy of Man with all things; with Works of Nature; with Works of Art; but chiefly with his own Kind, in the Energies of Benevolence, Friendship, and Love; exemplified in the celebrated Parting of Brutus *and* Portia, *taken from* Plutarch. *Conclusion.*

CONCORD

A POEM

The Deeds of Discord, or in Prose or Rhyme,
Let others tell. 'Tis mine (the better Theme)
Concord to sing; and thus begins the Song.

C O N C O R D.

Eliza A *Harris*

P O E M.

INSCRIBED

To the RIGHT HONOURABLE

John Robartes

~~the~~ EARL of *RADNOR*.

L O N D O N:

Printed in the Year M.DCC.LI.

11. Elizabeth Harris's signed copy of *Concord* (1751).

CONGENIAL THINGS TO THINGS CONGENIAL TEND:
So Rivulets their little Waters join,
To form one River's greater Stream: So haste
The Rivers, from their different Climes, to meet,
And kindly mix, in the vast Ocean's Bed:
So Earth to Earth down goes; and upward flies,
To Fires ethereal, each terrestrial Blaze.   10
Such *elemental* Concord. Yet not here
Confin'd the sacred Sympathy, but wide
Thro' *Plant* and *Animal* diffusely spread.
How many Myriads of the grassy Blade
Assemble, to create one verdant Plain?
How many Cedars tow'ring Heights conspire,
Thy Tops, O cloud-capt *Lebanon!* to deck?
Life-animal still more conspicuous gives
Her fair Examples. Here the social Tye
We trace, ascending from th' ignoble Swarms   20
Of Insects, up to Flocks, and grazing Herds;
Thence to the Polities of Bees and Ants,
And honest Beavers, bound by friendly League
Of mutual Help and Int'rest.—Cruel Man!
For Love of Gain, to persecute, to kill,
This gentle, social, and ingenious Race,
That never did you Wrong—But stop, my Muse,
Stop thy sad Song, nor deviate to recount
Man's more inhuman Deeds; for *Man* too feels
Benign Affection, nor dares disobey,   30
Tho' oft reluctant, Nature's mighty Voice,
That summons all to Harmony and Love.
Else would to Nature's Author foul Impute
Of Negligence accrue, while baser Things
He knits in holy Friendship, thus to leave
His chief and last Work void of sweet Attract,
And Tendence to its Fellow. But not so,
Not so, if truly sing the heav'n-born Muse:
And she can tell; for she the limpid Fount
Of Truth approaches; Rumours only reach   40
Our earth-born Ears. Then mark her Tale divine.

　　Ere yet Creation was, ere Sun, and Moon,
And Stars, bedeck'd the splendid Vault of Heav'n,
Was GOD; and GOD was MIND; and MIND was *Beauty*
And *Truth*, and *Form*, and *Order*: For all these
In Mind's profound Recess, and Union pure,

Together dwelt, involv'd, inexplicate.
Then Matter (if then Matter was) devoid,
Formless, indefinite, and passive lay;
Mysterious Being, in one Instant found   50
Nor any thing, nor nothing; but at once
Both all and none; none by *Privation*, all
By vast *Capacity*, and pregnant *Pow'r*.
This passive Nature th'active Almighty Mind
Deeming fit Subject for his Art, at once
Expell'd Privation, and pour'd forth Himself;
Himself pour'd forth thro' all the mighty Mass
Of Matter, now first bounded. Then was *Beauty*,
And *Truth*, and *Form*, and *Order*, all evolv'd;
Was open'd all, that lay enwrapt and hid   60
In the great Mind of Godhead. Forth it went,
Forth went the pure Quintessence far and wide,
Thro' the vast Whole; nor did its Force not feel
The last of minim Atoms. So (great things
If we compare with small) in sable Cloud
Invelop'd, lies the Lightning: Mortal Men
Look up, and dread th'Event: when, lo! illum'd
All in a Moment, the small nitrous`Seeds
Expanding, fill Heav'n's mighty Vault, and quick
From Pole to Pole the fiery Terror flies.   70

    Thus MIND *thro' all things pass'd*, Essence and Worth
Giving and limiting to each in Bounds
Proportion'd to its Kind. To Clods and Stones
It gave *Cohesion*; to things vegetant
Nutrition, and the Pow'r of Growth: To Brutes,
Sense, *Appetite*, and *Motion*: But to Man
All these it gave, and join'd to these the Grace,
The chosen Grace, of *Reason*, Beam divine!
Hence Man, ally'd to all, in all things meets
Congenial Being, Effluence of Mind.   80
And as the tuneful String spontaneous sounds
In Answer to its kindred Note; so He
The secret Harmony within him feels,
When aught of Beauty offers. This the Joy,
While verdant Plains, and grazing Herds we view;
Or Ocean's mighty Vastness; or the Stars,
In midnight Silence as along they roll.
Hence too the Rapture, while th'harmonious Bard
Attunes his vocal Song; and hence the Joy,

While what the Sculptor graves the Painter paints,   90
And all the pleasing Mimickries of Art
Strike our accordant Minds. Yet chief by far,
Chief is Man's Joy, when, mixt with human Kind,
He feels Affection melt the social Heart;
Feels Friendship, Love, and all the Charities
Of Father, Son, and Brother. Here the pure,
Sincere Congenial, free from all Alloy,
With Bliss he recognizes. For to Man
What dearer is than Man? Say ye, who prove
The kindly Call, this social Sympathy,   100
What but this Call, this social Sympathy,
Tempers to Standard due the vain Exult
Of prosp'rous Fortune? What but this refines
Soft Pity's Pain, and sweetens ev'ry Care,
Each friendly Care, we feel for human Kind?
O *Gomez!* gives thy Pelf such Bliss? Or ye,
Who wade thro' Blood to Fame, and, worse than Wolves,
Prey on your Kind, can your vain Triumphs give
Such solid Happiness? Like Giants old,
Ye fight 'gainst Nature, Nature's Order spurn,   110
And would o'erthrow. But she, be well assur'd,
Will baffle all your Efforts vain, and plant
Fell Daggers in your Hearts, Terror and Guilt,
Heart-burning Hate, and dreary black Remorse.

When *Rome* her last of Heroes lost (e'er since
The wretched Nurse of *Cæsars*, and of Monks),
When Brutus, urg'd by Faction, and a Mob
For basest Servitude now ripen'd, fled
From *Latian* Soil, then, to attend her Lord,
Fled too the faithful Partner of his Bed,   120
The wife, the virtuous Portia. Much she fear'd;
For much she lov'd. He, godlike Man, inspir'd
Not with less Love, tho' with superior Strength
Of Reason, thus her anxious Thoughts reliev'd.
'O Portia, best of Wives, grateful thy Sight,
'Grateful thy Converse. Yet, whene'er we part,
'(And soon we must) then do not, Portia, thou,
'Like other Women, sink; but bravely rouse
'Thy mighty Sire's Remembrance. His firm Deeds
'May steel thy Soul to Suff'rance. Me the Fates   130
'O'er distant Seas to hostile Arms compel.
'Should we succeed, then is thy Lot and mine

'Fortunate Virtue: Should we fail, 'tis still,
'Still, PORTIA, Virtue: Think on that; then turn
'Thy mental Eye to ev'ry worst Event;
'And, by premeditating, learn to bear
'Whate'er befalls of Ill. Joys will not come
'The less for this; and each Joy unforeseen
'With doubled Energy will bless thy Soul.'

   Thus he with balmy Words the lab'ring Pain   140
Within her Bosom sooth'd; and she was chear'd.
Stedfast she travell'd, stedfast she arriv'd
To the Sea-brink, where many a Vessel lay
With Sails expanded, BRUTUS to receive.
Now were they lodg'd in hospitable House,
The tender Scene of their long last Farewel:
Yet stedfast still she was; stedfast she saw
The Mariners prepare. When, lo! by Chance
A Picture meets her wand'ring Eye. It shew'd,
In living Lines, brave *Hector's* last Embrace,   150
When from his weeping long-lov'd Spouse he went,
Never to see her more. Ah, PORTIA! then
Where fled thy Courage? where thy stedfast Heart?
Thou lookst, thou feelest: The sad moving Scene
Too near Resemblance bears. Forth gush thy Tears,
Thy Spirits sink, thy Limbs forget their Strength,
And thou forgettest all thy BRUTUS said.
Yet he forgives. Forgives? Yet still he loves,
Loves Thee, that thou forgettest all he said;
For well he knows the Cause: 'Twas faithful Love,   160
By faithful Love affected; Like by Like;
CONGENIAL BY CONGENIAL.—

              But thy Song,
'Tis time, my Muse, to end. This Verse, O Thou,
RADNOR! who prov'st a secret Sympathy
With all that's fair; Patron and Judge of Arts;
Studious of Elegance in ev'ry Form;
RADNOR! this Verse be consecrate to Thee.   168

FINIS

Appendix II

'The History of the Life and Actions of Nobody'

HARRIS's 'History . . . of Nobody' (Harris Papers, vol. 7, part 10) is an example of the paradoxical encomium, dedicated to, and in the manner of, Fielding, whose 'An Essay on Nothing' had appeared as the fourth prose item in *Miscellanies*. Its date of composition may be set between the publication of the *Miscellanies* (12 April 1743) and May 1744, the publication date of *Three Treatises*. Harris uses the word 'caducous' at the beginning of the second paragraph of his dedication, punning on the sense of physical illness and the wand of Hermes, thus anticipating the eponymous hero of Harris's second book (1751). Additional allusions to Colley Cibber's autobiography, *An Apology for the Life of Mr Colley Cibber* (1740), and to the mocking of heroic genealogy in the second chapter of *Joseph Andrews*, confirm that Harris wrote this short piece to complement his friend's wit. Although it was not published, Fielding received the essay, and was delighted by it. Unfortunately, his letter of response is undated, but elaborates the Scriblerian joke by bringing Harris's attention to 'a curious Piece of Antiquity' which Harris had overlooked, i.e. a bust of Nobody in the collection of 'Jer. Pierce', Bath surgeon. A Bath provenance and a possible date in September of 1743 seem likely. Fielding apologises for his delayed response, and then goes on to discuss the contrast between Harris's *Three Treatises* (which he had read in manuscript), and the proposition that Wit and Philosophy are antithetical properties (Harris Papers, vol. 40, part 4):

> I have read yr History with great Pleasure. The Vein of Humour is rich and clear; and yo luxuriously throw away more Learning in yr Merriment than would set up a voluminous modern Author in the gravest Treatise. Others may be surprized to see such Excellence in the Ridiculous flowing from the same P[erso]n from whose Genius and Knowledge the World will shortly derive Treasures of so different a Kind, and of so much greater Value: but I am not of this Number: for to me Wit and Philosophy have always seemed to bear a closer Alliance than they are allowed by those who have little Acquaintance with either. Plato seems to have thought so w[he]n he introduced Socrates jesting in his Phaedon, and my Ld. Bacon expressly says Wit and Judgement are the same; nor would Mr Locke *have* asserted the contrary, had he not had a contrary Idea of the Word from the last mentioned

Philosopher, or Wit: for he hath Pretension to both Names. To say the Truth, the World have very wickedly abused these two Words; insomuch that I doubt whether more is generally meant by Wit now than a pert gay Folly or by Philosophy than a grave dull one; the former of wch may perhaps may be well personified by a Dancing-Master or Merry-Andrew, the latter by a Perriwig-pated Doctor or Presbyterian Parson. If this be so the vulgar Error and the Reason of it are apparent, and true Wit and Philosophy are out of the Question; and may, for all that hath been said of the false, be in Reality not only alike, but one & the same. In wch Light, as they appear to me, it can never surprize me to find them both united in the same P[erso]n. You will excuse me however if I am not so well able to account for yr Knowledge of such Variety of Facts concerning the only Hero of Antiquity with whose History I thought yo unacquainted.

Fielding's thoughts on the co-presence of Wit and Philosophy are the subject of another unpublished essay by Harris (see Appendix iv below, 'Upon Ridicule').

The text below is taken from Harris's 15-page holograph manuscript, preserving original spelling and punctuation, with original page numbers enclosed within square brackets.

THE HISTORY OF THE LIFE AND ACTIONS OF NOBODY

Dedication to H.F. Esqr.

Dear Sir,

You may remember, as we were once seriously discoursing on Time's being the general Demolisher of all sublunary Beings, we observed that Palaces, Temples, and spacious Cities he chose indiscriminately to attack at all Seasons; that the Human Body he assaulted with greatest vehemence about Autumn; but that the Works of human Wit, whether Verse or Prose, he was usually most prone to assail in Winter about Christmas. To these Observations on the seasons of Time we added others on his Weapons, which appeared no less remarkable. And here we both agreed, that whatever harsher [2] weapons he might use against grosser subjects, such as Human Heads, stone Walls, and the like, yet in destroying the more refined & delicate Works of Wit he chose for ye most part to employ a Pye. And hence the most judicious

Mythologists have esteemed the Pye, as much as the Scythe, to be one of his true & proper Symbols.

But here perhaps you may justly stop me, and demand in the Language of Horace Quotsum hac?—It is, Sir, from a consciousness of this caducous state both of me & mine, that I am ambitious to lay this my darling Offspring at your door. It contains the History of Nobody, collected from the best, and most authentic Memoirs. Should you ask, why as a Biographer I have preferred addressing myself to you, rather than to those other my most ingenious Contemporaries, whether male or female, who have written Lives under ye Names of Kouli Kan, Kouli Kibber, or the Duchess of Sheba &c; my Answer is— that as all these Authors have written the Life of Nobody as well as myself, they will rather consider me I fear as a Rival, than vouchsafe me their Patronage. But this is not ye case with the Author of the Life of Joseph Andrews. He [3] has writ the Life not of Nobody, but of Somebody; and is indeed almost the only one, who has written any such life, since Old Nepos & Plutarch. For this reason between you, Sir, & me there can be no Jealousy or Rivalship, as my Subject is so widely different, & of a kind so far inferiour. Permit me therefore to fly to you, as to a Patron, that when Time with his Pye shall have reduced this Work of mine to Nothing, & me its Author with his scythe to Nobody, I may still in tradition at least be remember'd to have been Somebody by having once been, Dear Sir,

Yr sincere Friend and Admirer
Y.Z.

[5] *The History of the Life and Action of Nobody*

Nobody, the illustrious Subject of these memoirs, is descended from a Branch of ye noble House of Nothing. From this aboriginal and primitive Stock have branched out, at different times, the several collateral and honourable Families of ye Privatives, Negatives, Non-Entities, Nullities, &, among the rest, that also of the Nobodies, of whom our present Hero is the Representative.

The great Dignity of the House of Nothing is allowed on all hands. Some modern Genealogists have gone so far, as to make it the Origin of all the Somethings & the Sombodies, which are now existing in the Universe. But the ancient Heralds seem to have been of a contraray opinion, allowing Nothing to have had no Children but Nothing, Nullity, Nobody, Non-Entity, & the rest of those worthy [6] Personages above mentioned with respect. That King at Arms, Lucretius, is positive as to this particular.

> Nullam Rem e Nihilo gigni————————[1]

says he; and Clarencieux. Aristotle affirms the same too in his Physics. It is farther urged on behalf of this Assertion, that so strong is the mutual Antipathy of Something & Nothing, that they cannot bear each other's Company, and are never seen together in the same place. But this with submission I think disproves not their consanguinity, since Experience shews us that the same may befall the nearest Relations among people of ye first rank. We shall however leave this arduous Point to the discussion of more subtle pens, assuring them that their labours, if not rewarded by Somebody, will be infallibly rewarded by Nobody, to whose History we now hasten.

We have already said that Nobody's Family was a younger Branch of the ancient House of Nothing. A younger Branch tis true they must be confessed, but yet of great Antiquity, as appears from the following Authorities. As long ago as the Trojan War we learn from Homer there was one ογτιε or Nobody, who was a particular Friend to the sage Ulysses. [7] So high indeed was he in the Estimation of this Hero, that when Ulysses was asked by Polypheme to tell his name, he thought the Name of ογτιε would carry more weight with it, than his own. Nobody (says Ulysses) is my Name; they call me Nobody. And that the same ογτιε was a prime Agent in ye ceremony of boring out the Cyclop's eye, is evident from the Words of the Cyclops himself, whose immense bellowings after that accident having brought his brother Monsters around him, upon their enquiry into his complaint, he thus answers them, *Gentlemen, this is Nobody's doing.*

And here indeed I cannot but by way of digression observe, that thro' all ages and Countries we meet in History of like Exploits, which have been done by the Nobodys, that

---

[1] '[That] nothing is born of nothing': Lucretius, *De Rerum Natura*, 1.150.

flourish'd in different Periods. Who for instance burnt such a Temple? Who broke such a China Jar, or robbed such an Old Woman's Orchard? We find the Books in a thousand instances all agree, 'twas done by Nobody. But to return [8] to our Subject.

The next Nobody of Note that occurs in Graecian History, was held in particular Estimation by the Stoic Philosophers. They even asserted that Nobody had attained to the highest pitch of their Philosophy; was utterly devoid of all Passion, & could treat the most exquisite Pain as no Evil. Chrysippus called one of his favourite Syllogisms after him, by the name of ογτιε; and Tully, in his Laelius, speaking of that Wisdom, which the Stoics supposed their perfect Character to possess, calls it a Wisdom, quam mortalis Nemo est consecutus.[2]

These are the two most eminent Greek Nobodys, which I have been able to discover in my reading. Among the Latins they were far more numerous. One of them was a perfect Orator, considered by Tully as his Model, & not Demosthenes, as some have falsely imagined. For this we have Tully's own Words in his Treatise called Orator, Sect 7.—Atque ego in summo Oratore fingendo talem informabo, qualis fortasse Nemo fuit.[3] But I in framing a perfect orator, shall form such a one, as perhaps Nobody was. There were many others, whom it would be needless here to enumerate. Yet did none of them ever shine with such transcendent lustre, as He who obtained that splendid Testimony of

[9] Nemo mortalium omnibus horis sapit.[4]

From the decay of the Roman Empire till about the time of Leo the tenth, I find that all parts of Europe were filled with Nobodys great and little. They abounded in all orders of Men, as well Laics as Ecclesiastics, but especially among the Laiety. There is nothing however recorded of them in this happy Interval, except it be the Invention of the Mariner's Compass. If this were not invented by Nobody (says one of the Irish

---

[2] 'Which no mortal has [ever] achieved': Cicero, *Laelius* (*De Amicitia*).

[3] 'And in fashioning the highest orator I shall create a person of such a sort as probably never existed': Cicero, *Orator VII*.

[4] 'No mortal is wise at all times': Pliny, *Natural History*, 7.40.131. One of Fielding's favourite tags (*Joseph Andrews*, Bk. III, ch. 5; *Tom Jones*, Bk. XII, ch. 13; *Jonathan Wild*, iii. 11).

Classics) 'tis at least certain, that Nobody knows very well, who was the Inventor, which in this Author's judgement comes to much the same.

Having premised thus much of Nobody's Ancestors, we come now to speak of our young Hero himself. The first remarkable Thing I find of him is, that by five years old he was a complete & perfect Master of all Arts & Sciences. The same it must be confessed has been also asserted of the Sons of many Kings, Princes, & Grandees. But in refutation of this, I believe if the candid & impartial of all Countries were to be interrogated—Who from his infancy possessed all Knowlege, as it were by Instinct? Was it young Prince A? Little Duke B? The Dauphin C? The Serene Infant D? They would all, I am well assured, answer No; [10] 'twas Nobody. And yet so far did our Hero excell those little Great Ones in the Modesty of his Demeanour, that 'tis a most certain fact that Nobody was at any time ready to own himself the greatest Dunce under the Heavens. When he came of age he obtained a seat in Parliament. Here his conduct was so upright & exemplary, that when any one talked of acting up to the highest pitch of Patriotism (a frequent Language in those virtuous days) 'twas a common saying that then he would act, as 'twas likely Nobody would act.

About the same time our Hero had a sister who then made her first appearance, a most beautiful young Lady, of great merit & fortune. Immediately the Hibernians & Caledonians modestly addressed her from every Quarter, making her immense Proffers of prodigious Settlements in the fertile Islands of Blask & Mull. Nobody (contrary to the practice of her whole Sex) refused them in preference to an English Gentleman of so small an Estate as £1500 per annum. Yet it is certain, that the abovesaid Caledonian & Hibernian Worthies were not only possessed of infinitely larger landed Estates in their own Country, but were likewise most of them Subalterns, either in whole pay or in half. And here by way of digression [11] I must intreat my Reader to observe that I relate this Fact of Nobody. Had I told it of Miss X, Lady Hariot Y, the Widow Z, or any other Miss, Lady, or Widow in the Universe, I should think myself justly ranged with Sr John Mandeville & Mendez Pinto. Nobody, beside this sister, had two

Brothers, John & Richard, the first of whom assumed the Name of Doe & the second that of Roe, but for what particular reason I have not been able to learn. All I could ever hear of them was, that they spent their whole lives in Law, & were seldom seen out of the Company of Attornies, Under-Sherifs, Bailiffs and their Followers.

Three years after this there were vast Ringing of Bells, Intoxications, Illuminations, Breaking of Windows, and all other suitable Demonstrations of Joy throughout every City & Town in the Kingdom. The pretext given out was, that Cambyses with six score ships only had taken old Carthage. But the truth is, twas in the event to the Praise & Glory of Nobody.

One fact I had like to have omitted. When young Miss Nobody came up first to Town, she had with her a Cousin one Miss Negative. This unhappy young Lady, being seduced by a specious Something, became unfortunately a Negative Pregnant. The Family [12] of the Nothings so much resented her misbehaviour, that they instantly discarded her, & would not own her as of Kin. She was therefore forced to take refuge among the Somethings, by whom she was willingly received & acknowledged as one of themselves.

About the same time our young hero Nobody had a Kinsman one Minus, a great Mathematician and Algebraist. This unlucky Rogue entered into an intrigue with his own Sister, and from the incestuous Congress of these two Minus's, instead of a Nothing, as was expected, came a certain Something, called Plus, to the great horrour of the whole Country.

And now in order of Chronology, I ought to enter into a detail of the great exploits of Nobody in India, Tartary, the Midland Parts of Africa and America, & at the two Poles. Here I should have a vast field, in shewing how many thousand Nations He converted, routed, civilized, pillaged, &c. &c. But as this laborious part of our Hero's History has been already fully handled by Quintus Curtius & the Father Missionaries, I shall to them refer the curious Reader, & pass myself to other Facts less known & celebrated.

As to the affair of Marriage, [13] we may I believe affirm that of Nobody, wch was never affirmed of Any-body, except

some serjeants & private Gentlemen in his Majesty's Infantry:
in short Nobody was married nineteen times, I am unable to
give a particular of his several Wives. All I can discover is that
Nine of them were old Women, not worth a Farthing, married
by Nobody for love.

It does not appear that he had Children by any of these
Wives. For tho I have met several in my time who have been
stiled Nullius Filii or Sons of Nobody, yet I have always had
too much reason to suspect their legitimacy.

And now I come to our Hero's Death. Nobody died after
the following manner, that is to say, by a Dislocation of his
little Finger. Immediately upon the accident use was made of
Cato's Remedy, mentioned in his Treatise de Re rustica Cap:
160. It consisted in singing every morning the following
Hymn—Huat, Huat, Huat, Ista, Sister, Sis, Ardannabou
dunnaustra.[5] But the Remedy, tho' never before known to fail,
did not in this instance meet with its desired success. This the
Physicians imputed to the Dislocation's being nervous.

[14] It is difficult to say certainly where Nobody was
buried; the most probable Conjecture is, along with Mahomet,
amid the Loadstones. But tho his Burial place be uncertain,
yet are the Monuments and Inscriptions erected to his honour
to be found in every Church & Churchyard throughout Chris-
tendom. The Erectors indeed of these Memorials have indulged
the wantonness of their own Fancies, in representing him
under different Characters, as well upon occasion under
different Sexes. Some of them have talked of Nobody, as tho'
he had been a beautifull young Virgin; others, as tho he had
been a Judge or a reverend Prelate. Some have supposed him
a King; others an old Woman; some, a great Scholar; & others
on the contrary a Page or General Officer. This variety at first
sight may appear to induce endless Perplexity, but the
difficulty is only in Imagination, & may be solved by the
attentive with the greatest facility. The Key to the whole in
short is no more than this—upon whatever Monuments we
read so much Virtue & Wisdom recorded, as common sense
tells us was never the Lot of Any Body, of all such Monuments

[5] The accepted text is 'huat huat huat istasis tarsis ardannabou damnaustra'
('words of a charm to cure a dislocated joint': Cato, *De Re Rustica*).

we may venture to affirm that they are clearly and undoubtedly the Monuments of Nobody.

[15] And now, as we have in all probability sufficiently obtained that End, at which all Historians sooner or later arrive, that of having related much more than 'tis likely any one will believe, it is time we think to conclude these most elaborate & curious Memoirs.

Finis

Appendix III
'An Essay on the Life and Genius of Henry Fielding Esqr.'

HARRIS'S memoir of Henry Fielding comprises 16 pages of hand-written, heavily revised, and frequently rewritten drafts numbered 1 to 16; a 3-page fair copy of the opening 6 paragraphs ending in mid-sentence (numbered 1 to 3); 12 pages of additional drafts which include several attempts at an opening; and an anecdote about Fielding's last days in England preserved on a separate leaf.

If the fair copy was made after the first 16-page sketch, comprehensive in its chronology, the fair-copy section on Fielding himself was kept by Sarah Fielding and has since disappeared.

The following edited version begins with this 3-page fair copy, and then reverts to the earlier 16-page version for the remainder of the essay. In a very few cases I have reproduced Harris's first phrasing rather than his revisions on the grounds that the former are more informative and particular, and I have conflated two drafts of a short section (pp. 2–4) introducing Fielding's early biography. Original punctuation has been retained almost completely. The process of revision led Harris to increasing generalization, a characteristic habit of seeking the causes rather than the particularity of things, and although this approach may be more 'philosophical', it inevitably downgrades the kind of anecdotal realism we might expect from one of Fielding's closest friends. But it is also clear from the essay that Harris regarded Fielding as a remarkable literary phenomenon, whom Harris fits adeptly to his theory of a sociable and modern talent working within the classical tradition. He was particularly informed about the working-conditions of Fielding's earlier career, had observed him closely in company, including his own, and thought of him after his death as an epitome of mankind itself. For an account of the composition of 'An Essay', see Chapter 4 above.

AN ESSAY ON THE LIFE AND GENIUS OF HENRY FIELDING
ESQR. BATH FEB. 5. 1758

[1] The Life of each particular Man is the Portion of Time during which he exists. Short indeed and fleeting this Portion is, and if he render it not memorable by some endeavours of

his own, it soon sinks into oblivion, and is irrecoverably lost.

> . . . sed Famam extendere factis
> Hoc virtutis Opus . . .[1]

Tis by Deeds, it seems, that our Fame is to be extended, and this the Poet tells us is the Work of Virtue. That it is the Work of Virtue we pretend not to deny, but we assert withal that tis the Work also of Vice, and of every other Conduct, whatever may be its Character, so that it render a Man conspicuous and distinguished above the rest. Socrates is not better known than his base accusers, Anytus and Melitus. Zoilus flourishes in Fame as vigorously as Homer, while Bavius and his Friend Maevius are [2] still remembred along with Virgil. The name of Ravillac is as familiar to our ears as that of Henry the Great. Nay even those paultry Friars and Fanatics, that did but meditate his Assassination, have obtained by the very design a Place in the records of Time, to which by their own obscurity they could never have had pretensions.

If we reflect on the Examples above mentioned, we shall find that some became famous by Action, others by Speculation, in short that the Active Life and the Contemplative appear to include them all. Nay with regard to those, who have signalized themselves by Writing, we may consider if we please even their Writings as their Actions, as those Works or Exploits, by which their Lives have been coloured, so that in a sense more humane than it was once applied at the trial of that worthy Patriot Algernon Sidney we may venture to affirm that *Scribere est agere*.

Upon these Principles it seems natural, when we inquire concerning Authors, to look upon their very Writings as the History of their lives, as the Lineaments and Portraiture of their internal Parts, whilst whatever they did else, is no farther of import, than as it tends to illustrate this literary Character.

The Genius of a Writer, when considered as a Seed, is that Divine Particle of Intellect, brought [3] into the World at his Birth. This, like other Seeds, if it meet not proper Nourishment, is either wholly destroyed, or exists in languid efforts.

---

[1] Virgil, *Aeneid*, 10. 468–9: 'But to prolong one's reputation with deeds, this is the work of excellence.'

There is no room to doubt that many a Horace, many a Swift have been lost behind a Compter, or in duty before the Mainmast. But the Genius of a Writer, when considered as something complete, is that Seed when carried to its full Maturity, that Seed when ripened by the proper concurrent Causes, by Education, by Countrey, by Fortune good or bad, and by other Incidents, that befall a Man during the Period of his Existence. Twas the free Countrey where he was born, and the liberal Sentiments in which he had been educated, that made Demosthenes attain that inimitable Sublime, which is so peculiarly the Character of that great Man's Rhetoric. Had he been bred at modern Constantinople, or even at the Court of his Contemporary, the Monarch of Persia, that Genius, which now we so justly admire, we might as probably have contemned for its servile Adulation.

Henry Fielding Esqr the subject of this Essay was the son of Lieutent General Fielding, the son of Mr Fielding a dignified Clergyman, who was himself the younger son to an Earl Denbigh. For his natural Accomplishments whether of Body or of Mind, his Genius was acute, lively, docile, capable equally both of the Serious and Ridiculous; his Passions vehement, and easily passing into excess; his person strong, large, and capable of great fatigues in every way; his Face not handsome, but with an Eye peculiarly penetrating and quick, and which during the Sallies of Wit or anger never failed to distinguish it self. His Education was at Eton School, where he became well grounded both in Latin and in Greek, and never after forsook ye Study and perusal of ye best Classics, particularly, among ye Greeks, of Homer, Aristotle, and Lucian, among ye Latins, of Terence, Virgil, and Horace and Ovid. Leaving School, he went to Leyden, whence returning soon to England, he fell into that Sort of Life, to which great Health, lively Witt, and yt flow of juvenile Spirits, so copious at this period, naturally lead every young man, unchecked by graver authority. His Company was highly pleasing, and his acquaintance of course became very extensive. He conversed not only with persons the first in fashion and quality, but with infinite others of indiscriminate rank and character, with whom either by chance or choice he was associated. Thus was he soon furnished with a wide and diversified view of Life and Manners, where his apt Genius did not fail to match those characteristics Strokes and morally distinctive Signs, which vulgar Minds behold witht. feeling or attention. His Prospects were enlarged even by his Adversities. Pleasures were expensive; Expenses produced debts; [5] creditors were importunate, and difficulties soon ensued. [4v] While a Small Family Estate was insensibly melting away, he married a Wife, for whom he had a great affection, and who of course became ye Partner both of his Prosperity, and his Adversity. Thus were not only his Joys but his anxietys also heightened, for his Passions, as we said before, were vehement in every kind, and ye Impressions on his Imagination were always durable and distinct. [5] Twas then he became acquainted with many of those Scenes of distress, which no Genius can describe, that has never seen, and which are for ever hid from those, whose Fortune is more equal, and whose Passions and Imagination are less vehement and prone to excess. If Necessity be not ye *natural*-Mother to Invention, she is at least ye *Step*-mother, who sets Invention to work; who commands her to toil and drudge on her account, and then appropriates to her self ye fruits of ye labour. Our Author now for the sake of a Subsistence began to apply those Talents, which had dawned long before, to Theatrical Compositions of a peculiar form and character. He might then have sayd with Horace.

> . . . Paupertas impulit audax
> Ut versus facerem . . .[2]

His Pasquin, his Historical Register, and other peices of the same sort resembled ye first Comedy of ye Greeks, that of Eupolis Aristophanes and others his Contemporaries; scenes of fancy and allegoric humour, pictures of human Life Extravagance and Nature, ye highest humour imaginable occasionally interspersed with a large mixture of bitter sarcasm and personal Satire, respecting ye leading Persons and measures of the times. How those Performances were received, those who saw them, may well remember. Never were houses so crowded, never applause so universal, nor the

---

[2] Horace, *Epistles*, 2. 2. 51–2: 'Reckless poverty drove me to write verses', adapted to 'drove *him*'.

same Peices so often repeated witht. interruption, or discontinuance. Tis enough to say that such was ye force of his comic humour and poignancy, that those in power in order to restrain him, thought proper by a Law to restrain the Stage in ye general, bearing even by this act of opposition and Suppression the highest testimony of Praise to ye force and influence of his abilities. The Legislature [6] made a Law, in order merely to curb one private man.

There is a Sort of Comedy called *genteel*, in which it has been observed that he seldom succeeded. The observation may be just, but if we consider ye Cause, the fault perhaps will be found rather in ye Subject, than in ye author. Could that Eye wch. was always turned to mark the distinguishing Strokes of Nature, could such a Mind succeed in Characters, where no Nature appears at all? Why do we admire in a beautiful Landskip of some able Painter ye Sheperdess and ye Plowman or ye Milkmaid, while ye Lady in her great hoop or ye Beau with his long Cue,[3] we are ready to confess w. render the picture ridiculous? Tis because, there is something Simple and natural in ye former, which we spontaneously applaud, while in ye latter we behold nothing, but what is capricious and forced. If we go from Landskip to Pastoral Poetry, for what reason are we charmed with ye simple Sentiments of a Thyestis or a Corydon, when from those of ye Beau and Lady before-mentioned we seek in vain for such delight? Tis owing to ye same natural Simplicity, which will at all times have its effect, and which once being supplanted there is nothing can supply its place. But to return to our Author.

During those theatrical Exhibitions, the face of his affairs was much altered. The Clouds of Fortune vanished, and were succeeded by her Sun-shine and genial Warmth. Yet here as in Nature the Scene was mutable, and as ye good succeeded ye bad, so ye bad again had its turn. To heighten his anxietys as well as his Joys, twas [7] during this interval that having buried his first Wife a few years before, he married a second, of whom he was extremely fond, and who of course became the Partner as well of his Adversity, as his Prosperity. Thus were not only his Joys, but his anxietys also heightened, for his

---

[3] A long roll of hair worn hanging down behind like a tail, from the head or from a wig (*OED*).

Passions (as we said. before) were vehement in every kind, and, thus surrounded with a family, he continued in ye administration of Justice.

Twas about this time, that, his theatrical Genius being thwarted by ye Servitude of ye Theatre to which he himself had reduced ye Stage, he took to ye Study of ye Law with indefatigable industry, and toiled like one of those drudges, to whom the fair, the elegant, the humorous and ye ludicrous are quite Non-entities, of wch. they have no conception. His Pains were not without Success. He made some figure on ye western Circuit, and had probably succeeded farther had not ye Gout rendred him incapable of performing those legal Journeys. By these however he obtained an insight into ye manners and passions of a variety of characters, not elsewhere to be seen, or at least not to be seen in so advantageous a light, ye natural Energies, and operations whether we begin from ye head, the Bench of Justice and ye Grand Jury, or descend to ye bottom, the Client the Attorney, ye Bailiff, ye Gaoler, and his fettered Followers. His real Inventive Genius was not inactive during these periods. At his Lodgings, upon ye Circuit he was often working on his Peices of Humour <Tracts>, which when Business was approaching, soon vanished out of Sight, while ye Law Books and the Briefs with their receptacle ye Green Bag lay on ye Table ready displayed, to inspire the Client with [8] proper Sentiments.

From ye Practice of ye Law, he became a Justice of Peace, and acted in that capactiy for ye County of Middlesex with much applause having frequent interviews with ye first <principal> Ministers of State on matters relating to ye civil Government, and ye administration of Justice. To the scenes, which pass before a civil Magistrate, where there is usually a mixture of resentment or distress, we may well apply those lines of Lucretius.

> Nam verae voces tum demum pectore ab imo
> Eliciuntur, et eripitur *Persona*, manet *Res*.[4]

. . . declining Health and a broken Constitution induced him to try in vain the better air of Lisbon, at wch. place he died,

---

[4] Lucretius, *De Rerum Natura*, 3. 57–8: 'For then a real cry is wrung from the bottom of his heart, and the mask is torn off, the truth remains.'

soon after his arrival. [7v, *marked for insertion as a note*: He died in Sepr. 1754 aged [blank] years, leaving behind him a Daughr. by his first Wife, and a Son and Daughr. by his Widow, who still survives him.]

[8] If we consider the various characters, which fill so vast a Metropolis as London, if we consider withal that Frays, Larcenys, and a thousd. other incidents bring some of every Sort, either as accusors or as partys accused, before a Magistrate in reputation we shall perceive that our Author in his last Employ had still farther opportunitys to increase his knowledge of human Life, and to behold it and ye Law moreover without yt. masque of Hypocrisy and complaisance, which in ordinary interviews keeps real Nature out of sight. Whoever will turn their Eyes back on what we have recorded, will find that few men had ever so various and diversified a prospect of human kind.

And now to reflect on the facts here recorded, it will appear, I should imagine, that what Homer said of Ulysses—*mores hominis multorum vidit*—was not more true of ye wandering Hero, than it was of our Author. He knew the world not as we now-adays [9] foolishly understand it, by knowing ye faces of a few eminent Persons, together with their Places of Resort (a Knowledge that descends even to Box-keepers and Door-keepers) but he knew ye World, by having viewed and conversed with all ye orders of Mankind, and seen the various operations of Reason, Habit, Opinion Passion Resentment and Appetite, those Springs, by whose power we are all of us actuated. As no one likewise was more strongly actuated by all those Springs than himself, so he had an internal and self Mirrour of his own Mind always at hand ready to consult, where precedents were not wanting for almost every given circumstance.

Such was our Author's Knowledge of Mankind, or (if I may be allowed the Phrase) his knowlege Experimental. For his Speculative or Litterary, it was derived from a sound Education in Classic Learning, never neglected (as too often happens) after he became a Man; from a diligent inquiry into our Laws and Constitution of his Countrey; from a perusal of the best authors as well as ye grave and ye gay, as well of his own Countrey, as well also as those of our ingenious neighbour the French, in whose Language he was knowing, and with whose best works not unacquainted, tho happy in this respect above most of his Contemporaries, that his Taste was not modelled after these alone. [10] Thus he had the finer languages, and Sentiments of Politer Nations to assist him, and thus he was not in his Compositions like one of those baser Statuaries, who never having it in their power to peruse the Grecian and Roman Sculpture have no other Models to follow than the Originals of Hyde Park Corner.

And now let us endeavour to collect an Idea of his Genius; not in its embryo State but in its Maturity and Vigour. It was in fact, when thus mature, an accute understanding, actuated by strong passions and a lively Imagination, and cultivated by sound Literature and an extensive intercourse with Mankind. From such a Mind so born, and so accomplish'd we need not be at a loss to explain from what principles arose his most celebrated performances and why we admire Joseph Andrews, and Tom Jones as two Master pieces in their kind. Had his learning been none, or of a narrow and vulgar kind, these works had been so far defficient in their elegance and taste. Had his Intercourse with Mankind been more limitted, and narrow (wch. is the case of many a Scholar, many a man of sober sense) his pictures of Man had not been drawn after the life, but had been so many imaginary beings, the insipid creatures of his own Brain. Had his passions and imagination been less vehement and strong, his Pictures had wanted force and the whole had been languid and inactive. [11] Lastly had his Understanding been wanting in native vigour and accuteness, nothing secondary, and adventitious could have supply'd its place, for Good Sense is the only Fountain, eternal and inexhaustible, from which alone whatever is genuine and meritorious can arise <basis on wch. every Structure of Science, every work of art, whatever is truly serious, or laudably truly humerous, alone can stand>.

. . . Sapere est principium et fons.[5]

[12] Among the various characters, wth. which his Works are adorned, there is none more admired than that of Mr Abraham Adams, a Portrait drawn after actual life, after ye

---

[5] Horace, *Ars Poetica*, l. 309: 'Wisdom is the beginning and font [of writing well].'

Revd. Wm. Young, a worthy Clergyman of Dorset. No character was ever more truly exhibited, perfectly drawn, with all its essential Attributes of Courage, Honesty, Integrity, Simplicity, and Learning, together with his perfect simplicity and ignorance of human life. We can only add that Mr. Young, after this public Exhibition in Print <ye publication of Joseph Andrews>, becoming more known, was called from his obscurity and made chaplain to ye Army Hospital, then abroad in ye late War. Here he had an opportunity to enlarge his acquaintance, and was much respected by such officers of distinction, as could relish his Learning and unaffected Wit. Here, too in ye places to which he moved at different times, many adventures befell him that were singular in their kind, tho none more remarkable than what happened to him near Frankfort. There on a day having crost the Bridge over ye Mayne, to indulge in what he loved, a solitary walk, he found [13] he had stroled insensibly into ye midst of ye French Camp, & that too at an unseasonable time, just after their defeat at ye Battle of Dettingen. The Incidents wch. he met with there, no one could relate but himself, as he always did it with a vein of humour which was perfectly original. Tis enough to say that as his character indangered him, so his character brought him off; that as his native Inattention led him into ye midst of Strangers and Enemies, so his native Intrepidity and honest Simplicity ushered him thro perills to his Countrymen and Friends. The War being over, by ye beneficence of Mr Ranby he found a retreat in Chelsea-College Hospital where his Valet de Chambre (as he used to call him) was a one-legged Corporal and where he lived amidst books and ye fumes of his own Pipe, till he peacably ended his days in ye yr. 1757 regretted and esteemed by all those who were so fortunate as to know him. But this by way of digression; to return therefore to our Subject.[6]

[14] There is a Race of Buffoons, whose Drollery is like a Raree-show, something prepard beforehand and of limited duration, something which when once exhibited, the whole is at an end, unless we can endure to see ye Show over again. Such was not our Author's case. His wit was native,

---

[6] For further discussion, see Probyn, 'James Harris to Parson Adams in Germany', pp. 130–40.

spontaneous, and ever new, derived instantly from those events that are arising every moment.

There are some too, whose Wit is said to be brightened by Wine; a fact indeed somewhat doubtful, as it commonly rests on doubtful testimony, that is, on those whose small Wit ye same Wine has already darkened. Such was not our Author's case; for tho' Wine could not suppress his natural Ingenuity, it never improved it, nor did he ever succeed so well whether in a Serious way or even a humorous one, as when he was sober temperate, cool, and free from ye loads of Intemperance.

There is a third Species of pleasant fellows, whose Wit is like a steel, it emits no fire, but when it has a flint to work upon. Provide them with their Butt, and they are sure to sparkle, but if the Wit be retorted [15] and they chance to meet an antagonist, their gallantry disappears, their Vivacity sinks, and the Dung-hill Heroes are glad to fly ye Pitt. No such Imputation ever befell our Author. He would parry with admirable humour a whole host of assailants, would maintain his ground and be a match for them all with invincible magnanimity and like Falstaff of old against his men in Buckram, wd. take all their Points upon his Target at once.

There is a fourth order of men, and those of no mean import, whose Wit and Humour keep pace with their condition. When Health and Fortune abound, then are Wit and Humour abundant also. If the Scene change, the Spring soon fails, and every attempt to exert themselves proves fruitless and vain. Nothing like this ever happened to our Author, whose Wit, though it might have had perhaps its intensions and remissions, yet never deserted him in his most unprosperous hours, nor even when Death itself openly lookt him in ye face.

Two Friends made him a visit in his last Illness on his leaving England, when his Constitution was so broken, that twas thought he could not survive a week. To explain to them his Indifference as to a protraction of Life, he with his usual humour related them the following story. A Man (sd. He) under condemnation at Newgate was just setting out for Tyburn, when there arrived a Reprieve. His Friends who recd. ye news with uncommon Joy, prest him instantly to be blooded; they were feard (they said) his Spirits on a change so unexpected must be agitated in the highest degree. Not in the least (replied the Hero) no agitation at all. If I am not hanged this Sessions, I know I shall ye next.

Such was the Man, the Subject of this short memorial, a Man, the Source of infinite entertainment to his Friends, whatever was ye Conversation, whether grave or gay, [16] for even in graver Conversation his particular Friends knew him to have excelled, a fact not to be admired, if we reflect on his natural good Understanding, that fundamental specific Principle, that Particle divine, which distinguishes real Wit from its Shadow, Buffoonery. As to ye public in general, they were obliged to him many years, as to an useful and able Magistrate; they were obliged to him for that Pleasure as well as for that Improvement and high Delight, which they derived from his various works, an obligation which still subsists and is not likely at all to cease.

How then shall we conclude, what is to be our final opinion?—As he is now no more, let our Judgment be candid; and yet not only candid, but also just at ye same time. Let his Faults be all forgot, which can now no longer offend; let his Virtues be all remembered, which still continue to give delight.

Appendix IV
'Upon Ridicule'

THE following unpublished essay on Ridicule is undated, but was probably written as early as 1750, shortly after the publication of *Tom Jones* (1749), and at the same time as Harris's *Concord* (1751), to which it bears a very close thematic similarity. Its chief inspiration is undoubtedly Shaftesbury's *The Moralists* (part 1, sect. 3), where Shaftesbury argues that the world's beauty is founded upon 'contrarieties' from which 'various and disagreeing principles a universal concord is established'. In Fielding's formulation this 'new Vein of Knowledge' (in fact a very ancient idea) becomes 'Contrast, which runs through all the Works of the Creation, and may probably have a large Share in constituting in us the Idea of all Beauty, as well natural as artificial' (*Tom Jones*, Bk. V, ch. 1).

The significance of Harris's essay lies in the fact that it applies a theory of contraries to comedy; methodologically, he tries to achieve for Ridicule or Satire what Longinus had done for the Sublime. The English translator of *On the Sublime* (1739), William Smith, inserted Shakespeare, Milton, and Pope into Longinus' treatise, and although there was no similar classical essay on comedy for Harris to start with, he nevertheless constructs an analogous comic tradition from Homer, Socrates, Menander, Virgil, Cicero, Samuel Butler, Swift, Fielding, and others. Ridicule is an air played on a logical substructure. The list of contrarieties, or 'co-ordinates', is, as far as one can see, original to Harris. The essay closes with an attack on Locke, who is named, and a commendation of Shaftesbury, using a quotation from one of Shaftesbury's unpublished 'epistles' which had criticized André Dacier's concept of irony in his translation and commentary on Aristotle's *Poetics* (Paris, 1692: discussed by Fielding in *Tom Jones*, Bk. VIII, ch. 1). Harris also alludes to, and paraphrases, parts of Shaftesbury's second treatise, 'Sensus Communis; An Essay on the Freedom of Wit and Humour in a Letter to a Friend' (1709). Whereas Shaftesbury designed his remarks for an aristocratic *club* of 'gentlemen and friends who know one another perfectly well', Harris both democratizes Ridicule and elevates it to the level of a general theory requiring a particular kind of intelligent understanding. As such, his essay offers a rare and very early theory of literary comedy, not least because of its analysis and demonstration of how the serious and the comic epic in Fielding's *Tom Jones* both relate to human universals.

Apart from the deletion of three Greek synonyms, and the addition of notes, the essay below preserves the spelling and punctuation of Harris's holograph manuscript of 30 numbered pages (Harris Papers, vol. 7, item 8, 'Concerning the Nature, Use, and Abuse of Ridicule', where this essay is grouped with the two 'Fielding' essays, i.e. 'History of Nobody' and 'Essay on . . . Henry Fielding'). In addition to specific analysis of *Tom Jones*, one senses that the particular example of Fielding's work has been a general influence on Harris throughout the essay. It is, of course, also possible that Harris's ideas may have influenced the novelist's discussion of the Ridiculous in the preface to *Joseph Andrews*, and on the co-existence of Tragedy and Comedy in classical literature in that novel (Bk. III, ch. 2).

UPON RIDICULE

The Effects of Art are pleasing, but ye Principles are painfull. We are all charm'd with a Picture of Raphael, an Oratorio of Handel, or a Part finely acted by Garrick, or a Quin. But when we come to enquire into those latent Causes, from which alone these Charms are derived, we immediately find ourselves involv'd in a Labour as devoid of Pleasure as the gravest Disquisition. Nowhere is this more true, than in the Subject of Ridicule. The Force of Wit & Humour is of universal Extent, being felt almost by all Men with a sensible Delight. Yet if we seek those Sources, whence the Streams of Humour flow, they prove as difficult to trace as the hidden sources of the Nile. Cicero's observation is not a little discouraging, who tells us that those who have attempted to explain Ridicule, have commonly written in such a Manner about it, as to exhibit nothing ridiculous, except their own Absurdity. Yet did not this deter him from entering himself into the Subject, which he has treated at large in his Book concerning the Orator. Aristotle too has mentioned it in his Treatise concerning Rhetoric, & gave it a fuller Examination (as he tells us himself) in that Part of his Poetics which are now lost.[1] Should we therefore venture to engage in the same Subject, we hope to shelter ourselves from ye Ridiculous, which we attempt to scrutinize under the Authority of two Geniuses so established

[1] Harris's marginal note cites Aristotle's *Rhetoric*, 1.2, as his authority, but for Aristotle's note on jests and irony in the lost *Poetics*, see *Rhetoric*, 3.18.

in reputation. Without farther preface, we begin, as follows. We only make one Request, as our inquiry is but short, that no one would censure, 'till he has read it thro', & bestowed sufficient Pains to understand the Whole, if he can bring himself to think, that such Pains are worth the Taking.

There is nothing ridiculous in what is clearly true; as when we assert two & two to make four; nor in what is clearly false, as when we assert two & two to make five, nor in what is admitted for true from the Strength of Appearance, as when we believe ye Moon to be bigger than any Star.

If therefore ye Ridiculous be neither seen in clear Truth, nor in clear Falsehood, nor in the Appearances of Truth assented to as real, it must needs be seen in some Mixture of these.

Now this Mixture is no other than that of Appearance with Reality when we recognize together, & under one Perception, an Appearance of Truth, & its opposite real Truth.

It must be observed by ye Way that every Falshood has an opposite Truth; if negative, then an affirmative Truth; if affirmative, then a negative one. So that as every appearance is it self but a Falshood, hence we see how every Appearance has of course an opposite Truth. But to return.

The Ridiculous is seen, when under one identical Perception we recognize a mere appearance, & its opposite Reality. Thus ye old Song, which proves that Drinking makes us immortal—

How can he turn to Dust, that still moistens his Clay?

In Death we turn to Dust; there can be no Dust, where there is continued moistening, such is continued Drinking, ergo *continued Drinking presents Death & makes us immortal.* Thus far ye Appearance. The opposite Truth (which needs no Comment, as being acknowledged by all) is, that *Drinking does not make us immortal.* This necessarily occurs to us, when its opposite is exhibited, being as it were in a kind of Jeopardy, from the Plausibility of its Antagonist. And hence the Wit, & the subsequent Delight, which is not unlike to that of Discord in Musick, where the Joy is from the Pain & the quick Return again to pleasure.

'Tis to be observed in this Union, that the Words themselves only express ye appearance, while the opposite

Truth is not exprest, but only felt. The Reason is plain. Were the words to express not ye appearance, but the Truth, the Hearer would rest *there*, without thinking of any Thing ridiculous, as we perceive to be the Case in all serious Assertions. But when ye opposite to any truth come to be asserted, with Plausibility, the Mind flies immediately to that particular Truth, & in viewing the two together forms the Union, which we have just mentioned. Here then we see the Reason why he who cannot recognize the two can have no conception of the Ridiculous but must needs be employed about the worst Part only, that is to say, the Falshood or mere Appearance, because the Words in themselves are expressive of nothing else. And hence the Cause of that splendid Puzzle, which sometimes spreads itself upon Baeotian Faces, when they see others laugh, & cannot find the Reason.

We may add to what has been said, that the stronger ye contrast between the apparent & the true, the stronger the Wit and Humour. This needs no proving, because here the very Essence of ye Ridiculous lies. Should it be asked, how then is this Contrast to be heighten'd—We may question, not by any heightening of its true Part, for it may be questioned, whether Truth admits any Intension. Appearances on ye contrary being many & various, are capable of intension in every degree. Besides, there is a more decisive Reason, which is, that the *Words* are expressive of the *Appearance only*, & cannot therefore be employed to heighten any Thing else.

And thus 'tis, that in ye strong contrast of False to True, thro' the Plausibility of Appearance; in ye Jeopardy of Truth from such Plausibility, in the Pleasure felt by ye Mind in being extricated from the Difficulty, & triumphing (as it were) in the Recognition of the Deceit—in this consists the Force of Wit & Ridicule.

It is to be observed in this Process of the Mind, that the Truth, which it recognizes <by a Sort of implication> suffers no Damage, ye Contempt, if any, being transferred to the Appearance, when we view it as it were vanquished, after all its specious Efforts to obtain our Assent.

And this leads us to find, in what manner Ridicule has its Effect in Reasoning. It cannot be said to prove any Thing directly, but 'tis a Branch of that Form, called *Reductio ad absurdum*, whose Force consists in shewing the Falsity of the Contrary. Thus for example, if I would recommend Liberty or Virtue, I endeavour to shew Tyranny or Vice to be ridiculous. It has likewise these peculiar Advantages, that 'tis mostly short, always joyous, & not painfull & lengthened, like the stricter Methods of Reasoning. This makes it, tho' seldom conclusive, yet nevertheless popular, inasmuch as vulgar Heads are not able to endure the extended connections of a more exact syllogizing, but faint under a Fatigue, to which they were never habituated.

As to the application of Ridicule, it may be applied either to Praise, or to Dispraise, or to Matters merely indifferent. To Praise, as when we censure a Man for Vices notoriously not his own—to Dispraise, as when we extoll him for Virtues which he as notoriously wants—to Matters indifferent, as in the Case of that Man, who being flea-bitten all Night, was asked ye next morning how he had rested, *Like the Great Scipio* (says he) *nunquam minus solus, quam cum solus.*[2]

It must be confest however that as all Ridicule implies a Mixture of Falshood, & that Falshood is in ye same Class with Defect, Ignorance, Folly & Vice, its more frequent Tendency is to lessen & vituperate. And hence it is, that all great Satirists have been likewise great Masters of the Ridiculous. Hence also, as we have *all* of us our Faults, few bear Ridicule with a patient Indifference, or care to trust themselves to become its Subject. Above all, none dread it so much, as those Imposters, who by Help of Visage, Gesture, Tone & Garb, maintain ye shew of Wisdom, without ye Reality. The Breath of Ridicule is so truly dangerous to those Houses of Cards, that they are often demolished by a single Puff.

Authors have been divided, as to the Use of ye Ridiculous. Cicero admits it into his Character of an Orator, & gives us both Precepts & Examples upon the occasion. Horace recommends it even in Matters of Importance

> *Ridiculum acri*
> *Fortius et melius magnas plerumque secat Res.*[3]

---

[2] 'Never less alone than when alone': the phrase occurs twice in Cicero, with reference to Scipio Africanus (maior) (*De Re Publica*, 1.27, and *De Officiis*, 3.1).

[3] 'Ridicule often decides important matters more effectively and better than severity': Horace, *Satires*, 1.10. 14–15.

Longinus seems to refer to a Test of this Nature, when he would distinguish between ye True & the False Sublime. *Nothing* (says he) *is great, which to contemn is great.* Now there is certainly a near Affinity between Contempt and Ridicule. Among ye Moderns none has more strenuously espoused the Cause of Ridicule, than the witty and learned Author of the Characteristics.

On the contrary, there have been a multitude of his Contemporaries, who have inveighed heavily against it, as a Thing subversive of every Thing serious. From what they say, & their Manner of saying it, one would think of Ridicule, as of some Weapon in War, that its Power must needs be more than commonly formidable, since Men so much dread to have it used against them. And yet we may observe in the Case of those very Censurers, that there is no weapon, when they can get it, that they use more unmercifully. Indeed wherever the Edge of Ridicule is successfully applied, the Sect or Party, which it happens to favour, whether Orthodox or Heterodox, for Court or for Country, are sure to support its Efforts with the strongest Applause. Witness the high Encomiums bestowed on the witty Authors of Hudibras, & the Rehearsal, on those celebrated writers, Dr. South and Mr Trenchard, not to mention Lestrange and others of inferiour Rank.[4]

However as to this Condemnation of Ridicule, 'tis not a Matter of much wonder. Even Reason itself has had no better Fortune, having been represented by some as a dangerous Thing, in which we must beware how much we confide too much. We may remark on this Occasion, as we may on Ridicule, that all men employ Reason, as long as they can have it; that they are sure to triumph, when they can prove their Adversaries to want it; & that they never renounce it, but in Cases of the last Extremity, that is to say in other Words, never 'till Reason renounced them. Nay, they do not

[4] Harris's first version was: 'Witness the high Encomiums bestowed on the witty Authors of Hudibras, & John Bull, on that celebrated Dr South, not to mention Warburton, and others of inferiour Rank.' John Trenchard (1662–1723) produced *The Independent Whig* (1720–1), subsequently reprinted as *Cato's Letters* (1721–47); Dr Robert South (1634–1716) employed ridicule in his *Sermons* (published from 1679–1744), and a collection of his *Maxims, Sayings, and Characters* appeared in 1717. Cf. Fielding's remark (*Covent-Garden Journal*, 3 Mar. 1752) that 'there is perhaps more Wit' in the Sermons of Robert South 'than in the Comedies of Congreve'.

give up the Cause too hastily. They would, if possible, demonstrate the greatest Monsters, even Transubstantiation & the whole stubborn Tribe of kindred Impossibilities have been attempted every way to be made tractable to Reason, while Logic has been ransacked thro' all its Subtleties & Distinctions with Subdistinctions multiplied like the Polypus.[5] Were one indeed ever to think of Reason, as contemptibly as these men would have us, there could be no stronger Inducement, than to behold it, as they use it. But not to digress.

The Truth is, Reason is a Power; so is Ridicule; so are our Senses; so are our Limbs; so are Swords, & Daggers;—All Powers capable of being applied either to Good, or to Bad; & the higher their Efficacy, the more dangerous their Abuse. Daggers may be employed to assassinate our Friends; our Hands may be employed to steal his Property; our Eyes to peep, & our Ears to hear at Times & Occasions, when their Use would be ungenerous. So too Ridicule & Reason may be employed by base Men, to answer ye worst & basest of Purposes. Thus the Objection at last meets that obvious Solution, that we are not to argue from the Abuse to ye Non-Use.

It may not be improper to try a few Examples of Wit, and Humour, that we may see, whether they may be accommodated to the above Hypothesis.

When the Poet Simonides was offered a trifling Sum to sing the Praise of certain victorious Mules, *he would have nothing to do* (he said) *with Demi-Asses*. When a larger Sum was promised, he then began his Song,

> Hail, Daughters of the generous Mare,
> That flies along, like fleeting Air.

The Wit of the Story is in the Noble & Ignoble united in the same Subject.

> *Qui Bavium non odit, amet tua Carmina, Maevi,*
> *atque idem jungat vulpes et mulgeat hircos.*[6]

---

[5] For Fielding's parody of Abraham Trembley's account of dissecting a polypus— 'each part becoming a new and whole polyp'—in the Royal Society's *Philosophical Transactions* (Jan. 1743), see *Miscellanies*, ed. H. K. Miller, pp. xxxix–xl, 191–204, 253–9.

[6] 'Let him who does not hate Bavius love your poems, Maevius, and let him harness foxes and milk billy-goats': Virgil, *Eclogues*, 3, 90–1.

Here the Wit or Satire is in ranging the Possible with the Impossible, and considering them as united in ye same Conduct. The above Verses have been thus parodied,

> The Man that hates not Ogilby the Great,
> May well admire thy Verses Nahum Tate,
> He too Jack Westley for a Saint may own,
> And seek for wealth in philosophic Stone.[7]

> Butler tells us,
> That when the Fight becomes a Chase,
> He wins the Day, that wins the Race.[8]

Here ye Wit is in the Union of Conquest with Defeat, that is to say, in the seeming Proof, how a man may be said to conquer by running away. There is the same Turn, with more Elegance (as ye Subject indeed required) in those Verses of Waller,

> In Love the Victors from the Vanquish'd fly,
> They run that wound, & they pursue that die.[9]

Swift tells us in his Tale of a Tub, that 'in all Assemblies, tho' you wedge them ever so close, we may observe this peculiar Property, that over their Heads there is Room enough; but how to reach it, is the difficult Point; it being as hard to get rid of Number, as of Hell;

> *Evadere ad auras,*
> *Hoc opus, hic labor est.*'[10]

He then proceeds to his three Methods, of the Pulpit, the Ladder, & the Stage itinerant, &c. Here the first Stroke of Wit lies in a seeming Discovery, which still leaves us no wiser than it found us.—*In the most crowded Assembly there is Room.—Where?—over their Heads.* The rest is a sample of continued Irony, where under the Appearance of arguing and saying every thing seriously, there is nothing seriously argued or said.

[7] Untraced.

[8] S. Butler, *Hudibras*, part 3, canto 3, 11. 291–2, slightly misquoted: 'And when the Fight becomes a Chace, | Those win the Day, that win the Race.'

[9] E. Waller, 'To A. H. of the different success of their Loves', *Poems* (1645), 11. 27–8 (p. 89). Harris misquotes the 2nd line: 'They fly that wound, and they pursue that die.'

[10] See *A Tale of a Tub* (1710), sect. 1; Dryden's trans. of Virgil, *Aeneid*, 6. 128–9 ('But to return, and view the cheerful Skies; | In this the Task and mighty Labour lies').

When we read in Tom Jones the Battle in ye Church-Yard, 'tis impossible not to feel the Force of the Ridicule, and where does it lie? In the happy Mixture of the Sublime in Appearance, with the Low in Reality; in seeing the Lofty, to raise us; the Pathetic, to melt us; the *Energeia*, to exhibit things, as tho' they were actually present, by an accurate Detail of the minutest Incidents; in short, the several Homeric Figures gravely introduced, & all to describe a Squabble among certain Barbarians of Somerset.

Thus we see that the Force of Ridicule lies in the plausible combination of things really inconsistent. But no where is this Force more striking, than in the Manners of Men, when we view those natural inconsistencies, that occur in the same Character. Here we have ye strangest Associations; Virtue & Vice, Wisdom & Folly, Reason & Passion; Stout Resolves & weak Executions; Serious Consultations & Determinations previously fixt; with infinite others too many to enumerate. This is what may be called Moral Humour, & is perhaps the highest & most excellent Species of the Ridicule. We may easily indeed perceive the Reason of its Merit. Wherever human Nature is faithfully described, whether in its regular Progressions, or its incidental Deviations, we are sure to sympathize at sight of the Picture, & whatever be the Event, whether serious or the contrary, we either weep for sorrow, or we weep for Mirth,

Sunt Lachrymae Rerum, et mentem mortalia tangunt.[11]

And hence 'tis, that the Moral Humour here mentioned has its latent use even in Tragedy & Epopy, but to every sort of Comic Work, tis so truly essential, that without some Sprinkling of this Ethic Salt, the whole never fails to be flat and insipid.

When Jones surprizes a Gallant with his Mistress in the Garret, had this Gallant been any common Character, a Groom or Lacquey belonging to Squire Western, there had been nothing to interest or strike us in the Event. But when this Gallant proves to be no less a Man than Mr Square, the Patron of eternal Fitness, the Assertor of ye Rule of Right, the

---

[11] 'There are tears for misfortune and mortal sorrows touch the heart': Virgil, *Aeneid* 1. 462.

Friend to rigid Virtue, & all that, we are immediately touched with the Moral Humour, to view so strange, & yet a natural Inconsistence; there being nothing incongruous (if we a little reflect) in deep Speculation with unruly Passion.

There is indeed the greatest care to be taken, that these Inconsistencies be natural. If ever the Wise lapse into Folly, or ye absurd rise out of it, it must be always by those Transitions, which are compatible with their several Characters; else the Performance instead of humorous becomes foolish & dull, & every Reader of Judgement is ready to say with Horace,

> Quodcumque ostendis mihi sic, incredulus odi.[12]

This Ethic Wit or Moral Humour shines no where more eminently, than in the celebrated Cervantes. The History of the Knight of Mancha from Beginning to End, is one continued Coalition of manly Sense, & gross Absurdity, which Coalition however strange, is still made natural by the artfull supposition of the Knights Romantic Madness. 'Tis in the Variety of Incidents, all referable to this original Idea, that the Beauty and Humour of ye Work consists.

Twas the same Moral Humour, which wrought so powerfully in Homer, which he knew how to infuse into his gravest Compositions, & which intitled him to be called the Father not only of Tragedy, but of Comedy. If we would be sensible of the Truth of what is here asserted, let us view the Dialogue in the last Iliad between Priam and Hecuba, divested of its Numbers & Heroic Stile.

Priam takes his Wife into an Apartment by herself, & then tells her with great seriousness, that he had received a Message from Jove, to set out for the Grecian Camp, & treat with Achilles about the Ransom of Hector's Body. He then asks her Advice, whether he should go, or no. Hecuba, upon merely hearing this adventure proposed, immediately begins a most lamentable Cry, & without the least Regard to Jove, or his Message, tells her Husband that he must needs have lost his Senses; that no Man in his Wits would think of such an Action; that Achilles was a cruel Wretch, & would certainly murder him; with much more to the same Purpose. Priam,

---

[12] 'Whatever you show me in this way, I disbelieve and hate'; Horace, *Ars Poetica* 1. 188.

after hearing her out, bids her not prove a Bird of ill-Omen, for that he was determined to go, let her say what she would, & accordingly leaves her, & sets forth directly.

Here in the Subject of Advice, we have a Sample of Ethic Humour, which has been verified in human Life a thousand Times since. Applied to Heroes & Heroines, with the proper Stile & Metre, it appears in all ye States of Tragedy or Epopy. Applied to Characters of lower Life, without such Ornaments, it descends to Comedy, or perhaps sinks into Farce.

Let us suppose the following Dialogue between Squire Tankard and his Wife.

SQUIRE. Dosn't remember, Bridget, how I lost my silver Tobacco-Box to Jack Sharp at Skittles?

WIFE. I do, my Dear, too well.

SQUIRE. Now I am come, Honey, for a crum of thy Counsel. Here's Sr Harry has sent a Message, that if I'll meet 'em at the Cat & Bag-Pipes, he'll engage that Jack shall give me my Box again. What sayst Thou, my Duck, shall I go, or no?

WIFE. Go! Mr Tankard? Why surely you are distemper'd—As mad as a March-Hare—Go to ye Cat and Bag-Pipes? In such Weather? Such Roads? Such a time of Day? And all to meet Jack Sharp? Why he's the greatest Rogue in all the Country; a very Pickpocket; an errant Gambler; he'll cheat you of every Penny; make you drunk, and then instead of bringing home your Box, 'tis well if you bring home yourself. I care not a Farth—

SQUIRE. Peace, sweet one, Peace!

WIFE. I say, Mr Tankard, I care not a Farthing.—

SQUIRE. Zounds! I say Peace, & leave off croaking this damneth Stuff. Mayst talk for a Month, I'm determined to go. Had the Landlord, or the Millar, nay had the Parson sent, I should not have minded it. But as the Message comes from Sr Harry himself, by G—I'll set out, tho' I break my Neck by ye Bargain.

Exit Squire, calling for his Boots

Here we have Priam & Hecuba in lower Life; the operating Causes, that is, the Passions the same; the Effects, if not the same, at least analogous; the difference arising only from

Accidents and Circumstances; as in Homer, we have Kings & Queens, in distant Countreys & remote Ages; while in ye other, we have Characters familiar & domestic.

These Accidents and Circumstances may be called the surface of human Life. To vulgar Eyes nothing else is visible. And hence it is, that being governed by appearances, they often make Distinctions, when Nature knows none. 'Tis ye diviner Eye, that can penetrate this Surface; that can view human Life in its inmost Recesses; & that has Strength to discern those latent Principles of Identity, which are common to Ranks and Characters apparently the most distant. He that has such an Eye, may be pronounced a great Genius, & 'tis in contemplating such a Genius under its different Operations, that we may comprehend the Force of that Paradox of Socrates, how 'twas the work of the same Capacity to write Tragedy or Comedy.

And this naturally leads to an Enquiry, what sort of Men those are, who are capable of Ridicule, either to exhibit it themselves, or to relish it in others. This we shall examine, when we have first given certain coordinations of things, out of which we conceive ye most of ye species of Ridicule may be derived. We ought perhaps to apologize for exhibiting any thing so dry, as these coordinations may possibly appear, yet may it be said in their Behalf, that they are of great antiquity, and not only this, but that they have their Foundation in Nature. We learn too in Scripture, that things are double, one against another.

> Co-ordinate ye first, which respects
> Being in general.
>
> True——False
> Certain——Doubtful
> Probable——Improbable
> Possible——Impossible
> Real &c——Apparent &c.
>
> Co-ordinate the second, which respects
> Natural Subjects & their Attributes.
>
> Substance——Nullity
> Great——Little

Many——Few
White——Black
Round——Square
Beautifull——Deformed
Admirable——Contemptible
Here——There
Above——Below
Standing &c——Sitting &c

Co-ordinate the third, which respects human Life.

Right——Wrong
Good——Bad
Virtuous——Vitious
Honourable——Dishonourable
Laudable——Shamefull or Base
Rational——Irrational
Reason——Appetite
Brave——Cowardly
Temperate &c——Intemperate &c.

We may call the first of these co-ordinates the Logical; the second, ye physical, & the third, the Ethical.

Now tis by taking some Couple or Pair out of one of these coordinates, & by artfully blending it, so that one opposite appear the other, whether it be that False appear True, or True False; Great appear Little, or Little Great, Honourable appear Base, or Base Honourable, 'tis (I say) in this artfull blending lies the Force of Ridicule, as the Pleasure may be said to lie in its Recognition & Resolution.

'Tis to be remark'd however in the Logical & Ethical Coordinates, that the Opposites are never in any Instance equipollent. True is better than False, Virtuous than Vicious, and so of the Rest. And hence 'tis that in all Ridicule deriv'd from these, there is always implied something laudable or faulty, so that 'tis never a Matter of Indifference to become this Way ridiculous. But in the Physical coordinate the Opposites are often equipollent; Round is often not better than Square, nor Great than Little, nor White than Black, nor Here than There, nor Standing than Sitting &c. and hence 'tis that the Ridicule derivd from hence may be endured without

Disparagement, & that this therefore constitutes what we call inoffensive Humour.

'Tis likewise to be remark'd in Ridicule, that it consists for the most part in Propositions after a manner analogous to ye being of Truth. Indeed as Ridicule itself may be called Truth's spurious Offspring, 'tis no Wonder it should retain certain Traces of its Parent. Truth is seen in bringing together certain Terms, & in exhibiting them when brought together, to the Mind's Eye. So is Ridicule. Truth brings these Terms together for two Purposes, & no others; one, to shew how Terms are connected; the other, to shew how Terms are dis-jointed: Ridicule does ye same. And hence 'tis there is no Ridicule, as well as no Truth, but what is of Necessity either Affirmative or Negative. Instances of the two Parts in Truth are obvious to all. In Ridicule, the Poet who sung that

> when the Fight becomes a Chace,
> He wins the Day, that wins the Race.

This Poet (I say) dealt in affirmative Ridicule, by connecting ye Ideas of Victory & Flight. On the contrary, the old Roman, who said of a thievish Servant, what was true of the most faithfull one, *that there was nothing lock't from him throughout the House*, this Roman might be said to deal in negative Ridicule by dis-joining from the Thief those Obstacles of Art, which Truth considers as chiefly made on his Account.

And now (as we promised) we are to enquire, what sort of Men those are who can either exhibit Ridicule themselves, or can relish it in others. They are not to be found among the heavy & dull; of this we may be assured from certain Experience. Yet dull Men may possess very acurate Sensations, & even so far enjoy the Use of their Understandings, as to comprehend the *Sameness* in things *nearly the same*, & discern ye Difference in things *widely different*. 'Tis plain then that the Power of Ridicule lies not in Sensation of any kind, nor yet in *ordinary* Intellect. It must lie then in that Intellect, which is better than the ordinary one, because excepting this, there is no Power left.

Now as every Intellect desires, as far as may be, to be conversant about Truth, & is unwillingly deprived of it, so is ye best Intellect of course the most conversant. And as all

Truth is affirmative or negative (the Affirmative in bringing things homogeneous together, ye negative in setting things heterogeneous apart) it follows that the best Intellect, & which for that Reason is most conversant in Truth, will be that, which recognizes not obvious Identities & obvious Diversities, but which can comprehend ye sameness in Things widely different, & discern the Difference in things nearly the same. To do this there must be an accurate Discernment in every Subject that occurs, of all its various Parts & Attributes; there must likewise be a strong Retention of every Subject so discerned, that whatever new Subject occurs, we may have Plenty of similar & dissimilar ones at hand, by which to try & scrutinize its Nature. Farther to enrich ye Mind with this Plenty of Subjects there are but two Ways; either various & well digested Literature, or a Life led in ye busy World, & conversant with a Variety of Men & Manners. The first is most profitable for Speculation, ye last, for Practice, but both united in one Mind form a far more perfect Character, than either of them singly.

Let us then suppose an Intellect, blest with the Faculties above, & enriched by every Method with a Multitude of Subjects. Now as there is diffused throughout ye Nature of things a Portion of *Appearance* as well as *Reality*, 'tis plain such an Intellect in its Survey of Being, will necessarily recognize both one & the other together with all their appendent Identities & Diversities. If in its Process, it wholly reject ye *Apparent* & adhere to the *Real*, such an Intellect forms the Character of a *Grave & wise Man*. If it choose in its Process to admit the *Apparent*, yet not with Design to supplant ye *Real*, but only to heighten each by reciprocal Opposition, such an Intellect forms ye Character of ye *Witty or facetious*. If it admit the *Apparent* not for ye Purpose just mentioned, but serious & deliberately to make it pass for ye *Real*, such an Intellect tis that constitutes the Character of a Sophist. There is no other Character left, except that which consists in the Privation of all these (I mean) the Character of ye Illiterate & Fool, whether it arise from Idleness, Intemperance, Age or natural Infirmity.

From what has been said it appears, that no Man can be a Wit, except he be a Man of Understanding. Each Character implies the same Sort of Mind; a Mind well stored with a Multitude of Ideas, which it can associate or dissociate, as Occasion requires. The Difference is, the Man of Understanding attends to ye *Real*, the Man of Wit attends to the *Apparent*. Now 'tis impossible for that mind to recognize the Apparent, which is not able in the first Place to recognize the Real.[13] 'Tis the same indeed here as in our other perceptive Faculties. There may be Palates, which cannot taste either the tainted, or the sweet, but the Palate, that tastes the Sweet, tastes its Opposite of course. There may be Ears perfectly indifferent as to Concord or Discord; but the Ear that knows Concord, cannot avoid knowing its contrary. 'Tis the same as to the Eye, in things beautifull & ugly; the same as to the mind in Matter of Truth & Falshood, or which is in fact the same, in things real & apparent.—the whole indeed may be reduced to this general assertion—that of Opposites, if they are sensible, there is but one Sensation; or if they happen to be intelligible, there is but one Intellection. This too depends on ye following necessary Truth,—that 'tis absurd that a Power, which perceives a thing to be *present*, should not be able to perceive, when the same is *not present*. And so much as to the Proof, that no man can be a Man of Wit, except he be a Man of Understanding.

This Affinity however between Wisdom & Wit depends not for its Certainty on Reasoning alone. Examples confirm it thro all Ages. Homer was the Father both of Tragedy & Comedy. The great Genius of Philosophy, Socrates was not only the wisest & best of Men, but perhaps the truest Master of Humour, that ever existed. His Disciples Xenophon & Plato followed the Example of their Master. To these may be added Aristotle & Theophrastus. The Ridicule of Euripides may be seen not only in his Character of Hercules in the Alcestis, but thro' his whole satyric Poem of the Cyclops. The Wit of Menander is allowed by all antiquity; yet his Fragments (the sententious Part of him having been principally preserved) give us far higher Ideas of his Wisdom, than of his Humour. The same may be said of his Contemporary Philemon. Among the Latins, Ennius & Lucilius were both of the wise & witty.

---

[13] Harris's note here is to Aristotle's *Rhetoric*, i.i.

So was Cicero; so were Virgil & Horace, tho' ye former as to Ridicule was more sparing & concealed. So was Augustus, & all his court. In our own Country we may reckon Chaucer, Shakespear, & Milton, who all excelled in both Characters, and gave us in their Writings many Samples of each. To these we may add Sr Thomas More, Sr Philip Sidney, the two Earls of Shaftesbury, the Statesman & the Author, & even among our Divines, Tillotson, Bentley, South, Middleton, Atterbury, Hoadly, Sherlock, together with others now living of deserved Repute.

From the foregoing Speculations we may be led to observe, that it is not so strange a Doctrine, as may at first perhaps appear, to make ye Risible a Peculiarity of Human Nature, if it be true that Ridicule is an Appendage to Reason.

Again, we may hence perceive the Force of that Argument of Socrates who compelled the Tragic Poet Agatho & the Comic Poet Aristophanes to confess, that it belonged to ye same Man to know how to make Tragedy & Comedy; & that he who was by his Art a Tragic Writer, was a Comic one too. Human Nature is clearly ye same in high Life as in low, & the Difference only lies in the merest of Accidents. Truth then of the same sort, is the Basis of both, for that Ethic Salt so essentiall to comedy, we can only assert from the Principles above, that he who knows the *True*, cannot miss of the *Apparent*. Homer who wrote the Iliad, wrote also ye Margites. Euripedes, who drew Hecuba, drew the Cyclops & Silenus; & the Genius that gave us Lear, gave us Falstaff & Shallow. Our later Countrymen, Dryden & Pope, are not wanting in many Passages highly sublime, and in many others of extreme Ridicule. Witness for Dryden Alexander's Feast & Macflecknoe; for Pope the Essay on Man & the Dunciad.

Again, we may hence perceive the Errour of ye celebrated Locke, who is for dis-joining the Natures of Wit & wisdom, by making Wit lie in *bringing together*. Now almost the whole of Euclid is Wisdom of this latter kind. The most valuable Part of all Science is clearly of this Kind, it being certainly more important, to know affirmatively what things *are*, than to know negatively what they *are not*. In every Syllogism there must be one Truth of this kind, since there can be no Conclusion from Negatives alone. To say therefore that Wit differs from Wisdom because Wit *brings together*, is to distinguish it from Wisdom by Wisdom's primary Attribute. On ye contrary to distinguish Wisdom from Wit, by Wisdom's *Power of separating*, is to distinguish it by a Power that is common to both, it being as much the End of Wit to make ludicrous Distinctions, as it is of Wisdom to make such as are serious.

What seems to have led Mr Locke and others into these Sentiments may possibly be this. They see many called Men of Wit, who are said to have no Wisdom; & many called Men of Wisdom, who are said to have no Wit. Hence they are induced to conclude that the two Characters are totally different, & explain the Difference, by referring them to opposite Principles. Now the Truth is, they do not differ, as Streams of the same Fountain, that pursue a different Course.—Whether there are many Sorts of Wit, we shall not here inquire; that there are of Wisdom is beyond Dispute. There is a Wisdom of Speculation, & a Wisdom of Practice. A Man may be excellent in ye one, & quite deficient in the other.

> Democriti edit pecus agellos,
> cultaque, dum peregre est animus . . .[14]

Here we have more than *Wit* without wisdom; we have even Wisdom without Wisdom, that is, the Speculative without the Practical. Yet who can doubt that a Mind, equal to the Survey of Nature, should not be equal, if applied, to so mean an Accomplishment, as that of cultivating or fencing a few Acres of land? And why should we not judge as favourably, as to the Faculties of ye Man of Wit? Let us help the facetious Character as we would the Philosophic. Let us borrow a Distinction even from ye Schools, to bring them off. Let us say, if they are not always the most worldly wise in *Act*, that in *Power* they are capable of all Discretion & Prudence. Let us compare them to those generous & high-bred Horses that are no way deficient in natural Strength, but are above the Drudgery of the Waggon or the Plough.

And this leads us insensibly to another Race of Men, who have confessedly no Wit, & yet are by some thought Men of Wisdom. These are those, who from constitutional Apathies

---

[14] 'The cattle of Democritus ate his little fields and crops while his mind was far away': Horace, *Epistles*, 1.12. 12–13.

have neither Passions to lead them astray, nor Intellect to exalt them; who jogg on in ye dirty Road of Lucre, with a perfect Insensibility to every thing scientific. But perhaps these are not worthy to be considered as Exceptions to our Hypothesis. There are others of a Rank superiour indeed to these, & yet we think our doctrine to be no Ways in danger, tho' what is called their Wisdom be often totally devoid of Wit. Such are Lexicographers, the Chronologers, the Index-makers, the Compilers, the painfull Calculators of long Tables, to shew Eclipses, Interest, Tides, & the like. These are framed by Nature for the Drudgery of the Brain, a serviceable Race as necessary to Science, as Ploughmen & Porters to Husbandry & Trade.

'Tis to the Men of Genius & Invention that we alone refer. Here if there were found Wisdom with Insensitivity to all Humour, we would freely confess our Hypothesis in Danger. But it may be justly questioned, whether this ever happened. Some Minds indeed of great Strength by severe Application to rigid Truth, may have possibly so suppresst their Turn to ye Facetious, as to acquire a Morosity even bordering on excess. Yet the very Sarcasms of such are a Proof of their Power, & shew what they might have been, had they cultivated the Graces. Again, there are others of great Capacity, who by totally cultivating Ridicule alone, come to consider nothing as serious, nothing as really valuable. With these not only Vices & the Vicious are derided, but even Virtue, Human Nature & Religion of all Kinds, Such of old were Aristophanes & Lucian, & in these latter Days Rabelais, and the Author of Gullivers Travels. I would not dissuade absolutely ye reading of these Writers, but I would caution against the Infection of unlimited Raillery, which insensibly begets within us a general Contempt, & that naturally leads to a general Malevolence, till in the End we lose every thing either social, or serious.

The celebrated Author of the Characteristics, tho' a Friend to Ridicule, was by no means (like the Authors just mentioned) a Friend to its excess. Speaking of Irony (one of its highest Species) he thus expresses himself in an Epistle yet unpublished, where he censures Daciere for supposing that celebrated Ode of Horace, *Parcus Deorum cultor et infrequens*, to be nothing more than a Banter on ye Stoical Philosophy.[15] 'Monsr Daciere (says the noble Author) knew little of ye Simplicity of Horace, or ye Measure of his Irony. There is a due Proportion in Irony well known to all polite Writers, especially Horace, who so well copied that noted Socratic Kind. Go but a little further with it, & strain it beyond a certain just Measure, and there is nothing so offensive, injurious, hypocritical, bitter and contrary to all true Simplicity, Honesty, or good Manners.'[16]

He indeed appears to have hit the Mark, that applies his Ridicule, as he would his Rhetoric (of which in fact Ridicule is a Part) not to supplant Truth, but to enforce it, not to darken the Brightness of Virtue, and Science, but where their natural Charms are too refined for vulgar Eyes, to recommend them by Charms more adapted & familiar; that has Ends like the following, when he deigns to be facetious—to gain the Attention of listless Readers; to reprove without Harshness; to avoid Ostentation; to cover a Sentiment better concealed for the present; to exhilerate the Severity of logical Scrutinies, which are often irksome from Perplexity & Length; in short that employs Wit as we do Salt (which is for that Reason in some Languages its Name) that it may give a Poignancy to other things, not itself be ye Repast. Rem tibi Socraticae poterunt ostendere chartae.[17]

And so much for ye Subject of Ridicule.

---

[15] Horace, *Odes* 1.34: [1] 'a sparing and infrequent worshipper of the gods.'

[16] Shaftesbury to Pierre Coste, Chelsea, 1 Oct. 1706 (PRO 30/24/45/iii/43), copied by Harris from the Shaftesbury papers into a MS vol. 'MS by 3ᵈ Lord Shaftesbury. Letters on Horace & upon other Philosophical Subjects', *c.*1737/8, p. 12. Shaftesbury read Dacier 'out of a kind of insulting Malice, to see how he, with his Court Models of Breeding & Friendship, would relish' this Horace poem (Monash University Library).

[17] 'Socratic pages will be able to set the matter before you': Horace, *Ars Poetica*, 1. 310.

Appendix V

Joseph Warton's 'Hymnus ad Pacem'

In September 1762 the young James Harris left Winchester, four years before Joseph Warton was internally promoted to headmaster, and spent the next six months with his father in London. As a raw youth of 17 he mixed with, in his own words, 'all the gaiety and dissipation of London . . . [which] gave me a knowledge of the world so much greater than I supposed my fellow-collegians could possess'. In March 1763 he proceeded to Merton College, Oxford, as a Gentleman Commoner, where his fellow collegians included Charles James Fox, Charles Marsham (later Lord Romney), Brownlow North (son of the Earl of Guildford and later Bishop of Winchester), Sir John Stepney (future British envoy to Dresden and Berlin), William Eden (Lord Auckland), and Robert Henley (Lord Northington). He saw his trigonometry tutor at most once a fortnight, and avoided lectures, chapel, and hall. Imitating what he called 'High Life in London', he drank claret, played cards, and generally wasted his time and finances (Malmesbury (ed.), *Diaries and Correspondence*, i, pp. xl–xli). That, at least, was how he himself saw his college years in later life. In spite of this unfavourable self-portrait of mild dissipation, the future Lord Malmesbury nevertheless made some attempt to study the classics: he asked his father to send him his observations on Aristotle's *Rhetoric*, which he had just begun to read (24 July 1763), and he continued to receive guidance from Joseph Warton (whose basic form of instruction at Winchester was verse translation), and from Thomas Warton, now into his sixth year as the Oxford Professor of Poetry.

Enclosed with this letter to his father of 24 July the young James Harris sent a transcript of a Latin poem by Joseph Warton celebrating the end of the Seven Years' War. The Peace had been signed on 10 February 1763, during the Pitt–Newcastle ministry, thus ending the 'German War' between Britain and the allied France, Austria, and Russia. In addition to European peace, the settlement also gained for Britain the whole of Canada, Louisiana east of the Mississippi, Florida (wrested from Spain), parts of the West Indies (the 'sugar Islands'), and access to the West African slave trade. France was forced to acknowledge the supremacy of the British East India Company, and effectively to remove herself from the British American mainland colonies.

As we have seen, the elder James Harris was given the task of moving the Address to the King on the Preliminaries of Peace which had been laid before the Commons on 1 December 1762. The speech was delivered eight days later. It was now the younger Harris's turn to represent the university constituency's sentiments on the same subject: the future British ambassador to Berlin and St Petersburg addressed national politics for the first time in his life by reciting a Latin poem written for him by the Oxford Professor of Poetry.

By 1760 the elder James Harris had already an extensive contact with both Joseph and Thomas Warton on scholarly matters. In April 1761 he was proof-reading Thomas's *Life and Literary Remains of Ralph Bathurst* (1761). He lent books to Joseph for the latter's edition of Theocritus (1770), advised him to incorporate useful Greek material in the Bodleian library, and discussed Kames's *Elements of Criticism* with him. The education of the young James involved all three of them.

Joseph Warton wrote the poem on the Peace for public performance in the Sheldonian Theatre at the annual commemoration. In the first week of July 1763, Thomas wrote to his brother: 'Harris has just brought me the Hymnus—they are absolutely the best Latin verses I ever read, both truly classical and poetical. He is to rehearse them to me tomorrow morning at seven, in the Theatre. You have done *him* and *us* infinite credit.' Joseph asked Harris senior for his opinion on the poem, and he thought it 'admirable; a happy temperature between the Lucretian rhyme and the Virgilian' (2 July). It is indeed an excellent Latin poem, and Harris could offer no suggestions for improvement.

The 'Hymnus ad Pacem' has not been previously published. An accurate copy was not available to John Wooll when he compiled *Biographical Memoirs of the Late Revd. Joseph Warton* (1806, p. 294), and was therefore omitted. The following is printed without change from a transcription of the poem in the younger Harris's hand, included with his letter to his father (Harris Papers, vol. 11, item 26, 'A Poem of Mr Warton's, spoken at Oxford in 1763 by Mr James Harris'). I am grateful to Gavin Betts for assistance in the transcription, and more particularly for the accompanying prose translation, which is his.

HYMNUS AD PACEM

Diva, Parens Frugum, Musarumque optima Nutrix
Optima Virtutum, Britonas jam Marte rubentes
Felici, et multâ satiatos Caede revisas,
Orbe triumphato—Tu, Pax Veneranda, per Arva

Europae viduata viris et diruta tecta
Incedas, placido vultu, gressuque decoro
Cuncta Coloribus egregiis et Odoribus opplens.
Quin Tecum properant faecundo Copia Cornu,
Flava Ceres, Spicam quassans, et plurima portans
Poma sinu, Bacchusque Pater cui Vitis obumbrat
Flexa genas; Aheni musto cui Crura madescunt.
Tum Charites aderunt, Zonis de more solutis
Quae Pedibus plaudent Choreas, dum ad nota vocabit
Prata Pales Tauros. Tecum alma Astraea redibit
Sacra Deum, Sanctas Leges et jura reportans
Heu violata Diu—Dum Tu facis, at fera Martis
Moenera per Maria ac Terras sopita quiescunt,
Ceu Jovis in Sceptrum truculentus decidit ales
Phoebeio Victus Modulamine. Cessat hiare
Os rapidum, assuetum nec spargunt lumina Fulgur.
Te, Dea, Te fugiunt atrae Formidinis ora
Te Dolor, insomnisque Fames, et pallida Egestas
De Casula reptans Combusta, Tegmina spinis
Subnectens Lacera. Effracto Bellona Flagello
Dentibus infrendens sequitur, Diraeque ululantes
Voce tua attonitae, et vellunt de Crinibus angues.
Interea Eoae Gentes, Indique remoti
Quique bibunt Nigrum vel praecipitem Niagaram
Danubium aut latum aut torrentia Flumina Volgae
Undique Amicitiam quaerunt, et Foedera jungunt.
Ergo rite Tibi meritum instauremus Honorem;
Munera Pieridum ferimus, Phoebique Corollas
Isiaci Juvenes, Patriae nam Tempore iniquo
Belli inter rabiem haud Musas sectarier almas
Possumus aequo animo; languentia Pectora Fervor
Deserit Aonius. Quin Te redeunte virescunt
Ingenium atque Artes. Tibi Laurus Olivaque surgunt
Cecropia ut quondam e Terra cum Graecia magnos,
Xerxi debellato, egit servata Triumphos.
Salve, Diva iterum, O dudum exoptata Britannis
Ipsa inter Spolia et repetitos Martis Honores.
Extremae Gentes, atque ultima gestiet aetas,
Pax, Laudes celebrare tuas, Nomenque Georgi.

'Goddess, mother of crops and best nurse of the Muses, best nurse of virtues, return to the Britons who are now red with successful war and satiated with much slaughter after the conquest of the globe. Go, revered Peace, through Europe's fields, denuded of men, and through her ruined buildings; go with calm face and becoming gait and fill all with your colours and scents. Yes, and with you Plenty hastens with bounteous horn, and golden Ceres shaking her ears of grain and carrying abundant fruit in her bosom, and father Bacchus, his cheeks shaded with bent vine and his legs dripping with must from the vat. Then will the Graces approach, their girdles loosened according to custom. They will strike up the dance with their feet while Pales calls the bulls to familiar pastures. With you Astraea will return, bringing back the ceremonies of the gods, holy laws and rights, alas! too long violated. This you do, but the fierce works of Mars are lulled and grow quiet through land and sea, just as the fierce bird of Jove falls down on to his sceptre when overcome by the music of Phoebus; his quick beak ceases to gape and his eyes do not scatter their normal gleam. You it is, goddess, who turn to flight Fear's black face, and Grief and sleepless Hunger and pallid Need as she crawls from the burnt hut fastening her torn garment with thorns. Bellona follows with broken whip and gnashing teeth, and the howling Furies aghast at your voice and tearing the snakes from their hair. Meanwhile the peoples of the East and the distant Indians and those who drink the Niger or headlong Niagara or the broad Danube or the gushing streams of the Volga, everywhere men seek friendship and make treaties. So let us duly restore to you your deserved honour. For you do we, the youth of the Isis, bring the Muses' gifts and the wreaths of Phoebus. Amid the rage of war in our country's danger we cannot pursue the benign Muses with calm mind; the Aonian fervour deserts our languishing breasts. But on your return talent and arts revive. For you the laurel and Cecropian [i.e. Athenian] olives rise, just as formerly they rose from the earth when Greece, safe after the defeat of Xerxes, conducted mighty triumphs. Hail once more, O goddess long desired by Britons amid the very spoils and repeated honours of war. Distant peoples and the most remote age will desire to celebrate your praises, O Peace, and the name of George.'

Composite Chronology: 1736–1980

THIS lists all the known published and significant unpublished works of James Harris, including revised editions, translations, collaborative works, works dedicated to Harris, significant reviews, and critical studies in English, German, and French. It does not include unpublished letters nor the frequent performances of his musical arrangements, such as *The Spring* (the pastoral musical drama which Garrick invited Harris to adapt from the original 'Daphnis and Amaryllis'). A query indicates putative authorship, and quotations are included from some of the more significant unpublished works, which are marked with an asterisk. Some shorter unpublished poems are printed in full from Harris's manuscripts.

1734

'The Fable of the two Weather Cocks'*: 4-page poem.

'Of the Nature and Genius of Poetry'*: 10-page prose essay on the topic that ''Tis by Fiction or Invention only, by which a Poet is constituted'.

'Ode to Common Sense'* (complete text):

> Source of Science & of Arts
> Unerring Guide of Human Hearts,
> Parent of Truth, who first uprear'd
> By Thee, to Anxious Man appeard:
> O! march the num'rous, barb'rous Host,
> To Thee, and thy kind influence lost.
> See Purple Tyranny advance,
> Grasping ye sharp, relentless Launce.
> With sullen Brow, and solemn Walk,
> See mitred Superstition stalk.
> Lo! Myriads wait these monsters dread,
> Of Creatures Black, and Creatures Red;
> And dire the clashing of their Swords,
> And dark the Jargon of their Words.
> O stretch thy glorious Aegis out,
> And drive far hence ye Savage Rout.
> Dazzled with Intellectual Light,
> Abash'd they'll fear, though fly the Light
> So, while the Realms of Night endure,

And beauteous Nature lies obscure,
Batts, Wolves, and Owls, a gloomy Train,
Nocturnal Anarchy maintain.
But soon as e're ye God of Day
Dispells ye Dusk with joyous Ray,
The Sons of Night all scout away.

1736

Second treatise, 'A Discourse on Music, Painting, and Poetry',
finished.

1738

First treatise, 'Concerning Art, A Dialogue', finished.

1739

5 Jan. 'L'Allegro'*: extracts from Milton's poem divided into two
Acts, with parts for accompanied singing and Chorus (10
pages).

T. Birch, 'Life of Shaftesbury', in vol. ix of *A General Dictionary
Historical and Critical* (1734–41), 179–86; based on Harris's
research in the Shaftesbury papers, if not wholly on his writing.

1741

15 Dec. Third treatise, 'Concerning Happiness, A Dialogue',
finished.

[?] 2 articles in *The Westminster Journal*, 4 (19 Dec.), and 15 (6 Mar.
1742) (see 1747 below).

1742

19 Oct. 'The Plumbeians a Vision'*: 13 pages of octosyllabic
couplets about Salisbury intellectual ennui, in particular the
tedium of Edward Goldwyre's oft-told story of eating fresh
sturgeon, concluding:

At that my Hair began to stand,
A sudden Tremour seiz'd each hand.
Gods! am I doomed to such Perdition,
To hear a Week without remission?
To hear a Week the Sturgeon Story?
Ah! Traytor! This is Purgatory;
Tis Hell, tis Worse—the sudden Fright

Dispell'd the Fantoms of the Night,
And who can 'ere the raptures frame
When wak'd, I found it but a Dream?

'Polonius and Ophelia, a True and Sorrowful Tale wch happened in
the Ninth Century'*: 12-line parody of 'Ophelia's' elopement
and pursuit by her father.

'A Hymn to Hymen, being a Fragment of Chaucer', later printed in
Dodsley's *Collection* (1758), v. 296. The holograph draft is as
follows:

Full well of lerned Clerkes it is said,
That Woman-hood for Man'is use was made.
But naughty Man liketh not one, or so,
He lusteth aye, unthriftily for mo.
And whom He whilom cherished, when ty'd
By Holy Church He cannot her abide.
Like unto Dog, which lighteth of a Bone,
His Tail he waggeth, glad thereof ygrown.
But th'ilk Bone if unto his Tail Thou Tye,
Pardie He, fearing it, away doth flye.

'Caelia'*: whole poem given below from holograph in bound vol. of
miscellaneous poems:

Gaudia dum furtiva petit, vetitosque Hymenaeos,
Nocte fugit caeca celeri pede Caelia Patrem.
Quid Pater? incassum furit, et magno intonat ore,
Unde mihi Currus? Sequar hanc per tela per ignes
Tartareas usq <ue> ad fauces. Nec plura locutum
Nox manet in praesens placido dare membra Sopori.
    Cras surgit, jentat, provisa viatica fidis
Demandat loculis; tum cunctis rite peractis
Inscendit lente Currum; revocare fugacem,
Heu! longe elapsam, et sibi jam mala nulla timentem,
Binorum tarda nitens ope Iumentorum.
    Felix, qui potuit rerum cognoscere causas,
Atque procul nexas, et ineluctabile Vinclum
Conjugii fugit, strepitusque metusque jugales.
Fortunatus et Ille, ausus qui tollere prolem,
Natarum cum jam pubescere corpora cernit,
Ocius explorans dignos parat ipse Maritos,
Nec miseras ultro suadet cunctator iniquus,

Quod presto est, captare avide, discrimine nullo,
Si Scotus, Hibernusve, aut si venit advena Miles,
Candenti Vestes conspersus pulvere et auro.

'Seeking furtive pleasures and forbidden marriage Celia flees
from her father with swift foot in the blind night. What is her
father to do? In vain he rages and thunders with mighty voice,
"Where is my carriage? I shall follow her with bolts of fire right
to the gates of Hell!" He says no more. For the present, night
waits for him to give his limbs to gentle sleep. On the morrow,
he rises, takes breakfast, attends to provisions and puts them
in his trusty trunks. Then, everything duly arranged, he slowly
mounts his carriage and strives with the tardy help of twin
beasts to recall the fleeing girl, who, alas!, has long since
escaped and no longer fears any misfortune.

'Happy the man who was able to discover the causes of
things and who fled far from the bonds and the inescapable
chain of marriage and the noise and fears of matrimony.
Happy also is he who, after daring to raise children, makes a
quick search when he sees the bodies of his daughters already
approaching maturity and himself finds them worthy
husbands. He does not make unfair delay and encourage the
unhappy creatures to snatch indiscriminately whatever is at
hand, whether a Scotchman or Irishman comes, or a foreign
soldier with clothes besprinkled with shining dust and gold.'

(Trans. G. Betts; ll. 12 ff. are modelled on Virgil, *Georgics*, 2.
490 ff.; there are other Virgilian echoes)

1743

7 Apr. publication of H. Fielding's *Miscellanies*, 3 vols., including
Harris's revision of 'The First Olynthiac of Demosthenes', in
the first vol. First advertised 5 June 1742.

Notes to *Three Treatises* finished.

Sept. 'The History of the Life and Action of Nobody'* written:
dedicated to Henry Fielding.

1744

May. *Three Treatises* published.

William Collins, MS poem [Lines addressed to James Harris].

Oct. 'Much Ado: A Dialogue', written.

'Upon the Species of Folly'* written.

1746

Tuesday, 7 Jan. *The True Patriot: and The History of Our Own Times*, no. 10; Harris's lead article on the Jacobite rebellion (dated 14 Dec. 1745).

C. Batteux's trans. of first and second of the *Three Treatises*, in *Les Beaux Arts reduits à un même principe* (Paris, 5 edns. by 1774). Harris saw the Leiden edition of 1753.

'Fashion: A Dialogue', written.

'A Short Map of Human Knowledge'* written.

Dec. 'Knowledge of the World, or Good Company: a Dialogue', written.

1747

20 Aug. Writing of *Hermes* begins, with the original title *Hermes or A Philosophical Inquiry concerning Universal Grammar Logic & Poetry*.

[?] Articles 4 and 5 (dated 19 and 26 Dec. 1741) repr. in *Letters from the Westminster Journal*.

1750

'Upon Ridicule'* written.

J. Petvin, *Letters Concerning Mind. To which is added, A Sketch of Universal Arithmetic, Comprehending the Differential Calculus, and the Doctrine of Fluxions*. Rev., co-ed. with G. W. Harris, and prepared for the press by 'Author of a Book called Three Treatises' (p. iii).

1751

Hermes: or, A Philosophical Inquiry concerning Language and Universal Grammar.

Concord: A Poem, dedicated to [Sir John Robartes, fourth] earl of Radnor.

'To the pious memory of Canon Malvolio, who died 1751'* (i.e. Canon Bampton, residentiary of Salisbury, who, in Harris's words, 'left half his Estate to a Woman, whom he had kept, half for Lecture Sermons to defend the Trinity'):

> Malvolio dies, & leaves his Store
> Half to the Church, half to his Whore.
> Is not Malvolio wondrous civil,
> To trim so nice, 'twixt God & Devil?

1752

'The Gardens of St Giles's inscribed to ye Countess of Shaftesbury by J.H.'* Sixteen prose pages, and Harris's only extensive writing on landscape, this is written in the form of a prose epistle. It describes the St Giles estate as possessing 'at once the four capital sources of natural Beauty, Lawn, Water, Wood, & uneven Ground, from the artful diversification of which depends all that is truly great, or elegant in Gardening' (p. 4). Harris picks out 'Shakespeare's Cell' (a small building containing a bust of the dramatist, an urn, and an edition of his works), a Gothic hermitage made from tree roots, lined with moss, with a couch and rush floor and an hourglass, a Grotto lined with 'Sparrs, Minerals, & Fossils', the 'Towers of Salvator, the Castle in the Island, the thatched House, the Pavilion'.

Upon the Rise and Progress of Criticism.

'Knowledge of the World, or Good Company. A Dialogue', printed 'from the MSS. of J.H. of S, in the County of W'.

[?] 14 Apr. 'A Dialogue at Tunbridge Wells, between A Philosopher and a Fine Lady. After the Manner of PLATO', *The Covent-Garden Journal*, 30.

A Short Account of the Four Parts of Speech according to Aristotle, as explained in Hermes (probably by J. Upton).

1753

'A Tour from St Giles's in Dorsetshire, thro' Yeovil, Bridgewater, Wells, Bath, Bristol, &c. performed July 1753'*: 4-page poem.

1754

26 Jan. 'Papers relating to Mrs Carter's Translation of Arrian'*: 7 pages in Harris's hand, followed by 16 in Mrs Carter's hand, followed by 16 in Harris's hand. Harris saw the first of the 2 trans. chapters of Carter's translation of Arrian's *Epictetus*, and thought it 'a very fine one . . . done with great accuracy & care; with an excellent choice of words; & with a close attachment to the Letter of the Text', lacking only a 'natural ease and appearance of an Original, while tis withal an accurate & faithfull Copy' (p. 4). Carter addressed Harris with questions about trans. of the more difficult abstract Greek concepts such as *Imagination, Principles, preconceived Ideas, Qualifications, Materials of Action*. Harris replied with detailed etymological commentaries and added a final remark: 'As to any Emendations, Notes, or Explanations, respecting this Treatise of Arrian, I have

made none worth notice, since Mr Upton's Edition, & those I had made before, were inserted by him into that' (p. 15).

15 July. Verse epistle from St Giles to the Revd John Upton, prebendary of Rochester*: 3-page poem, including the lines:

> While you my friend, at Rochester employ'd
> Review yr Spenser with a Critick's eye . . .

advising Upton to 'shun broils & discord'. Harris says of himself:

> Health I gain without a nauseous draught,
> And blend the sweet & useful.

J. Collier, *The Cry: A New Dramatic Fable*, 3 vols., published after Harris's revisions and editorial assistance.

1755

Charles Batteux's trans. of 2 treatises reappears in *Principes de littérature* (Paris).

1756

German trans. of *Three Treatises* (Danzig).

8 Aug. 'The Travellers, a Heroic Poem in Two Books inscribed to ye Honble Susan Countess of Shaftesbury'*: 27-page blank verse poem describing a tour (July 1755) from the seat of Sir Wyndham Knatchbull near Plymouth to Exeter via Saltram, Blandford, Dorchester, Lyme Regis, Charmouth, and Axminster, by the fourth earl, Susannah, Countess of Shaftesbury, William, Thomas, and James Harris, among others. The poem opens:

> Sights and the Men I sing, that left their home
> Thro' distant Towns & Provinces to stray.
> Long did they labour, many a storm indure,
> Steep hills ascend, & cross impetuous floods,
> Fir'd with the love of objects great & fair.
>
> O Shaftesbury, Judge & Patroness of Arts!
> Mistress of all that's elegant! to Thee
> The Poet sings; do Thou protect the Song,
> Or else in vain he tunes his Epic lay.

1758

5 Feb. MS of 'An Essay on the Life and Genius of Henry Fielding Esqr'*, completed at Bath.

Eleven 'Epigrams from Martial' [by John Hoadly] dedicated 'To James Harris, Esq;', in R. Dodsley (ed.), *A Collection of Poems*, 6 vols., v. 285–8.

E. Carter, *Epictetus . . . consisting of his Discourses preserved by Arrian . . . the Enchiridion and Fragments*, trans. from the original Greek. With Harris's assistance on 'that wicked logical chapter . . . I am greatly obliged to Mr Harris' (Carter to Catherine Talbot, Sept. 1753).

1759

Schlegel's trans. of Batteux, as *Einschränkung der schönen Künste* (Leipzig).

1760

Harris 'Catalogue of the Works of George Frederic Handel' included (pp. 147–55) in J. Mainwaring, *Memoirs of the Late George Frederic Handel. To which is added. A Catalogue of his Works, and Observations upon them.*

1761

Jan. 'The Complainers: a Sentimental Tale'*: 3-page octosyllabic poem beginning 'For years did mortal man complain | Of bad success to Jove in vain.'

27 Jan. 'The Burcomb Hero, or Pitts the Great'*, a verse satire on Salisbury personalities, beginning 'Immortal Homer! cried the Grecian Witts | But those of Burcomb cry, Immortal Pitts!'

1762

French trans. (extract) of 'Concerning Happiness: A Dialogue' in *Pensées angloises sur divers sujets de religion & de morale* (Paris): reviewed in *L'Année littéraire* (1762), letter 3, pp. 45–62.

S. Fielding, *Xenophon's Memoir of Socrates with the Defence of Socrates before his Judges. Translated from the Original Greek* (Bath and London); with Harris's acknowledged assistance.

Oct. 'The Maceiad: An Heroi-Comic Poem'*: 7-page poem about the visiting magistrates on the Western Circuit sitting at Sarum, 'but chief . . . the dreadful *Dowdy*': 'I sing the Prelate, whose victorious Arm | Diffus'd thro' Mobs & Magistrates th' Alarm.'

25 Oct. Drury Lane, 'The Spring' (i.e. 'Daphnis and Amaryllis') performed.

1763

Concord repr. in F. Fawkes and W. Woty, *The Poetical Calendar*, xii.
53–9.

1764

25 July. 'New Pastoral of Menalcas'*: 9-page draft for performance
of musical pastoral.

Undated 2-stanza fragment of privately performed verse drama
'The Druids'*, 'The Speech of one after the Romans had taken
Mona', written out for 'Warton Sr' (i.e. Joseph Warton), not in
Harris's hand, beginning:

> Gorg'd is the ravenous throat of War,
> Slaughter stops her scythed car,
> While o'er the bloody field of prey,
> The greedy Eagle wings his way.
> Mona! are thy Children Slain?
> Sounded the sacred trump in vain?
> Could not thy robed Bards thy altars save?
> Is all thy Glory sunk into the Grave?

1765

2nd, rev., and corrected edn. of *Three Treatises* and *Hermes* (the latter
retitled *Hermes, or a Philosophical Inquiry concerning Universal
Grammar*.

'To R. O. Cambridge of Twickenham'*; verse epistle beginning:

> Various the pleasures you dispense;
> You treat us, while we wake, with sense.

June. Poem on Pitt. In May 1762 Harris described William Pitt as
'an Inigo Jones in Politics, a man of great Ideas, a Projector of
noble and magnificent Plans—But Architects, tho they find the
Plan, never consider themselves as concerned to find the
Means.' Three years later, Harris wrote the following untitled
and undated attack on William Pitt, now Earl of Chatham,
probably shortly after Rockingham left office in July 1765,
when Pitt rather than Harris's mentor Grenville was asked to
take his place. Pitt appointed only 4 of Grenville's followers to
office:

> Tis vain to murmur, vain to point;
> We once were in, but now we're out.
> Great Pitt, with his almighty Broom

Has swept thro' ev'ry Courtly Room,
Thro' ev'ry Office, great & small,
First Lords & Seconds, one & all.
Yet Lords remov'd, He soon may find
Another Sett, to please his mind.
Ah! could he but as soon supply
The Sums he sent to Germany,
The Millions yt have sunk this Isle,
And made his Prussian Hero smile.
   Strive, orator, with all thy might
To shew that this black Scheme was white,
The more improbable the story,
To prove it thine the greater glory.
If Logic no support affords,
Sesquipedalian, Gallic words,
From long experience thou canst tell,
Will do the bus'ness full as well.

1767

20 Apr. 'Society: A Satire or Miscellaneous Poem'*:
   I care not what the rigid say,
Three hours with pleasure I can stay,
To hear an Opera, or a Play.
When Sunday comes, I ne'er refuse
To go to Court, to hear the News,
Altho I'm satisfied full well,
That those, who know it, never tell:
But chat & Laughter have their merits;
I feel they help to raise the Spirits.
Sometimes at Ranelagh I'm found,
Viewing the merry Crowd go round,
That tread for hours one Circle still,
Like horses grinding in a Mill.
When Friends invite me to Vaux-hall
I readily obey the Call;
Nor do I blame the man, that loves
Intire Butt, Beer, and Shady Groves.
Let him with caution take the Beer,
There's nothing from the Groves to fear.
   .   .   .   .

> But above all there's none so sweet,
> As when choice Friends together meet;
> The Banquet, where the God-like Spark
> Of Reason helps illume the dark.

S. Fielding, *Familiar Letters between the Principal Characters in David Simple*, 2 vols. Harris's 2 dialogues, 'Much Ado' and 'Fashion' included, ii. 276–93.

Sept. 'The Procession: An Entertainment'*: a 7-page verse play-script wholly drafted on the back of used letters, performed by the mayor of Salisbury, the recorder, the canons, Walter Long, James Harris, Nurse Causway, Tom the Porter, and others of 'the Salisbury Company'.

1769

May: 'Dialogue between K[ing] &c.'*, beginning:

K. What are my Ministers about?
M. Striving to keep those Patriots out.
K. Why keep my Patriots such a din?
P. Because we think twill help us in.

1770

'Classis Russiaca contra Turcos missa'*: R. Lowth's verses, spoken by himself at Winchester school, 6 July.

1771

3rd edn. of *Hermes*.

1772

3rd edn. of *Three Treatises*.

1773

Dublin 'edn.' of *Hermes* (Dublin).

1774

Sept. 'Musaeus: An Eclogue'*: 4-page comic poem celebrating the musical skills of the Vicars Choral at Salisbury, and describing the high spirits of one who fell asleep on his own doorstep after a night spent carousing.

Undated 36-page MS (not in Harris's hand), entitled 'Adam: An Oratorio', an 'historical drama' compiled from *Paradise Lost* and adapted to music by 'R.J.'

1775

Philosophical Arrangements.

1775–86

Miscellanies by J. Harris, 5 vols. (collected works from original edns. except for vols. i and ii, which are 4th edn., rev. and corrected: printed by C. Nourse).

Miscellanies by James Harris, 4 vols: vol. i *Hermes* (3rd edn. of 1771 rev. and corrected); vol. ii *Three Treatises* (4th edn. of 1782 rev. and corrected); vol. iii *Philosophical Arrangements* (1775); vol. iv *Philological Inquiries* (1781).

1776

Review by A. von Haller of *Philosophical Arrangements* in *Göttingische Anzeigen*, 14 (1 Feb.), 110–11.

1777–86

Miscellanies, 5 vols. (made up of previous edns.).

22 June 1777. 'Meditation upon Cheese in the manner of the Honble Mr Boyle F.R.S.'*, including: 'What is a Dilettante, a Man of Taste, what but a Parmesan Cheese, whose merit is all *exotic*? The plain, honest Englishman is a fine Cheese from North-Wiltshire, devoid of Faults, and free from blemish.'

1778

3 Dec. 'Winter Amusements: an Ode'*; 5-page poem beginning:
> Ye beauteous Nymphs, and jovial Swains
> Who deck'd with youthful bloom

and ending,
> Perpetual charms, unfading Spring
> In sweet Reflexions find,
> While Innocence and Virtue bring
> A Sunshine o'er the Mind.

1779

27 Apr. 'After Horace'*:

> By the Banks of my Avon while pensive I stray,
> For what d'ye believe I honestly pray?
> May I keep what I have—so much for my Pelf—
> And the time that I live, may I live to my Self:
> May my Study ne'er want a good Book, that I prize,
> Nor my Kitchen nor Cellar their proper Supplies,
> To receive my good Friends, & prevent a Surprize.
> —But perhaps tis enough from kind Heav'n to implore,
> What it gives, and it takes, a competent Store
> Of Riches & Health—For a calm equal Mind,
> Tis a Blessing I know *for my Self* I must find;
> If without my own choice I'm compell'd to be just,
> My Virtue & Merit soon sink in the Dust.

1780

German trans. of 3rd London edn. of *Three Treatises, Abhandlungen über Künst, Musik, Dichtkunst und Glückseligkeit* (Halle).

Nov. 'Master and Man: A Dialogue', 2-page dialogue written 1 month before Harris's death, on the subject of the Gordon riots in London:

MAN. Sir, can yr Honour time afford,
　To hear yr Servant speak a Word?
MASTER. Tom, speak thy mind, and thou shalt know
　Thy Master's mind as on we go.
MAN. I never, Sir, to Alehouse went.
MASTER. You never were to Bridewell sent.
MAN. I never bid Postilion stand.
MASTER. You ne'er were bid, Hold up your hand.
MAN. I've robb'd no Mail, the Office knows.
MASTER. You're not hung up to feast the Crows.
MAN. And don't this prove I'm good thro'out?
MASTER. Not quite so fast Tom, there I doubt
　No Kite was ever taken in,
　When he beheld an open Gin.
　Yet for that reason who can say
　Kites are not reckon'd Birds of Prey?
　Nor Fear, nor Pain, 'tis understood,

But Love of Virtue guides the Good.
If that be wanting, give but hope
To Rogues that they shall scape the Rope,
They dare to every Crime aspire;
Dare Newgate or the Bank to fire,
Dare speed thro every street the blaze,
Dare the devoted Temple raze,
Confound, consume——
MAN.   Stay good Sir stay
I'm terrified at what you say;
At what you say confounded quite—
Upon my word Sr, I'm no Kite;
You have no cause to sound th' alarm;
I never meditated harm
Against mens houses or their lives—
Pray lett me go & whet my knives

Exit Man Exit Master

1781

Philological Inquiries, 2 vols.

1782

C. Heyne's review of *Philological Inquiries* in *Göttingische Anzeigen*, 58 (13 May 1782), 466–70.

1783

4th edn. of *Three Treatises*.

1785

French trans. of part 3 of *Philological Inquiries*, as *Histoire littéraire du moyen age*, by A. M. H. Boulard (Paris). Boulard (the trans. of Adam Smith) notes (p. vi) that the medieval period 'est peut-être la partie la moins connue de l'Histoire de l'Esprit humain', and inserts 6 pages (pp. 92–8) from La Harpe's *Discours sur les Grêcs anciens et modernes* into his trans. of Harris. For a review, see *L'Année littéraire*, 8 (1785), letter 6, pp. 145–68.

1786

4th edn. of *Hermes*.

1788

German trans. of *Hermes* by C. G. Ewerbeck: *Hermes oder philosophische Untersuchung über die Allgemeine Grammatik* (Halle).

1794

5th edn. of *Three Treatises* and *Hermes*.

1796

French Directory ordered trans. of the 2nd, rev., and corrected edn. of *Hermes*, by F. Thurot: *Hérmès, ou recherches philosophiques sur la grammaire universelle*.

[c.] 1800

J. Corfe, arranger and publisher, *Sacred Music: Dedicated by Permission to The Right Honble. Earl of Malmesbury . . . Consisting of a Selection of the most admired Pieces of Vocal Music from the Te Deum, Jubilate, Anthems, & Milton's Hymn, Adapted to some of the Choicest Music of the Greatest Italian and other Foreign Composers . . . By the Late James Harris, Esq.*, 2 vols. ('London Printed for the Editor, & to be had of him at Salisbury, & the principal Music dealers in Town and Country.')

1801

Malmesbury's 2-vol. folio edn. of *The Works of James Harris*.

1802

New edn. of *Philological Inquiries*.

1803

Works, 5 vols.

1806

6th edn. of *Hermes*.

1816

New edn. of *Hermes*.

1825

7th edn. of *Hermes*.

1841

New 1-vol. edn. of Malmesbury's *The Works of James Harris*.

1849

Sir John Stoddart, *The Philosophy of Language; Comprehending Universal Grammar* (first division of Coleridge's projected *Encyclopaedia Metropolitana*): 2nd rev. edn. ed. W. Hazlitt, 1854.

1870

A Series of Letters of the First Earl of Malmesbury His Family and Friends from 1745 to 1820, 2 vols, ed. the Earl of Malmesbury.

1929

G. ten Hoor, 'James Harris and the Influence of his Aesthetic Theories in Germany', Ph.D. thesis (Ann Arbor, Mich.).

1934

O. Funke, *Englische Sprachphilosophie im späteren 18. Jahrhundert* (Berne): on Harris and Horne Tooke.

1968

Scolar Press facsimile of the unrev. and uncorrected 1st edn. of *Hermes*.

1971

Upon the Rise and Progress of Criticism (1752) and *Knowledge of the World, or Good Company*, facsimile repr., Garland Publishing, Inc., New York.

1972

Reissue of Thurot's French trans. of *Hermes*, ed. with an extensive introd. A. Joly (Geneva).

1975

Facsimile repr. of *Works by James Harris* (*Hermes* and *Philological Inquiries* only), 2 vols., AMS Press, New York.

1980

Facsimile repr. of [J. Mainwaring] *Memoirs of the Life of the Late George Frederic Handel* (1760). Harris's catalogue of Handel's works appears on pp. 147–55, and is attributed to him by O. Deutsch, *Handel: A Documentary Biography* (London, 1955), 842.

Bibliography
Manuscript Sources

The largest single source is the manuscript collection referred to here as 'Harris Papers', which is in the private possesion of the Earl of Malmesbury. This extensive collection of 48 volumes, covering the period 1588 to 1780, has yet to be fully catalogued, but a manuscript 'Index to the Contents of the Catalogue' was drawn up by the first earl in the early years of the nineteenth century, and this has provided me with the references to volume and item numbers used in the present study. The collection contains several volumes of parliamentary memoranda, letters, Harris's unpublished poems, and prose works (including the manuscript of *Hermes*, and the 'Memoir' of Henry Fielding). In 1958 a very small selection (from the parliamentary memoranda) was photocopied for the History of Parliament Trust and used by Sir Lewis Namier and John Brooke in the preparation of their 3-volume *The House of Commons 1754–1790* (1964). I have referred to these photocopies as Malmesbury Papers.

In the Public Record Office (Kew), the Lowry Cole Papers (30/43–56) contain a fair copy of the first earl's manuscript memoir of his father, James Harris (30/43/1/2: dated 1800), another version written as a preface to a new edition of James Harris's works (30/43/1/3: dated 1801), Catherine Gertrude Robinson's undated and unfinished 'Memoir of J. Harris Author of Hermes' (30/43/1/4), and Gertrude's 'Portrait of her parents, James and Elizabeth Harris' (30/43/1/5: dated 1806), together with a collection of her childhood exercise books (GD 30/43/1), letters to her from James and Elizabeth Harris in London (30/43/2: dated 1767–81), and accounts of her own journey to St Petersburg (10–25 Oct. 1777), and of her visits to Moscow, Scotland, and Bath.

The library of Merton College, Oxford (F33, series a), has the first earl's letters to his family from Paris, Madrid, Brunswick, The Hague, Berlin, and St Petersburg (from 1762 to 1781).

The British Library includes James Harris manuscripts in the Andrew Mitchell Papers, vol. 55 (BM Add. MSS 6858, pp. 21–4), and the P. A. Taylor Papers, vol. 2 (BM Add. MSS 37683). The collection known as 'James Harris Autograph Papers 1768–9' (BM Add. MSS 18729) includes Harris's letter to Lyttelton of 28 Dec. 1769, and various manuscript poems; the J. Mitford Note Books, vol. 7 (Add. MSS 32565) include letters to Jonathan Toup from Harris from 1747 to 1776; letters from Thomas and James Harris to

Dr Thomas Birch (1738–57: BM Add. MSS 4308–9); and among the Hardwicke Papers, vol. 785 (BM Add. MSS 36133), there are Royal Warrants signed by James Harris. BM Add. MSS 18728 is Harris's own proof copy of *Upon the Rise and Progress of Criticism*, heavily corrected and annotated.

The Wiltshire Public Record Office (Trowbridge) contains James Harris's release and marriage settlement (Sa. 34: 212B/5950: dated 1 June 1745), and a dozen sets of papers relating to the lease and release of land in the Salisbury area (212B/91 and 8M51 coffer 52). There are many letters (dating from 1758 to 1771) from James Harris to Thomas Robinson, second Lord Grantham, in the West Yorkshire Archive, Leeds (2072–3a), and Mr John Jacob of Durrington Manor made available to me copies of letters (1779–80) written by James Harris to the son of his physician, Dr Jacob, then at The Hague.

The Baillieu Library of the University of Melbourne holds many annotated volumes from 'Hermes' Harris's library, dispersed at the Christie's sale of the Hurn Court library, 9–10 and 30–1 March 1950; and the library of University College, London, holds Harris's own annotated copies of the first edition of *Hermes*, and of *Philosophical Arrangements*. Monash University holds the 130-page folio vol. 'MS by 3^d Lord Shaftesbury. Letters on Horace & upon other Philosophical Subjects' (*c*.1737) in Harris's hand.

PRIMARY PRINTED SOURCES

(*Except where stated, London is the place of publication*)

MALMESBURY, first earl of (ed.), *The Works of James Harris, Esq.*, 2 vols. (1801), which includes a prefatory memoir of James Harris by his son (pp. i, ix–xxvii).

—— *The Works of James Harris*, in 1 vol. (Oxford, 1841).

MALMESBURY, third earl of (ed.), *Diaries and Correspondence of James Harris, First Earl of Malmesbury*, 4 vols. (2nd edn., 1845).

—— (ed.), *A Series of Letters of the First Earl of Malmesbury His Family and Friends from 1745 to 1820*, 2 vols. (1870).

MALMESBURY, fifth earl of, 'Some Anecdotes of the Harris Family', *The Ancestor*, 1 (Apr. 1902), 1–27.

SECONDARY PRINTED SOURCES

Manuscript collections containing Harris material published by the Historical Manuscripts Commission include:

Fifteenth Report, appendix, part 2, *The Manuscripts of J. Eliot Hodgkin, Esq., F.S.A. of Richmond Surrey* (1897).

REDINGTON, J. (ed.), *Calendar of Home Office Papers of George III, 1760–1765* (1878).

Report on Manuscripts in Various Collections, iv. *The Manuscripts of Bishop of Salisbury . . . Corporations of Salisbury, Orford and Aldeburgh* (Dublin, 1907).

ROBERTS, R. A. (ed.), *Calendar of Home Office Papers of the Reign of George III*, 1770–72 (1881).

ROUTLEDGE, F. J. (ed.), *Calendar of the Clarendon State Papers Preserved in the Bodleian Library* (Oxford, 1970).

SELECTED CRITICAL AND HISTORICAL SOURCES

AARSLEF, H., *The Study of Language in England, 1780–1860* (Princeton, 1967).

ACWORTH, R., *The Philosophy of John Norris of Bemerton (1657–1712)* (Hildesheim and New York, 1979).

ALGAROTTI, F., *An Essay on Opera* (1762).

Anon., *Magna Britannia et Hibernia, Antiqua & Nova, or, A New Survey of Great Britain*, 6 vols. (1720–31).

Anon., *A Description of England and Wales*, 10 vols. (1769–70), x (1770).

'Anti-CHUBBIUS' [J. Horler], *Memoirs of the Life and Writings of Mr Thomas Chubb* (1747).

D'ARBLAY, Mme, *Memoirs of Dr Burney*, 3 vols. (1832).

ASHLEY, M., *Cyropaedia: or, the Institution of Cyrus, by Xenophon*, 2 vols. (Dublin, 1728).

AVISON, C., *An Essay on Musical Expression*, (2nd edn., 1753).

BALDERSTON, K. C. (ed.), *Thraliana: The Diary of Hester Lynch Thrale (later Mrs Piozzi)*, 2 vols. (Oxford, 1942).

BARBAULD, A. L. (ed.), *The Correspondence of Samuel Richardson*, 6 vols. (1804).

BATTESTIN, M. C., *The Providence of Wit: Aspects of Form in Augustan Literature and the Arts* (Oxford, 1974).

BEATTIE, J., *Dissertations Moral and Critical* (1783).

BENSON, R., *Memoirs of the Life and Writings of the Rev. Arthur Collier . . . with Some Account of his Family* (1837).

—— and HATCHER, H. (eds.), *The History of Modern Wiltshire by Sir Richard Colt Hoare: Old and New Sarum* (1843).

BIRCH, T., *A General Dictionary, Historical and Critical*, 10 vols. (1734–41).

BIRKBECK HILL, G. (ed.), *Johnsonian Miscellanies*, 2 vols. (Oxford, 1897).

BLAKEY, R., *Historical Sketch of Logic, from the Earliest Times to the Present Day* (London and Edinburgh, 1851).

BLUNT, R., *Mrs Montagu: 'Queen of the Blues' Her Letters and Friendships from 1762 to 1800*, 2 vols. (n.d.).

BOULARD, A. M. H., *Histoire littéraire du moyen age* (Paris, 1785).

BOWRING, J. (ed.), *The Works of Jeremy Bentham*, 8 vols. (1838–43).

BREVA-CLARAMONTE, M., *Sanctius' Theory of Language: A Contribution to the History of Renaissance Linguistics* (Amsterdam, 1983).

BRYDGES, Sir Egerton (ed.), *Collins's Peerage of England*, 9 vols. (1812).

BURNET, J., Lord Monboddo, *Of the Origin and Progress of Language*, 6 vols. (Edinburgh, 1773–92).

—— *Ancient Metaphysics: Or the Science of Universals* (1779–99).

BURNEY, C., *A General History of Music*, 4 vols. (1776–89), ed. F. Mercer, 2 vols. (1935).

CAMPBELL, G., *The Philosophy of Rhetoric* (Edinburgh, 1776).

CARLILE, N., *Description of Endowed Grammar Schools* (1818).

CHOMSKY, N., *Cartesian Linguistics: A Chapter in the History of Rationalist Thought* (New York and London, 1966).

CHRISTIE, I. R., *Wilkes, Wyville and Reform* (1962).

CHUBB, T., *A Collection of Tracts* (1730).

COLLIER, A., *Clavis Universalis* (1713), ed. E. Bowman (Chicago and London, 1909).

COLLIER, J., *Essay on the Art of Ingeniously Tormenting; With Rules for the Exercise of that Pleasant Art* (1753).

—— *The Cry: A New Dramatic Fable* (1754).

CORFE, J. (ed.), *Sacred Music . . . by the Late James Harris*, 2 vols. (c.1800).

COXE, W., *Anecdotes of George Frederick Handel, and John Christopher Smith* (London and Salisbury, 1799).

CUMBERLAND, G., *Thoughts on Outline, Sculpture, and the System that Guided the Antient Philosophers* (1796).

CUMBERLAND, R., *Philosophical Inquiry into the Laws of Nature* (1750).

DAVIS, R. M., *The Good Lord Lyttelton: A Study in Eighteenth-Century Politics and Culture* (Bethlehem, Penn., 1939).

DEAN, W., *Handel and the Opera Seria: The Ernest Bloch Lectures* (1970).

DE MADARIAGA, I., *Britain, Russia, and the Armed Neutrality of 1780* (New Haven, 1962).

DEUTSCH, O. E., *Handel: A Documentary Biography* (1955).

DOBSON, A., 'Hermes Harris', in *Later Essays* (Oxford, 1921), 46–69.

DU BOS, Abbé, *Critical Reflections on Poetry, Painting and Music*, 3 vols., trans. T. Nugent, (1748).

EASTON, J., *The Salisbury Guide, giving an account of the antiquities of Old Sarum, and the ancient and present state of New Sarum or Salisbury* (22nd edn., Salisbury, 1801).

ENGLAND, M. W., 'The Satiric Blake: Apprenticeship at the Haymarket?', *Bulletin of the New York Public Library*, 73 (1969), 440–64, 531–50.

FAWKES, F., and WOTY, W. (eds.), *The Poetical Calendar* (1763).

FEARN, J., *Anti-Tooke; Or an Analysis of The Principles of Language*, 2 vols. (1824–7).

FERDINAND, C., 'Benjamin Collins: Salisbury Printer', in *Searching the Eighteenth Century*, ed. M. Crump and M. Harris (1983), 74–92.

FIELDING, S., *Xenophon's Memoir of Socrates with the Defence of Socrates before his Judges* (Bath and London, 1762).

FOUCAULT, M., *The Order of Things: An Archaeology of the Human Sciences* (New York, 1970); 1st pub. in French as *Les Mots et les choses* (Paris, 1966).

—— *The Archaeology of Knowledge*, trans. A. M. Sheridan Smith (1974).

FUNKE, O., *Englische Sprachphilosophie im späteren 18. Jahrhundert* (Berne 1934).

GARDINER, R. B. (ed.), *The Registers of Wadham College, Oxford, Part One, 1613–1719* (1889).

HABERMAS, J., *Knowledge and Human Interests*, trans. J. J. Shapiro (Boston, 1968).

HAMILTON, Sir William, *Discussions on Philosophy and Literature, Education and University Reform* (New York, 1868).

—— *Lectures on Metaphysics and Logic*, 4 vols. (Edinburgh 1870).

HARE, A., *The Georgian Theatre in Wessex* (1958).

HAYWARD, A. (ed.), *Autobiography, Letters, and Remains of Mrs Piozzi*, 2 vols. (1861).

HEPWORTH, B., *Robert Lowth* (Boston, 1978).

HERBERT, E., 1st Baron Herbert of Cherbury, *De Veritate* (1624, 1625), trans. M. A. Carré (Bristol, 1937).

HERBERT, Lord (ed.), *Pembroke Papers (1780–94): Letters and Diaries of Henry, Tenth Earl of Pembroke and his Circle* (1950).

HIGHFILL, P. H., *et al.* (eds.), *A Biographical Dictionary of Actors, Actresses, Musicians, Dancers, Managers and other Stage Personnel in London 1660–1800*, 12 vols. (Carbondale, Ill., 1973–87).

HILLES, F. W., *The Literary Career of Joshua Reynolds* (Cambridge, 1936).

HORNBY, E., 'Aspects of Social, Intellectual and Leisured Life in Salisbury in the Eighteenth Century', MA thesis, Leicester University, (1979).

—— 'Some Aspects of the Musical Festivals in Salisbury in the Eighteenth Century', *The Hatcher Review*, 12 (Autumn 1981), 78–85.

HOWELL, W. S., *Logic and Rhetoric in England, 1500–1700* (Princeton, 1956).

—— *Eighteenth-Century British Logic and Rhetoric* (Princeton, 1971).

HULL, R. F. C. (ed.), *The Collected Works of C. G. Jung*, xiii (1968).

HUME, D., *Dialogues Concerning Natural Religion* (1779).

HUTCHINS, J., *The History and Antiquities of the County of Dorset*, 4 vols. (3rd edn., 1861–74).

JACKSON-STOPS, G., *Malmesbury House* [Salisbury, 1971].

JANSSENS, U., *Matthieu Maty and the Journal Britannique 1750–55* (Amsterdam, 1975).

JOLY, A. (ed.), *Hermès ou recherches philosophiques sur la grammaire universelle*, trans. F. Thurot (Geneva, 1972).

KEMBLE, F. A., *Record of a Childhood*, 2 vols. (1878).

KITTREDGE, G. L., *Some Landmarks in the History of English Grammar* (1906).

KNEALE, W., and KNEALE, M., *The Development of Logic* (Oxford, 1962).

KNIGHT, W., *Lord Monboddo and Some of his Contemporaries* (1900).

LANSDOWNE, Marquis of, 'Wiltshire Politicians (c.1700)', *Wiltshire Archaeological and Natural History Magazine* 46:157 (Dec. 1932), 64–85.

LAWSON, P., *George Grenville: A Political Life* (Oxford, 1984).

LELAND, J., *A View of the Principal Deistical Writers*, 2 vols. (5th edn., 1798).

LIPKING, L., *The Ordering of the Arts in Eighteenth-Century England* (Princeton, 1970).

LITTLE, D. M., and KAHRL, G. M. (eds.), *The Letters of David Garrick*, 3 vols. (Cambridge, Mass., 1963).

LYTTELTON, G., *History of King Henry II* (1761–71).

McKENZIE, G., *Critical Responsiveness: A Study of the Psychological Current in Later Eighteenth-Century Criticism* (Berkeley and Los Angeles, 1949).

[MAINWARING, J.], *Memoirs of the Life and Writings of George Frederic Handel* (1760).

MALEK, J., 'Art as Mind Shaped by Medium: The Significance of James Harris' "A Discourse on Music, Painting, and Poetry" in Eighteenth-Century Aesthetics', *Texas Studies in Literature and Language* 12:1 (1970), 231–9.

MARSH, R., *Four Dialectical Theories of Poetry: An Aspect of English Neoclassical Criticism* (Chicago, 1965).

MATTHEWS, B., 'Handel—More Unpublished Letters', *Music and Letters*, 42 (1961), 127–31.

METTRIE, J. O. de La, *Man a Machine* (Paris, 1748), trans. G. C. Bussey (La Salle, Ill., 1912; 1961 edn.).

MICHAEL, I., *English Grammatical Categories and the Tradition to 1800*, part 2 (Cambridge, 1970).

MILBURN, G. E. (ed.), *The Diary of John Young 1841–43* (Surtees Society, vol. 195; Leamington Spa, 1983).

MILLER, H. K., *Essays on Fielding's Miscellanies: A Commentary on Volume One* (Princeton, 1961).

—— (ed.), *Miscellanies by Henry Fielding, Esq., Volume One* (Oxford, 1972).

MITFORD, W., *Inquiry into the Principles of Harmony in Language, and of the Mechanisms of Verse* (1774).

NAISH, T., *A Sermon Preach'd at the Cathedral Church of Sarum, November 22 1700 Before a Society of Lovers of Musick* (Salisbury, 1701).

—— *Sermon Preach'd at the Cathedral Church of Sarum, November the 30th, 1726. Being the Anniversary Day Appointed for the Meeting of the Society of Lovers of Musick* (London, Sherborne, and Sarum, 1726).

—— *Sermon Preach'd at the Cathedral Church of Sarum, November the 30th 1727, Being the Anniversary Day Appointed for the Meeting of the Society of Lovers of Musick* (London, Sherborne, and Sarum, 1727).

NAMIER, Sir Lewis, and BROOKE, J., *The House of Commons 1754–1790*, 3 vols. (1964).

—— and —— *Charles Townshend* (1964).

NEWMAN, E., *Gluck and the Opera: A Study in Musical History* (1964).

NICHOLS, J., 'James Harris', in *Illustrations of the Literary History of the Eighteenth Century*, 8 vols. (1828), v. 345–8.

NORRIS, J., *An Idea of Happiness, in a Letter to a Friend* (1691).

—— *A Collection of Miscellanies: Consisting of Poems, Essays, Discourses and Letters* (2nd edn., 1692).

—— *Treatises upon Several Subjects* (1698).

—— *An Essay Towards the Theory of the Ideal or Intelligible World*, part 1 (1701).

NUGENT, T. (trans.), *An Essay on the Origin of Human Knowledge, being a Supplement to Mr. Locke's Essay on Human Understanding, Translated from the French of the Abbé de Condillac* (1756).

OLSON, E. (ed.), *Aristotle's 'Poetics' and English Literature: A Collection of Critical Essays* (Chicago, 1965).

PADLEY, G. A., *Grammatical Theory in Western Europe 1500–1700: The Latin Tradition* (Cambridge, 1976).

—— *Grammatical Theory in Western Europe 1500–1799: Trends in Vernacular Grammar I* (Cambridge, 1985).

PARR, S. (ed.), *Metaphysical Tracts of the Eighteenth Century* (1837).

PENNINGTON, M., *Memoirs of the Life of Elizabeth Carter with a New Edition of her Poems* (1807).

PETVIN, J., *Letters Concerning Mind* (1750).

PETVIN., J. *Letters Concerning the Use and Method of History* (1753).

PROBYN, C. T., 'Johnson, James Harris, and the Logic of Happiness', *Modern Language Review*, 73:2 (Apr. 1978), 256–66.

—— 'James Harris to Parson Adams in Germany: Some Light on Fielding's Salisbury Set', *Philological Quarterly*, 64:1 (Winter 1985), 130–9.

—— 'James Harris: Salisbury Philosophe 1709–1780', *The Hatcher Review*, 2:19 (1985), 421–35.

PUGH, R. B., and CRITTAL, E., *A History of Wiltshire (The Victoria History of the Counties of England*, vol. 5; 1957).

RAINE, K., and HARPER, G. M. (eds.), *Thomas Taylor the Platonist: Selected Writings* (Princeton, 1969).

REID, D. J., with PRITCHARD, B., 'Some Festival Programmes of the Eighteenth and Nineteenth Centuries: 1. Salisbury and Winchester', *R.M.A. Research Chronicle* 5 (1965), 51–67 (addenda by A. D. Walker in 6 (1966), 23, and again by B. Matthews, 8 (1970), 23–33).

REID, T., *Essays on the Intellectual Powers of Man* (Edinburgh, 1785).

REYNOLDS, Sir Joshua, *Discourses on Art*, ed. R. R. Wark (New York and London, 1959).

RICHARDSON, Mrs Herbert, 'Wiltshire Newspapers—Past and Present: Part III, The Newspapers of South Wilts', *Wiltshire Archaeological and Natural History Magazine*, 40 (June 1920), 53–69.

RISSE, W., *Bibliographica Logica*, 2 vols. (Hildesheim, 1973).

ROBERTSON, D. H., *Sarum Close: A History of the Life and Education of the Cathedral Choristers for 700 Years* (1938).

ROBERTSON, J. M. (ed.), [Shaftesbury's] *Characteristics of Men, Manners, Opinions, Times*, 2 vols. (Gloucester, Mass., 1900; repr. 1963).

ROBINS, R. H., *A Short History of Linguistics* (1967).

ROBINSON, I., *The New Grammarians Funeral: A Critique of Noam Chomsky's Linguistics* (Cambridge, 1975).

RUDÉ, G., *Wilkes and Liberty* (Oxford, 1962).

SADIE, S. (ed.), *The New Grove Dictionary of Music and Musicians*, 20 vols. (1980).

SAINTSBURY, G., *History of English Criticism* (Edinburgh, 1930).

Salisbury Journal, 1729– .

SEDGWICK, R., *The House of Commons, 1715–54*, part 1 (1970).

SHARPE, G., *Origin and Structure of the Greek Tongue* (1767).

SHEPHERD, C., *A Tour through Wales and the Central Parts of England* (1799).

SHERBO, A., 'Some Early Readers in the British Museum', *Transactions of the Cambridge Bibliographical Society*, 6:1 (1972), 56–64.

SLATTER, D. (ed.), *The Diary of Thomas Naish* (Devizes, 1965; vol. 20 of *WANHS*, 1964).

SPENCE, J., *Essay on Pope's Odyssey* (1726–7).

—— *Polymetis* (1747).

—— *Observations, Anecdotes, and Characters of Books and Men* (1820), ed. J. M. Osborn, 2 vols. (Oxford, 1966).

—— *Letters from the Grand Tour*, ed. S. Klima (1975).

STEWART, D., *Elements of the Philosophy of the Human Mind* (Edinburgh, 1792).

STODDART, Sir John, *The Philosophy of Language; Comprehending Universal Grammar, or Pure Science of Language* (1830, 3rd rev. edn., 1854).

STONE, G. W., *The London Stage, 1660–1800, Part 4: 1747–1776* (Carbondale, Ill., 1962).

SUPHAN, B. (ed.), *Herders Sämtliche Werke* (Berlin, 1877–1913).

SUTHERLAND, L. S., and MITCHELL, L. G., *The History of the University of Oxford* (Oxford, 1986).

TOMLINSON, J. R. G., 'The Grenville Papers, 1763–5', MA thesis in 3 vols., Manchester University (1956).

TOOKE, J. H., *A Letter to John Dunning, Esq.* (1778).

—— *The Diversions of Purley* (1798).

TRAPP, J., *Lectures on Poetry Read in the Schools of Natural Philosophy At Oxford . . . Translated from the Latin* [by W. Bowyer and W. Clarke] (1742): the original, *Praelectiones Poeticae*, appeared in 1711, 1715, 1719, with a 2nd edn. in 1722, and a 3rd edn. in 1736.

TYERS, T., *A Biographical Sketch of Dr. Johnson* (1784).

VOITLE, R., *The Third Earl of Shaftesbury, 1671–1713* (Baton Rouge and London, 1984).

WARBURTON, W., *The Divine Legation of Moses Demonstrated on the Principles of a Religious Deist*, 6 books (1738).

WARTON, J., *Life . . . of Ralph Bathurst* (1761).

WEBSTER, N., *Dissertations on the English Language* (Boston, 1789).

WHEELER, W. A., *Sarum Chronology* (1889).

WIMSATT, W. K., *Philosophic Words* (New Haven, 1948).

WOOLL, J. (ed.), *Biographical Memoirs of the Late Revd. Joseph Warton D. D.* (1806).

YARBOROUGH, M. C., *John Horne Tooke* (New York, 1926).

Index